Signal Processing and Biomedical Engineering Research

Ammar Ahmed · Joseph Picone
Editors

Signal Processing and Biomedical Engineering Research

Applications of Machine Learning Based on Big Data Principles

Springer

Editors
Ammar Ahmed
Advanced Safety and User Experience
Aptiv
Agoura Hills, CA, USA

Joseph Picone
Department of Electrical and Computer Engineering
Temple University
Philadelphia, PA, USA

ISBN 978-3-032-18508-2 ISBN 978-3-032-18509-9 (eBook)
https://doi.org/10.1007/978-3-032-18509-9

This Springer imprint is published by the registered company Springer Nature Switzerland AG
The registered company address is: Gewerbestrasse 11, 6330 Cham, Switzerland

Contents

Enhancing Computational Diagnostic Tools for Major Depressive Disorder in Children and Adolescents: Deep Learning Analysis of Resting-State and Functional Connectivity from High-Density EEG

Amir Jahanian Najafabadi and Khaled Bagh

Abstract This chapter explores the application of deep learning techniques specifically convolutional neural networks to identify neurological biomarkers associated with major depressive disorder in children and adolescents. It reviews foundational studies and highlights recent advancements in the field, with a focus on using resting-state functional brain connectivity derived from electroencephalography data to distinguish individuals diagnosed with major depressive disorder from healthy controls matched by age and sex. A convolutional neural network model based on the VGG16 architecture is employed for data analysis, following a comprehensive preprocessing workflow and the estimation of brain connectivity using a multivariate auto-regressive model combined with independent component analysis. The model is designed to identify meaningful neurological biomarkers across different frequency ranges and specific regions of interest within the brain. The chapter also explores various methods for interpreting the deep learning model's decision-making process, and compares the insights obtained from the convolutional neural network with results from conventional statistical approaches based on power spectral density and functional connectivity measurements analyzed separately. Furthermore, the potential use of this model as a diagnostic support tool is discussed, including its ability to classify individual participants based on short segments of their resting-state electroencephalography recordings. To enhance the model's accuracy and reliability, the chapter addresses advanced strategies such as refined evaluation methods, robust cross-validation procedures, and ensemble learning approaches.

A. Jahanian Najafabadi (✉) · K. Bagh
Department of Cognitive Neuroscience, Bielefeld University, Bielefeld, Germany
e-mail: amir.jahanian@uni-bielefeld.de

K. Bagh
e-mail: khaled.bagh@uni-bielefeld.de

K. Bagh
Department of Technology, Bielefeld University, Bielefeld, Germany

A. Ahmed and J. Picone (eds.), *Signal Processing and Biomedical Engineering Research*, https://doi.org/10.1007/978-3-032-18509-9_1

Ultimately, this work contributes to the growing body of research on the neurobiological underpinnings of major depressive disorder and supports the development of electroencephalography-based diagnostic tools tailored for pediatric and adolescent populations.

Keywords Electroencephalography · Children · Adolescents · Major depressive disorder · Resting-state functional connectivity · Power spectra density · MVARICA · CNN · Deep learning

1 Introduction

The early detection of depression has become a key focus in the scientific and medical communities, as this mental health condition affects over 350 million people worldwide—a number that continues to rise due to factors such as age [139, 146] gender [96, 100], and global events, e.g., COVID-19, war [58]. Depression is a complex heterogenous psychiatric disorder with multiple contributing factors, including genetic, biological, environmental, and psychological influences (for review, [6, 33]). It is classified in the *Diagnostic and Statistical Manual of Mental Disorders* (American Psychiatric Association: DSM-5-TR, 2022 [10]) into several distinct categories, such as major depressive disorder (MDD), postpartum depression, disruptive mood dysregulation disorder, persistent depressive disorder (dysthymia), premenstrual dysphoric disorder, and depressive disorder caused by another medical condition.

MDD is characterized by a wide range of symptoms that go beyond just feeling sad or down. Individuals with MDD often experience a persistent low mood, feelings of worthlessness or excessive guilt, and a loss of interest or pleasure in activities they once enjoyed (anhedonia) [108]. Cognitive difficulties such as trouble concentrating or making decisions, poor working memory, and slowed information processing are common in children and adolescents, along with physical symptoms like disrupted sleep patterns, chronic fatigue, and changes in appetite [24, 144, 160, 161]. These major symptoms are strongly correlated with brain and physiological activities which emphasize multidimensionality and complexity of the disorder [94, 116, 123, 156].

Additionally, MDD can cause significant emotional distress, leading to suicidal thoughts or behaviors in severe cases (for review, [94, 127, 129]). Recent studies (e.g., [27, 129]) indicated that adolescents experiencing depression often have difficulty directing or shifting their attention when processing emotional stimuli. They are also more likely to perceive situations negatively and recall past experiences in a broad, less detailed manner. However, there is limited evidence to suggest a widespread decline in overall executive function. In addition, while the biological roots of suicide in MDD are still being explored, research suggests that it's likely the result of a combination of different influences. These include genetic vulnerability, environmental and internal stress factors like relationship issues, work or financial problems, and co-occurring mental health conditions. Other contributing elements may involve changes at the epigenetic level, dysfunction in the brain's stress response system (the Hypothalamic–Pituitary–Adrenal axis), imbalances in key neurotransmitters especially serotonin as well as alterations in cholesterol levels, immune system markers, and brain chemicals like brain-derived neurotrophic factor and other neuromodulators (e.g., for review; [130]).

Research consistently highlights irregularities in various brain networks linked to adolescent suffering from depression, particularly involving the amygdala, medial prefrontal cortex, and ventral striatum, along with disruptions in their connectivity. Moreover, as reported by Bienek et al. [24], the neurocognitive deficits linked to the first episode of MDD (recent-onset depression) in adolescents compared to adults. Results of their study indicated that the impact of recent-onset depression on neurocognitive functioning is similar in both age groups, as lower scores in adolescents can be attributed to differences in age and education suggesting that neurocognitive impairment remains an overlooked aspect in clinical treatment guidelines for adolescent MDD (for review; [24]). Schumacher et al. [144] further reported lower performance in areas like working memory, long-term memory, attention, executive function, and language in children and adolescents with depression. Their results underscore the importance of evaluating cognitive functioning in depressed young age population and the need for further research into how treatment affects these cognitive skills. Due to its complexity and diverse presentation, diagnosing and treating MDD remains challenging, necessitating a comprehensive and personalized approach to care [22, 114, 123, 154, 166].

In this chapter, we focus specifically on an American children and adolescents diagnosed with MDD, one of the most severe and debilitating forms of depression. Unlike short-term or situational depressive episodes, MDD is a chronic condition that profoundly affects an individual's ability to function in daily life. It is classified as a unipolar disorder, meaning it does not involve manic or hypomanic episodes. Due to its persistent nature, MDD requires prediction, early diagnoses, long-term clinical care, often involving individualized treatment plans that may include psychotherapy, medication, or a combination of both (e.g., [86, 157]). While pharmacotherapy and psychotherapy remain the primary treatment approaches in many cases, they are not universally effective. Research indicates that nearly one-third of individuals diagnosed with MDD do not respond to these standard treatments, highlighting the urgent need for alternative or supplementary therapeutic strategies (for review, e.g., [34–36, 43, 82, 87, 88]). The challenge becomes even more complex when diagnosing and treating MDD in children and adolescents [35, 111]. While the prevalence of MDD in children and adolescents are major concern (e.g., see [102]), unlike adults, younger individuals undergo continuous cognitive, emotional, and physiological development, making it difficult to distinguish typical developmental changes from early signs of depression [46]. Factors such as hormonal fluctuations, puberty, and varying stages of emotional maturity further complicate the diagnostic process, increasing the risk of misdiagnosis or delayed intervention [8]. It is worth noticing that these mental health changes will later lead to further challenges as those prevent them from many typical experiences.

Given these challenges and health concerns, it is essential to develop advanced tools that complement existing diagnostic methods, allowing for more precise and reliable clinical evaluations. Enhancing diagnostic accuracy in younger populations will not only improve early intervention efforts but also contribute to more effective, personalized treatment strategies, ultimately improving patient outcomes and long-term well-being and other individuals' functionality.

1.1 The Rising Economic and Healthcare Impact of Major Depressive Disorder

Using the U.S. as a case study, the economic impact of MDD has increased significantly over the past three decades. In 1990 the economic burden of depression was $77.4 billion and later increased to $83.1 billion in 2000 (based on year 2000 values, [59]). During this time, the proportion of MDD cases among adults aged 18–34 grew significantly, jumping from 34.6 to 47.5% [59]. All components of this financial burden such as direct medical costs, suicide-related costs, and workplace-related expenses experienced growth, with workplace costs showing the most dramatic rise at 73.2% [59]. Expenses further increased significantly over the time and an amount of $173.2 billion was reported by 2005 and reaching $210.5 billion in 2010 (also in 2012 values). Direct medical expenses accounted for 35% of the total cost (down from 47% in 2010), while suicide-related costs made up 4% (a slight decrease from 5% in 2010). Workplace costs, on the other hand, rose to 61% of the total financial burden, up from 48% in 2010. This increase in workplace-related costs reflects improved employment opportunities for individuals with MDD. Despite these shifts, the proportion of overall costs directly attributed to MDD, rather than coexisting health conditions, remained relatively stable at 37%, compared to 38% in 2010.

The 2010 cost breakdown showed that 47% was attributed to direct medical expenses, 5% to suicide-related costs, and 48% to workplace-related losses [57]. Prior research [58] has reported that between 2010 and 2018, the number of adults in the U.S. diagnosed with MDD increased by 12.9%, rising from 15.5 million to 17.5 million. This rise in MDD cases was accompanied by a substantial increase in its economic burden. The total cost associated with MDD in adults increased by 37.9%, climbing from $236.6 billion to $326.2 billion (adjusted to 2020 values). The significant economic burden of MDD has far-reaching effects on policy decisions, society, and overall psychological or societal well-being. This growing impact highlights the urgent need for politicians, scientists and healthcare professionals to develop effective solutions in funding relevant research and in scientific implementations. Their efforts should focus on prevention, early detection, public education, and personalized treatment strategies. By improving monitoring and tailoring therapeutic plans to individual needs, scientists can help enhance productivity, quality of life, and overall mental well-being.

1.2 Current Measurement and Status of Major Depressive Disorder in Clinical Settings

Over the past two decades, researchers have increasingly turned to computational modeling and advanced algorithms to identify neuro-biomarkers of MDD using Electroencephalography (EEG). These efforts aim to improve diagnostic models and tools, either as standalone methods or in combination with traditional psychiatric evaluations, which typically rely on reported symptoms, life events, and psychological assessments [29]. These include instruments such as the Beck Depression Inventory (BDI; Beck et al. [20, 21]), the Reasons for Living Inventory (RFL; Pirani et al. [137]), the Self-Disgust Questionnaire (SDS; Olatunju et al. [128]), and the 17-item Hamilton Rating Scale, which is the most frequently used observer-rated measure for depression. These tools are used in accordance with DSM/ICD criteria [132, 143, 147].

2 Background

2.1 Resting-State Electroencephalography in Major Depressive Disorder

EEG is a widely used, non-invasive neuroimaging technique that measures electrical activity in the brain with exceptional temporal resolution. Due to its ability to track brain activity in real time, EEG has become an essential tool in both clinical and research settings, aiding in the diagnosis and study of various neurological, developmental, and psychiatric disorders. It plays a crucial role in understanding brain function by detecting patterns of electrical activity associated with different mental states, cognitive and motor processes in healthy and unhealthy individuals [71, 72, 77, 78]. EEG has several advantages, not only because of its high temporal resolution and real-time data, but also due to its cost-effectiveness, wide availability, and relatively simple setup requirements [117].

Unlike other neuroimaging methods such as fMRI or PET, which often require expensive equipment and controlled environments, EEG can be performed in more flexible settings, including outpatient clinics, schools, and even mobile research labs Banaschewski and Brandeis [14]. This makes it particularly valuable for large-scale studies or in contexts where access to advanced imaging facilities is limited. Furthermore, because EEG is non-invasive and well tolerated by children, it is especially suited for pediatric populations, where cooperation and minimal discomfort are critical [39]. These features make EEG a practical and powerful tool for investigating neural mechanisms across a wide range of conditions and age groups, and for potentially guiding diagnosis and treatment planning in real-world clinical practice [105].

One of the key advantages of EEG is its sensitivity to subtle changes in brain activity, allowing researchers and clinicians to differentiate between typical and abnormal neural functioning. By analyzing EEG recordings under specific conditions such as at rest, with eyes closed, with eyes open, or during cognitive tasks, it is possible to identify distinctive neural signatures associated with various disorders. This makes EEG particularly useful in exploring neuro-biomarkers for conditions like epilepsy, Autism Spectrume Disorder, ADHD, schizophrenia, and depression, ultimately contributing to the development of more effective diagnostic and therapeutic approaches [3, 23, 31, 103]. Several studies [38, 69, 100, 131, 140] have indicated that the brain's electrical activity, as measured by EEG during rest, is closely related to the symptoms observed in individuals with MDD. Using a technique called Fast Fourier Transform (FFT), which translates EEG signals from the time domain into the frequency domain, researchers can examine various brainwave frequencies namely delta (0.5–4 Hz), theta (4–8 Hz), alpha (8–13 Hz), beta (13–30 Hz), and gamma (above 30 Hz). Changes in the strength and pattern of these frequency bands have been linked to core features of depression, such as persistent sadness, loss of interest, difficulties with thinking and concentration, and altered levels of alertness. For example, lower alpha activity or increased theta waves in the frontal brain areas have been repeatedly observed in people with depressive symptoms. These findings suggest that resting-state EEG patterns may hold promise as indicators for diagnosing or tracking the progression of MDD.

For example, increased delta activity has been suggested to indicate a shift toward internal mental processing, especially during tasks that demand sustained cognitive effort [65]. Resting state Delta is often associated with the brain's default mode network (DMN), which plays a role in regulating cognitive control [83–85]. Some research has shown that delta rhythms can aid in internal monitoring and regulating attention, particularly when external stimuli are minimal [19]. For instance, studies involving meditation have demonstrated that delta activity increases during the suppression of distracting thoughts and external interference. Elevated delta power, especially in the medial prefrontal cortex during eyes-closed resting states, has been linked to reduced emotional and cognitive engagement [153]. Increased delta power has also been observed in children while performing semantic tasks suggesting that the rise in delta activity during mental tasks may be linked to a decrease in sensory input to the cortex, helping to reduce distractions and allowing for better focus on internal processes [64]. This pattern is particularly relevant to MDD, where individuals often experience heightened internal focus, such as rumination, and struggle with cognitive control [4, 79]. Abnormal delta activity may, therefore, reflect a brain that is less responsive to external stimuli, contributing to the cognitive and emotional disengagement frequently observed in depressive states [28, 121].

A recent study by McVoy et al. [115] revealed that in comparison with adults diagnosed with MDD, adolescents with depression exhibited lower overall brain coherence in the theta frequency band when compared to healthy controls. Additionally, researchers observed a reduction in functional connectivity within the frontal cortex at the theta frequency in the depressed adolescents' group [115]. Their study reported certain brain regions, such as the right prefrontal cortex and left parietal cortex, showed reduced connectivity across multiple frequency bands (delta, theta, and alpha) in individuals with MDD compared to healthy controls. Theta activity, as a slow brain wave, is closely linked to emotional responses and cognitive functions such as attention and memory. In individuals with MDD, disruptions in theta rhythms are often observed, as these patients frequently experience difficulties in regulating emotions [142], and struggle with cognitive tasks, including attention and memory [47]. Over time, these deficits can worsen, making it more challenging for individuals diagnosed with MDD, especially children and adolescents, to manage emotions and maintain focus when learning, which are essential for daily functioning. The altered theta activity in MDD may reflect underlying neural mechanisms that contribute to these cognitive and emotional struggles [53, 125].

Alpha waves have been a fundamental focus in EEG research ever since Hans Berger first identified them in 1929 [12]. These rhythmic patterns are most commonly seen when a person is awake but relaxed, especially with their eyes closed, and they remain one of the most easily recognizable features in EEG readings [17–19, 101, 170, 172]. Alpha activity is often used as an indicator of cortical excitability and is thought to reflect how well the brain maintains a healthy balance between excitatory and inhibitory signals [93].

Functionally, alpha rhythms have been tied to a range of mental processes. Studies have consistently linked them to attention, memory, and learning [26, 40, 113]. Alpha oscillations also seem to play a role in how emotional memories are stored and recalled [145]. In particular, alpha activity in the prefrontal brain regions has been associated with key cognitive abilities like emotional regulation, decision-making, and overall executive function—all of which are often disrupted in mood-related disorders. When it comes to MDD, changes in alpha wave patterns have been proposed as a potential neurophysiological marker. One area that's received a lot of attention is alpha asymmetry, which refers to the uneven distribution of alpha power between the two hemispheres of the brain. This imbalance, especially between the left and right frontal lobes, has been linked to emotional processing and is thought to reflect a vulnerability to depressed mood [120]. In individuals with depression, alpha activity tends to be elevated in regions tied to internal thought and self-reflection. This may interfere with the brain's ability to redirect attention away from negative thoughts, which could contribute to persistent rumination and emotional "stuckness" [119, 171]. Some researchers believe that this might be partly explained by increased activity in the default mode network (DMN)—a brain system involved in self-referential thinking—which is often more active than usual in people with MDD [28]. Although similar DMN disruptions have been noted in bipolar disorder [55], the nature of alpha oscillations in bipolar patients appears to be more inconsistent, suggesting that the underlying brain mechanisms differ somewhat from those seen in unipolar depression [150]. This distinction highlights the potential of Alpha wave research not just in understanding depression, but also in helping differentiate between various mood disorders. Altogether, alpha oscillations offer a valuable perspective on the brain's role in depression. Their involvement in attention, emotion, and cognitive regulation makes them a meaningful area of study for both understanding symptoms and developing new clinical tools.

Beta oscillations are sensitive to motor task components related to top-down processing and sensorimotor behavior [16, 44, 78]. Also, Beta oscillations are integral to motor learning, particularly through its involvement in the primary motor area (M1) and their functional connectivity with other brain regions. Additionally, beta activity is associated with active thinking, problem-solving, and focused mental tasks, all of which individuals with MDD often struggle with. When people engage in these types of activities, beta waves become particularly prominent in the brain. Reduction of beta power may reflect cognitive slowing, difficulties in attention and increased beta activity in certain brain regions can be indicative of overactive brain states that may contribute to anxiety or hyperarousal [125]. Prior research that revealed lower beta activity linked with MDD may indicate a compromised ability for cortical inhibition. Furthermore, the abnormal increase in beta phase synchronization seen in MDD might represent the brain's attempt to establish a new functional network to compensate for deficits in attention control, memory retention, and executive processing that are crucial during demanding working memory tasks [99]. Supporting this, another study [141] found that beta (12–30 Hz) and low gamma (30–50 Hz) activity in the fronto-central regions of the brain was inversely related to inattention. In simple terms, higher power in these frequency bands was associated with lower inattention scores. Additionally, overall symptom severity emerged as a strong predictor of inattention, with the BDI accounting for 23.9% of the variability in both inattention and anxiety scores.

Gamma rhythms are closely linked to heightened neuronal activity, particularly during the processing of sensory stimuli and the generation of action potentials [92, 162]. These oscillations are particularly important in the hippocampus, where they are strongly associated with moments of information encoding [48]. Gamma activity is integral to various cognitive and emotional processes [97], and disruptions in these rhythms may contribute to conditions like MDD, where altered neural processing is often observed. Research has highlighted the importance of gamma activity in the context of MDD (for review; [32]). When comparing MDD patients to healthy controls, significant reductions in gamma power were observed in the left temporal and both occipital regions of the brain. Additionally, gamma connectivity between the left hemisphere and the right frontal region was weakened in MDD patients. In the left temporal regions, there was also a noticeable decrease in beta-gamma phase-amplitude coupling. In patients with MDD, gamma-band connectivity between regions of the left hemisphere particularly the temporal and parietal cortices and the right frontal cortex was notably reduced [63]. For example, previous EEG work has documented decreased low-gamma coherence between regions such as the left pre-supplementary motor area (pre-SMA) and the right inferior frontal gyrus in MDD patients [63]. This disruption in long-range communication suggests a breakdown in the coordination between hemispheres, which may interfere with the brain's ability to regulate emotions and maintain cognitive control. The right frontal cortex, known for its involvement in executive processes and inhibitory control, typically depends on well-synchronized input from the opposite hemisphere. When this input is weakened, as observed in the MDD group, it can undermine the brain's capacity to manage mood and behavior effectively. This loss of synchrony in gamma activity may help explain some of the cognitive and emotional symptoms commonly seen in depression, including persistent negative thought patterns, difficulty shifting attention, and reduced mental flexibility.

Gamma waves are associated with high-level cognitive processes such as attention, memory integration, emotional regulation, and conscious perception (for review; [32]). Weakened gamma connectivity in this context suggests a disruption in the coordination and communication between brain regions that are crucial for emotional and executive functioning. Specifically, diminished interaction between the left hemisphere often linked to language, logical reasoning, and positive affect and the right frontal cortex which is known for its role in regulating emotions and social behavior may underlie some of the cognitive and emotional disturbances seen in MDD. This impairment in functional connectivity could contribute to symptoms such as indecisiveness, reduced emotional responsiveness, and impaired cognitive control that are frequently reported by individuals with depression.

Furthermore, lower gamma activity in the left occipital region was found to be negatively correlated with sleep disturbances, while disruptions in beta-mid-gamma and beta-high gamma phase-amplitude coupling in the left temporal regions were associated with cognitive difficulties [103]. These findings suggested a crucial link between gamma activity and the neural dysfunctions observed in MDD [103]. An EEG study by Pizzagalli et al. [138] found that individuals with higher depression scores, as measured by the BDI and the Mood and Anxiety Symptom Questionnaire, had reduced resting gamma activity in the anterior cingulate cortex. In contrast, another study reported an increase in gamma activity in the frontal and temporal regions when individuals with depression performed spatial and arithmetic tasks [149]. These contradictions have potentially arisen by the difference in study subjects and their level of depression severity.

To our knowledge there are a limited number of studies focusing on the differences in power spectra density (PSD) and resting-state functional connectivity (rsFC) in childhood and adolescents diagnosed with MDD. Due to the heterogeneity of MDD and the nonlinear, dynamic nature of EEG signals, capturing their complex, constantly shifting, and chaotic patterns using simple linear methods is highly challenging [1, 2]. Moreover, in many cases MDD symptoms overlap with other psychological disorders, e.g., anxiety [112, 168] and sleep disorder [124, 126].

2.2 Resting-State Functional Connectivity in Major Depressive Disorder

In the context of neuroscience, connectivity refers to dynamic interactions between different brain regions and is typically categorized into several key types, each offering unique insights into brain function. Among these, functional connectivity, spectral connectivity, phase-based connectivity, and effective connectivity have been particularly valuable in identifying biomarkers for psychiatric disorders such as

MDD [50, 95]. Functional connectivity captures statistical dependencies between EEG signals from distinct brain areas, often using measures like coherence or phase-locking value. Spectral connectivity focuses on these interactions within specific frequency bands, such as alpha and theta. Phase-based connectivity, using metrics like the Phase Lag Index, helps detect abnormal synchrony patterns while minimizing the confounding effects of volume conduction, thereby improving the specificity of potential biomarkers. Effective connectivity refers to the directional influence between brain regions and is typically quantified using approaches such as Granger causality. Collectively, these connectivity measures have deepened our understanding of neural mechanisms and hold significant promises as non-invasive biomarkers for diagnosis, prognosis, and treatment monitoring.

In recent years, researchers have increasingly turned to rsFC analysis using EEG to investigate how depression impacts brain networks [25, 51, 52, 54]. Unlike traditional EEG methods that concentrate on individual brain regions or specific event-related responses, rsFC examines how neural activity synchronizes across various brain areas over time. This broader approach offers a more complete picture of the brain's functional networks and helps identify the disruptions associated with MDD. By detecting patterns of either increased or decreased connectivity, rsFC has the potential to uncover EEG biomarkers that can distinguish individuals with MDD from healthy controls, ultimately enhancing diagnostic accuracy and paving the way for more personalized treatment options. Rigorous EEG-based classification models for MDD increasingly recognize the importance of balancing accuracy with generalization. Such models balance accuracy and required generalization, with functional connectivity in the theta frequency band revealed as a key role in improving robustness. Previous research has identified distinct connectivity patterns associated with depression (dysphoria). For instance, a study by Ghiasi et al. [52] demonstrated that combining spectral and functional connectivity measures yields the highest classification accuracy (83.91%), emphasizing the value of integrating multiple features in data sources to assess neural dynamics. However, the strength of a diagnostic model lies not just in peak accuracy but in its stability across varied samples and conditions. Prior research [50] revealed that in individuals diagnosed with depression, there was a noticeable imbalance in brain functional connectivity between the two hemispheres. Specifically, the left hemisphere showed a greater number and stronger short-range cortical functional connectivity, whereas the right hemisphere had more and stronger long-range of such connections. Some of these neural connections—particularly those in the alpha frequency range located in the front, back, and left regions of the brain, as well as short-range frontal connections in the theta range—were found to be positively linked to how severe the depressive symptoms were. This suggests they could potentially serve as indicators of depression severity. Overall, the study found that people with MDD showed a unique pattern of brain connectivity, especially in the right front and left back areas of the brain. Increased theta-band connectivity was also observed in frontal brain regions, with the right frontal and left posterior areas being the most affected in individuals with MDD. Similarly, Whitton et al. [166] reported stronger beta-band connectivity (12.5–18 Hz and 18.5–21 Hz) between the default mode network and the fronto-parietal network, suggesting that elevated high-frequency connectivity between these regions may serve as a neural marker for recurrent depression.

Using high-density EEG, Damborská et al. [37] reported that patients with moderate to severe depression showed significant neural alterations compared to healthy controls. Directed functional connectivity was notably higher in the right amygdala and caudate, suggesting increased neural communication in regions linked to emotional regulation and cognition. The study also reported elevated theta and alpha power across all examined brain structures, which may reflect disrupted neural oscillations associated with mood disturbances, potentially indicating compensatory or maladaptive brain activity.

There has also been significant work on analyzing the type of connectivity neuro-biomarkers important for MDD [60]. Their study demonstrates that MDD is associated with unstable, abnormal functional connectivity patterns. This was also corroborated by later research [81], where meta-analysis identified consistent, large-scale network dysfunctions in MDD. Moreover, this was agreed upon by later research [62], where the meta-analysis identified stable neural activation and connectivity patterns in MDD. These findings suggest that MDD is characterized by enduring neural alterations.

Connectivity and brain oscillation patterns, providing valuable insights into the neural mechanisms underlying the disorder. Recent research employed the phase lag index to construct functional connectivity matrices and used an altered Kendall rank correlation coefficient to select features with high discriminative power [136]. A range of classifiers were applied to distinguish between 27 MDD patients and 28 demographically matched healthy controls [136]. Permutation tests were utilized to evaluate the performance of these classifiers. The binary linear support vector machine (SVM) classifier achieved the highest classification accuracy, correctly classifying over 92% of participants using leave-one-out cross-validation for the full frequency band, with an impressive area under the curve (AUC) of 0.98. This strong classification performance suggests that functional connectivity features, combined with the appropriate machine learning models, hold significant potential for distinguishing individuals with depression from healthy controls. Additionally, the study revealed that depression leads to widespread alterations in brain activity, affecting nearly the entire cortex. The most significant changes in brain oscillation patterns were found in the delta, theta, and beta frequency bands, while changes in the alpha frequency band were less pronounced. These findings indicated that depression is associated with complex, widespread dysregulation in neural oscillations, with specific frequency bands playing a more prominent role in these alterations. Such findings highlight the growing interest in developing objective, data-driven approaches for detecting MDD-related brain alterations.

2.3 Study Objectives

In this study, we aim to build upon previous research to better understand the neural signatures underlying MDD in children and adolescents, comparing these findings to those of healthy controls. To enhance our knowledge in computational diagnostic models, in this chapter, we present the latest advancements in our research as well as comparison with prior studies conducted by colleagues in literature, which is dedicated to the classification of MDD in children and adolescents. Our study specifically compares U.S individuals diagnosed with MDD to sex and age-matched healthy controls, utilizing a multidimensional analytical approach. By applying advanced techniques to resting-state EEG data alongside detailed patient phenotypic and depression index information, we aim to enhance the accuracy and reliability of MDD classification and shed light for further research and implementation in clinical settings. This comprehensive analysis not only deepens our understanding of the neural and behavioral markers of MDD but also contributes to the development of more effective diagnostic tools for early intervention and personalized treatment strategies.

This chapter further builds upon our previous large-scale EEG studies aimed at enhancing classification-based models for identifying neuropsychiatric disorders, including MDD, in children and adolescents [74, 75, 78]. Understanding the neuro-biomarkers linked to MDD is crucial for improving both diagnosis and treatment, as these biological markers can provide valuable insights into the underlying mechanisms of the disorder. By integrating neurophysiological data with clinical assessments, medical professionals can use these biomarkers as a complementary or confirmatory tool alongside traditional diagnostic methods. This approach not only refines current diagnostic models but also paves the way for future applications in personalized and precision medicine, early intervention strategies, and more targeted therapeutic approaches for individuals affected by MDD [78].

Moreover, despite numerous studies, including our own, a reliable, multidimensional, and optimized model for accurately identifying abnormal biomarkers that distinguish MDD from healthy individuals especially in children and adolescents remains elusive. Variations in analytical methods, participant age groups, and especially sample sizes and challenges in generalizability and interpretation of results contribute to inconsistencies in the literature. To address this gap, we integrate rsFC analysis with deep learning models to classify normal and abnormal neuro-biomarkers using EEG data from young individuals diagnosed with MDD. By detecting abnormal rsFC patterns in specific frequency bands, our findings offer deeper insights into the neurophysiological disruptions associated with MDD, paving the way for improved diagnostic and clinical applications.

In the context of this chapter, we focus on Auto-Regressive (AR) models, where AR models have gained prominence for their ability to capture temporal dependencies and provide interpretable connectivity measures. AR models describe the current value of a signal as a linear combination of its past values. We use AR models to extract connectivity metrics like Partial Directed Coherence (PDC), where PDC is considered to be a combination of several connectivity types, including Effective Connectivity (as it utilizes Granger Causality) and Spectral Connectivity (as it in the frequency domain). For multivariate data, such as multi-channel EEG recordings, a MultiVariate Auto-Regressive (MVAR) model is used. The MVAR model for a system of N time series is defined in Eq. 1.

$$X(t) = \sum_{k=1}^{p} A(k)X(t-k) + E(t) \tag{1}$$

where $X(t)$ is an N-dimensional vector representing the observed signal at time, $A(k)$ is an N x N matrix of AR coefficients at lag k, p is the model order (number of past time points used for prediction), $E(t)$ is the residual error vector, assumed to be white noise.

The AR coefficients $A(k)$ are typically estimated using methods such as Yule-Walker equations or lease squares optimization [107]. These coefficients encode the temporal dependencies between the signals and serve as the foundation for computing connectivity measures.

A significant advancement in the application of MVAR models is the Multi-Variate Auto-Regressive Independent Component Analysis (MVARICA) framework [54], MVARICA integrates MVAR modeling with Independent Component Analysis (ICA) to address the limitation of MVAR. As the residual $E(t)$ is assumed to contain two components, Gaussian noise and uncaptured patterns by the MVAR model. These uncaptured patterns are assumed to be non-Gaussian and statistically independent, making ICA a natural choice for extracting independent sources. The MVAR coefficients are then transformed to characterize these independent sources in the residual errors. This coefficient transformation is illustrated in Eq. 2.

$$B(k) = M.A(k).U \tag{2}$$

where M and U are mixing and unmixing matrices from the ICA model. $B(k)$ is the transformed MVAR coefficients. By separating the mixing effects of volume conduction, MVARICA provides a more robust framework for estimating directed connectivity between brain regions.

Once the MVARICA model is fitted to the data, its coefficients can be used to compute various connectivity measures in the frequency domain, by applying Fourier transform to the coefficients. These measures differ in their sensitivity to directionality, frequency specificity, and robustness to noise. In this work, we focused on three different measures: Partial Directed Coherence (PDC), full-frequency PDC (ffPDC), and PDC factor. PDC and its related measures proved to be the most compatible with ML models [74, 75, 78].

PDC quantifies the directed influence of one signal on another while accounting for the influence of all other signals in the system [13]. It is defined in Eq. 3.

$$PDC_{ij}(f) = \frac{|A_{ij}(f)|}{\sqrt{\sum_{k=1}^{N} |A_{kj}(f)|^2}} \tag{3}$$

where i and j are two EEG channels, N is the number of signals (EEG channels or brain regions) in the multivariate system being analyzed, $A_{ij}(f)$ is the Fourier transform of the AR coefficients:

$$A(f) = I - \sum_{k=1}^{p} A(k)e^{-i2\pi fk} \tag{4}$$

where I is the identify matrix and p is the model order. Note that Eq. 4 is derived from rewriting the MVAR model in the time domain equation into the frequency domain, where the time-domain model is rewritten fro Eqs. 1 to 5.

*

$$X(t) - \sum_{k=1}^{p} A(k)X(t-k) = E(t) \tag{5}$$

where taking the Fourier transform of both sides gives Eq. 6.

$$(I - \sum_{k=1}^{p} A(k)e^{-i2\pi fk}).X(f) = E(f) \tag{6}$$

where $X(f)$ and $E(f)$ are the Fourier transforms of the EEG time-domain signal and the residual respectively. PDC is normalized to range between 0 and 1, with higher values indicating stronger directed connectivity. However, PDC is sensitive to the scale of the signal, which can limit its interpretability. To address the scale sensitivity of PDC, the ffPDC was introduced, ffPDC normalizes the PDC scale across all frequencies, making it scale-invariant:

$$ffPDC_{ij}(f) = \frac{|A_{ij}(f)|}{\sqrt{\sum_{f'} |A_{ij}(f')|^2}} \tag{7}$$

wheref' represent all possible frequency combinations (i.e., from 0 Hz to the Nyquist frequency).

The PDC factor incorporates the residual error covariance matrix Σ to account for the noise structure in the system, providing a more robust estimate of the directed connectivity:

$$PDC\,factor_{ij}(f) = \frac{|A_{ij}(f)|}{\sqrt{A_j^H(f)\Sigma A_j(f)}} \tag{8}$$

where $A_j^H(f)$ is the j-th column of $A(f)$, and Σ is the covariance matrix of the residuals $E(t)$. The PDC factor is less sensitive to noise correlations and provides a more accurate representation of true connectivity.

3 Materials and Methods

3.1 Dataset

For this study, we analyzed resting-state EEG data from 329 children and adolescents aged 5 to 21. Of these, 88 had been diagnosed with MDD, while the remaining 241 were not given any diagnosis, classified as healthy and labeled as HBN by Child Mind Institute. The data came from the publicly available Healthy Brain Network (HBN) dataset [5]. The demographic data of the participants were also provided by the HBN institute, as shown in Table 1, where the average age of healthy dataset (labeled as HBN) is 9.94 ± 3.50, while MDD's average was 15. ± 2.77. While there were other phenotype data available, such as medications and Child Depression Index surveys, most of the participants opted out of doing either, therefore, that information is not included in this chapter.

The recordings were taken using a 128-channel EEG (Table 2) HydroCel Geodesic system (Electrical Geodesics Inc.), with a sampling rate of 500 Hz and a frequency range of 0.1 to 100 Hz. However, after removing the outer channels—typically excluded due to their susceptibility to noise, poor signal quality, and limited relevance to cortical activity—only 109 channels were retained for analysis (cf., Table 3). The arrangement of EEG electrodes on the scalp is shown in Fig. 1a, divided based on Regions of Interest (ROI, Fig. 1b), and a summary of the dataset is depicted in Table 4.

The seven brain regions used for grouping EEG channels in Table 1—frontal right and left, temporal right and left, central, and occipital right and left—were chosen based on their established relevance to the neurobiology of MDD, particularly in pediatric and adolescent populations. Frontal regions, especially the left and right

Table 1 Demographic information across healthy (HBN) and MDD, showing the number of samples per group, and showing the mean and standard deviation of the age group

Class	Children	Adolescent	Male	Female	Height	Weight	BMI
HBN	178 (8.57 ± 2.14)	63 (15.5 ± 2.21)	134	107	136.4 (±18.7)	37.6 (±19.0)	18.7 (±4.7)
MDD	10 (10.1 ± 1.21)	78 (16.1 ± 2.15)	33	55	160.4 (±12.0)	63.2 (±21.4)	23.4 (±6.3)

Table 2 128 Channels grouped into 7 clusters based on the 7 regions of the brain

Region	Channels
Frontal right lobe	E1, E2, E3, E4, E5, E8, E9, E10, E14, E118, E123, E124
Frontal left lobe	E11, E12, E15, E16, E17, E18, E19, E20, E21, E22, E23, E24, E25, E26, E27, E32
Temporal right lobe	E86, E92, E93, E97, E98, E100, E101, E102, E103, E104, E108, E109, E110, E111, E114, E115, E116, E117, E121, E122
Temporal left lobe	E28, E29, E33, E34, E35, E36, E38, E39, E40, E41, E42, E44, E45, E46, E47, E50, E51, E52, E53, E57
Center	E5, E6, E13, E30, E31, E37, E54, E55, E79, E80, E87, E105, E106, E112
Occipital right lobe	E76, E77, E78, E82, E83, E84, E85, E89, E90, E91, E95, E96
Occipital left lobe	E58, E59, E60, E61, E62, E64, E65, E66, E67, E69, E70, E71, E72, E74, E75

frontal lobes, have long been associated with emotional regulation, executive functioning, and cognitive control domains frequently disrupted in MDD. For instance, frontal alpha asymmetry, a well-replicated EEG marker of depression, often reflects reduced left frontal activity relative to the right, which is associated with anhedonia and withdrawal [68, 155]. The temporal lobes play a key role in emotional memory, language, and social cognition. Altered activity in the right and left temporal regions has been linked to negative emotional processing and rumination, which are commonly observed in individuals with depression [42, 73]. The central region includes sensorimotor areas and is involved in integrating internal and external information. Dysregulation here may reflect disruptions in the balance between cognitive and emotional networks, particularly in younger populations whose brain systems are still developing [106, 169]. Lastly, the occipital lobes are primarily responsible for visual processing. While not traditionally emphasized in depression research, several studies have found abnormal occipital EEG patterns in MDD, possibly reflecting changes in attention and perceptual bias [61]. Grouping electrodes by these functional ROIs provides a neurobiologically meaningful way to interpret connectivity differences and aligns with both developmental neuroscience and clinical research in mood disorders.

3.2 Pre-processing

The data pre-processing was divided into two main stages: (1) EEG signal pre-processing and (2) computation of the connectivity model and extraction of connectivity measurements. The EEG pre-processing steps are illustrated in Fig. 2. The process begins with loading the raw EEG data using the MNE library in Python,

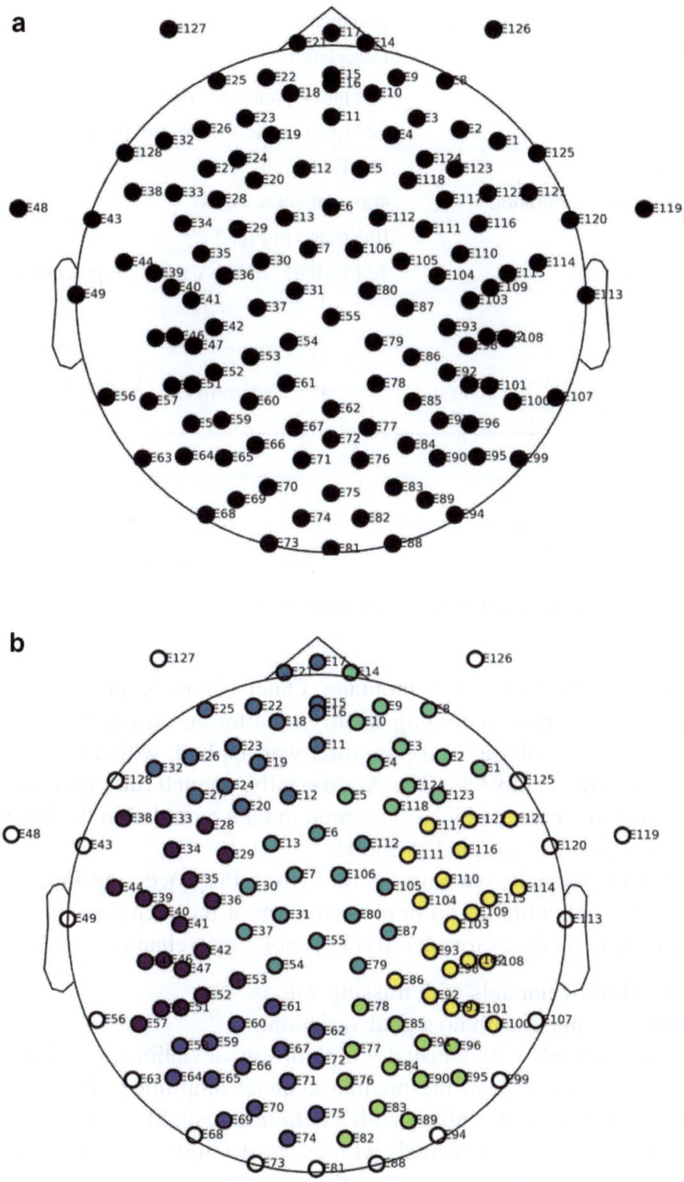

Fig. 1 **a** Channel location of 128-channels in the HBN dataset; **b** Regions of Interest (ROI)

Table 3 Excluded outer channels

Outer channels
E43, E48, E49, E56, E63, E68, E73, E81, E88, E94, E99, E107, E113, E119, E120, E125, E126, E127, E128

Table 4 The healthy brain network dataset parameters

Parameter	Description
Subjects	329 participants (88 MDD, 241 HBN)
Channels	128 raw channels (109 after pre-processing)
Sampling rate	500 Hz
Avg. Dataset recording duration	400 s (about 6.7 min)
Pre-processing	Illustrated in Fig. 2
Dropped datasets	12 (5 MDD, 7 HBN), after pre-processing

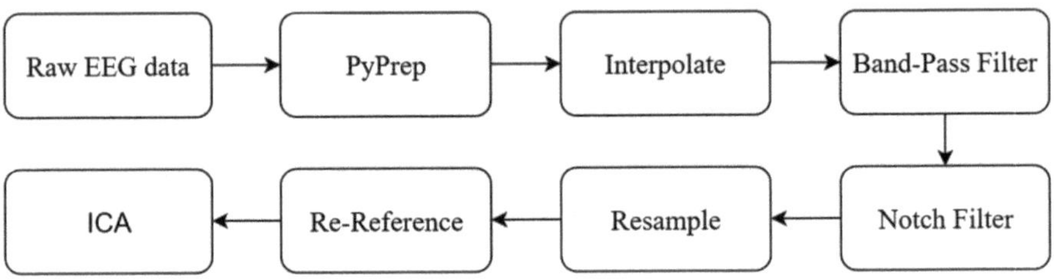

Fig. 2 Flowchart of the pre-processing and cleaning of EEG data

applying the GSN-HydroCel-128 montage. Outer channels and any zeroed channels—resulting from poor recording quality or filtering during data acquisition—were removed. A zero-phase band-pass filter was applied, with a low-pass cutoff at 70 Hz and a high-pass cutoff at 1 Hz. Additionally, a notch filter at 60 Hz is used to eliminate power line noise, which is common in data recorded in the USA. The data was subsequently down-sampled to 256 Hz.

To detect bad channels, we employ the Python Preprocessing Pipeline for EEG Data (PyPREP) [25], which is an implementation of the prep pipeline in Python. It integrates multiple methods for identifying problematic channels, including:

- **Missing Values**: Channels with missing values.
- **Flat Lines**: Channels with no signal variation.
- **Deviation**: Channels with unusually high or low deviations, i.e., if its amplitude deviates significantly from the median channel amplitude. This is determined using a robust Z-scoring method with a default threshold of 5.0
- **High-Frequency Noise**: Channels with excessive high-frequency activity, i.e., if its high-frequency content is significantly greater than the median across channels. This is also determined using a robust Z-score, with a default threshold of 5.0.
- **Correlation**: Channels with low correlation to other channels, indicating poor skin conductance or malfunction.
- **Signal-to-Noise Ratio (SNR)**: Channels with low SNR.
- **Dropouts**: Channels with intermittent signal loss, i.e., channels that frequently lost signal were marked as bad.

- **Random Sample Consensus (RANSAC)**: RANSAC is a robust method for detecting outliers. It identifies channels that consistently deviate from the expected signal pattern based on a consensus of other channels [25].

The PyPREP pipeline was executed using three iterations of the RANSAC algorithm, applied in a sample-wise manner (i.e., across time samples for each channel, instead of across channels for specific time). During each iteration, RANSAC identified channels whose signals consistently deviated from predictions based on subsets of other channels, flagging them as outliers. Once the bad channels were detected from the pipeline, they were interpolated using data from the surrounding clean channels to restore their signals. Following interpolation, the entire dataset was re-referenced by computing the average signal across the remaining non-bad channels.

Finally, ICA is applied using the FastICA model, with the number of components computed as shown in the equation below

$$Components_n = Channels_n - Channels_{bad} - 1 \tag{10}$$

where $Components_n$ is the number of possible components, $Channels_n$ is the number of channels and $Channels_{bad}$ is the number of bad channels that could not be interpolated. The ICA model is then fitted to the EEG data and used to identify and remove artifacts, where bad artifacts in this context refer to unwanted signals that can distort the EEG signal. These can include signals from muscle movements, heartbeats (ECG), eye movement (EOG), and technical or environmental noise (e.g., poor electrode contact). These artifacts are considered "bad" because they can obscure the true EEG signal and reduce the signal-to-noise ratio (SNR). Artifacts related to EOG were detected by Magnetoencephalography and Electroencephalography in Python (MNE-Python) detection methods [56], using Pearson correlation, based on adaptive z-scoring in the frequency range of 1–10 Hz. ECG related artifacts were also detected using MNE based function for detecting such artifacts, through Cross-Trial Phase Statistics, between frequencies 8–16 Hz.

3.3 Power Spectral Density

Following the pre-processing and artifact removal steps, we performed Power Spectral Density (PSD) analysis on the cleaned EEG data to examine the distribution of signal power across different frequency bands, across ROIs. PSD was computed using Welch's method [165], as implemented in MNE library. The number of points for the Fast Fourier Transform (FFT) is 4096, which gives a frequency resolution of 0.0625 Hz. with segments overlap of 50%.

The PSD computation was then used to compute the absolute and relative power, for the different frequency bands (Delta: 1–4 Hz, Theta: 4–8 Hz, Alpha: 8–12 Hz, Beta: 12–30, Gamma: 30–70 Hz), with the absolute power defined as

$$ABS = \sum_{i=f_0}^{f_n} PSD_{welch}(x_t) \tag{11}$$

where $PSD_{welch}(x_t)$ is the PSD Welch power, computed on time series EEG data x_t, with f_0 and f_n being the borders of the frequency range.

Relative power is computed as

$$REL = \frac{\sum_{i=f_0}^{f_n} PSD_{welch}(x_t)}{\sum_{i=1}^{70} PSD_{welch}(x_t)} \tag{12}$$

Which is the ratio of PSD of the specific frequency range over the full frequency range (0–70 Hz).

We also further computed the absolute and relative power and averaged over channels, to compute the ROI-based PSD values. The PSD values then underwent statistical analysis; using R-based Jamovi (v 2.3.28) software. The latter, we fitted a Linear Mixed Model (LMM) into the data, with relative power as the dependent variable, and with frequency band, disorder label (HBN/MDD) and ROI as independent variables and participants (cf., section IV-C).

3.4 Resting-State Functional Connectivity

After pre-processing and cleaning the EEG data, we segmented it into non-overlapping intervals using distinct time thresholds of 15 s. For each segment, a MVARICA model was applied to estimate connectivity patterns. The model coefficients derived from these segments were then used to calculate functional connectivity measures.

The hyperparameters of the connectivity model were first optimized to ensure robust performance. To achieve this, the segmented EEG data was further divided into one-second epochs, resulting in 15 epochs per segment. The model order parameter was optimized by minimizing the mean squared generalization error using leave-one-out cross-validation (LOOCV). The search range for the model order was set between 10 and 50, where this range allows MVARICA to focus more on capturing longer and more stable connectivity patterns. Another parameter of interest is the delta parameter, used in ridge regularization [135], as MVAR models fitting is ill-posed with high-dimensional/colinear data (e.g., EEG), leading to unstable coefficient estimates. Here, Tikhonov (L2) regularization is applied to the MVAR coefficient matrix to prevent overfitting. The delta parameter was optimized using the bisection search method [158].

This optimization process was performed independently for each EEG segment to account for the non-stationary nature of the EEG signal. Once optimized, the MVARICA model was fitted to the original EEG segments [54], and a variety of connectivity measures were computed from the fitted model [45]. These measures included PDC, ffPDC, and PDC factor, all calculated bidirectionally between all channel pairs.

The connectivity measures were computed with a frequency resolution of 0.1024 Hz, i.e., 2500 points, covering the range from 0 Hz to half the sampling rate (0–128 Hz), as shown in Fig. 3a. The results were then divided into the frequency bands: Delta, Theta, Alpha, Beta, and Gamma. Each connectivity measure in a frequency band was used to generate an image (as shown in Fig. 3b), where rows and columns correspond to different EEG channels, and their intersections represent the connectivity strength between those channels. Channels were organized according to ROIs: FL, FR, TL, TR, OL, OR, and Center.

The 15-s threshold, which is relatively long compared to the more commonly used 2–5 s segments, was selected based on findings from our prior research [15], Jahanian [75]. These findings indicated that longer segments enhance the performance of rsFC and auto-regressive models in classifying MDD; since the longer segments provide more data for the MVARICA model, it enhances the resolution, which is necessary when focusing on lower frequency bands (e.g., delta).

Additionally, we opted for non-overlapping segments to ensure the independence of the connectivity measurements [30, 118]. This approach is particularly advantageous for machine learning applications, as it reduces redundancy and enhances the generalizability of the derived features. That's especially the case when talking about CNNs, which are designed to learn hierarchical features from the data, and highly correlated inputs may hinder the model's ability to identify distinct biomarkers.

3.5 Architecture of CNN

The CNN models utilized here are based on VGG16, pretrained on ImageNet dataset, the model setup and hyper-parameter are based on our previous work [75], which is a basic VGG16 model, similar to the architecture illustrated in Fig. 4, with top layers removed (up to block L.11). The top layers are replaced with a flattening layer, followed by a batch normalization layer, then two fully connected layers with 512 units and ReLU used as activation function. Then a dropout layer of 0.2 was added, followed by another fully connected layer of 512 units, finally, a decision layer with sigmoid as activation function. The input of the model is of shape (218, 327, 3), with the image duplicated across the three channels.

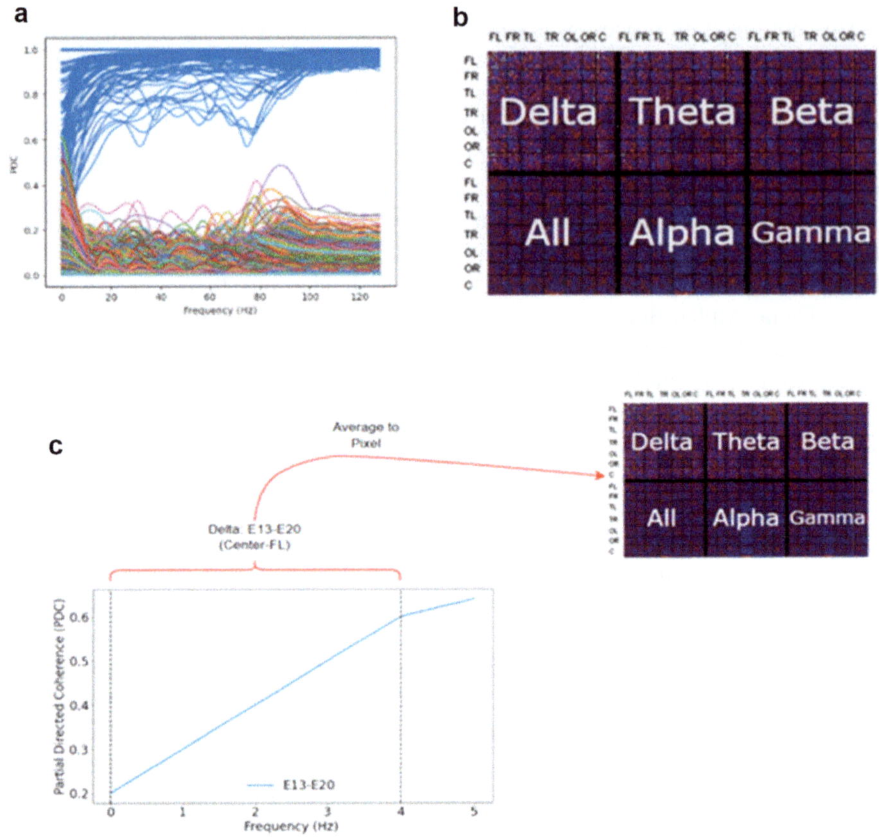

Fig. 3 **a** Example of computed PDC connectivity for a single segment. **b** Example of processed connectivity measure images, which is used as input to the CNN modules. **c** Example of how each pixel is generated in the image

3.6 ROI Analysis

When it comes to extracting features importance from the models, a method that focuses on global feature and another that focuses on instance-level explanations were utilized, mainly Permutation Feature Importance (PFI) [7] and Integrated Gradient [152].

PFI quantifies the contribution of each feature in a trained machine learning model by assessing the impact of feature perturbation on model performance. In this context, the values of a single ROI interaction are randomly shuffled while all other features remain unchanged. The resulting decrease in model performance—measured using the F1-score—indicates the importance of the shuffled feature. This procedure is repeated ten times to ensure robustness and reliability in the estimation of feature importance.

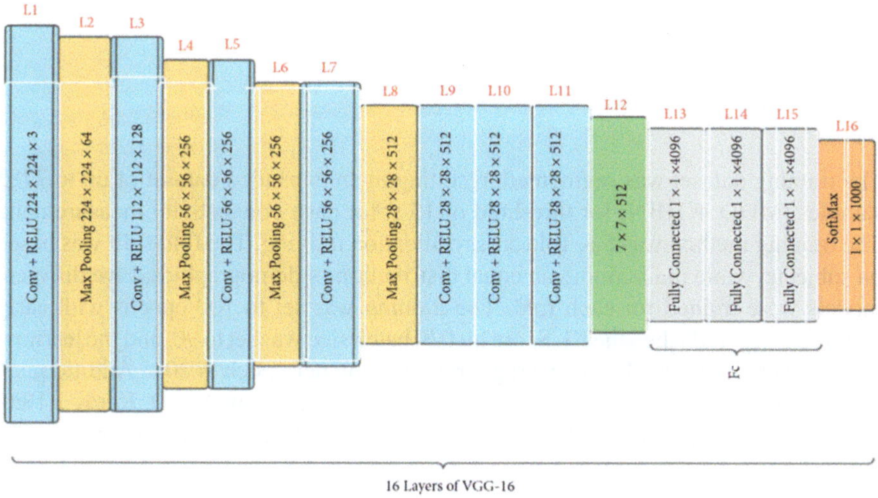

Fig. 4 Convolutional Neural Networks' (CNNs) VGG16 architecture [66]

Integrated Gradients is a popular explainability method, which was used in this work to interpret the predictions of a CNN model on the test data. The method computes the gradients of the model's predictions with respect to interpolated inputs between a baseline (zero tensor here) and the actual input. The integrated gradient is defined as

$$IG_i(x) = \left(x_i - x_i'\right)\frac{1}{m}\sum_{k=1}^{m}\frac{\vartheta F\left(x' + \frac{k}{m}\left(x - x'\right)\right)}{\vartheta x_i} \tag{13}$$

where $IG_i(x)$ is the integrated gradient of input i, x_i is the value of input i, x_i' is the value of input i in the baseline input x', $F(.)$ is the model output function, m is the number of steps used to approximate the integral (50 steps). The gradients are averaged across interpolation steps and multiplied by the input to produce attribution maps, which quantify the contribution of each input feature to the model's output. The results are aggregated across test data used to produce a single score, that expresses the average importance of a single pixel (connectivity channel interaction), this is then averaged across channels to get a score for each ROI, in each frequency band.

Both techniques provide insights into the model's decision-making process, with PFI focusing on global feature rankings (e.g., for feature selection), which is also a model agnostic method. Meanwhile, Integrated Gradients quantify the contribution of specific channels to the model's predictions. The advantage of using such method is that it also provides information on the influence direction, where a positive score of a specific feature, indicates that higher value of the feature pushes the model decision-making process toward the positive class, while a negative score, indicates the opposite. This method, however, is limited to differentiable models, as it relies on computing gradients to attribute feature importance.

4 Experimental Results

4.1 Setup

The training dataset was segmented, resulting in an expanded dataset of up to 7783 samples (MDD + HBN) for threshold of 15 s, for each connectivity measurement. The training method was 5 K-fold cross-validation training. The fold split was based on subjects, while maintaining an equal distribution of demographics. Five different models were trained for each fold. The training was set to 100 epochs with early stopping triggered when the F1-Score hit 0.8, batch size was set to 90, and the learning rate was set to 0.000004. The average training time per epoch is 20 s. Additionally, the training of CNN was done in Python 3.10.12, Tensorflow 2.18.0, Keras 2.18.0, scipy 1.10.1. Weights for each class were added according to the equation below

$$W_j = \frac{n_{samples}}{n_{classes} \cdot n_{jsamples}} \tag{14}$$

where W_j is the weight of class j, $n_{samples}$ is the number of training samples available in the data, $n_{classes}$ is the number of classes (2 classes), $n_{jsamples}$ is the number of samples in the dataset, belonging to class j.

4.2 Results

The results of training the CNN module are presented in Table 5, with the best test performance achieved using FFPDC, yielding an F1-score of 0.815, an MCC of 0.753, and Cohen's Kappa of 0.746. These results indicate that the model was successfully learned despite the dataset's imbalanced nature. The high performance of FFPDC aligns with our previous findings [75], where FFPDC effectively preserves relative amplitude differences across frequencies—features that the CNN can leverage for classification. Additionally, FFPDC is robust to scale, making it particularly suitable for resting-state EEG, where low brain activity produces low connectivity values. PDC Factor emerged as the second-best performing metric, with an F1-score of 0.782, an MCC of 0.713, and Cohen's Kappa of 0.703. This performance may be attributed to PDC Factor's ability to suppress noise effectively. Notably, the fact that PDC Factor outperformed PDC suggests that the MVARICA model did not overfit, even with an increased model order optimization search range of 10–50, instead of restricting it to lower values.

For our further analysis, we primarily focus on testing the model based on FFPDC. Table 6 presents the CNN module's performance across demographic groups of males and females, where test scores are expected to remain largely consistent. However, we opted not to conduct testing based on age groups, as only 10 MDD datasets belonged to children. This limited sample size could prevent the model from detecting potential

Table 5 CNN testing score on PDC, FFPDC, and PDC factor

Connectivity measurement	Loss	F1-Score	Sensitivity	Precision	Specificity	MCC	Cohen Kappa
PDC	0.334	0.705	0.706	0.734	0.913	0.620	0.609
FFPDC	**0.262**	**0.815**	**0.848**	**0.797**	**0.922**	**0.753**	**0.746**
PDC factor	0.293	0.782	0.843	0.746	0.903	0.713	0.703

biomarkers and would also provide insufficient data for the explanation method to effectively highlight relevant biomarkers.

Figure 5a presents heatmaps of the ROIs derived from the Integrated Gradients method across different frequency bands. The importance score for each ROI was calculated by summing the Integrated Gradients scores for all connectivity interactions involving that ROI across frequencies. Higher scores in the heatmap (red) indicate that an increase in FFPDC values in the corresponding ROI contributes more to the positive class (MDD), whereas lower scores (blue) suggest that an increase in FFPDC values contributes to the negative class (HBN). The heatmap reveals that MDD exhibits higher FFPDC values in the Center and OL ROIs, while HBN shows higher connectivity values in the FR ROI. Other ROIs display a slight bias toward HBN, with higher FFPDC connectivity values.

Table 6 Test score of the CNN module on the test data on specific demographic groups (Male and Female)

Group	Data size	F1-score	Sensitivity	Precision	Specificity	MCC	Cohen Kappa
Male	24 HBN + 8 MDD	0.824	0.848	0.813	0.848	0.759	0.754
Female	23 HBN + 8 MDD	0.800	0.849	0.776	0.918	0.740	0.730

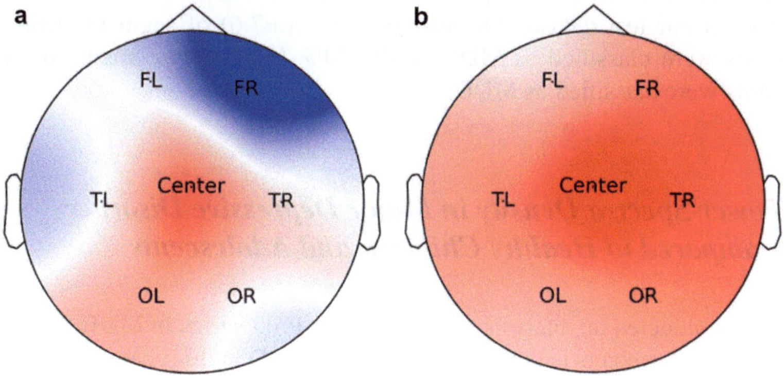

Fig. 5 **a** Summed Weights of Batch Integrated Gradient across frequency bands and ROI interactions. **b** Averaged Weights of PFI across frequency bands and ROI interactions

In Fig. 5b, a, heatmap generated using the PFI method illustrates the impact of each ROI on model performance. The importance score for each ROI was determined by computing the mean drop in the F1-score across all interactions in which the ROI was involved across different frequencies. The heatmap clearly highlights that the Center ROI had the most significant impact on the F1-score, indicating its crucial role in classification performance.

Figure 6 provides a detailed visualization of the normalized inter-ROI interactions across frequencies using the integrated gradient method. The results reveal distinct frequency and ROI-dependent patterns: Delta is characterized by Center-FL and Center-OR HBN biased interactions, with Center-FR, Center-TL interactions, that are MDD biased. Theta has the highest inter-ROI interaction, with the strongest interaction being OR-TL, followed by OR-FR, which are HBN biased. OL-FL, FR-FL, and FL-Center are MDD biased. Alpha shows OR-FR and OR-FL interactions that are HBN biased, with FL-TL, FL-OL, FL-FR, FL-Center and Center-TL interactions that are MDD biased. Beta does not exhibit any strong bias, except for FL-FR interaction, which is HBN biased. Gamma also shows no strong interaction, except for FL-TL toward the MDD class. Finally, in the combined frequency range, no strong bias is detected, with the strongest being OL-FR with MDD bias.

Figure 7 illustrates the normalized inter-ROI interactions using the PFI method across different frequency bands. Notably, within each frequency band, certain ROI interactions prominently contribute to the drop in the F1-score, indicating that the CNN model assigns particular importance to these specific ROIs. Additionally, interactions involving the Center ROI consistently appear across all frequency bands, highlighting its critical role in the model's decision-making process. In Delta, Center-TR, Center-OL, Center-OR, Center-FL, TL-OR, and TR-FR are highlighted. In Theta, Center-TL, Center-OL, Center-OR, OR-FR, OR-TR, OR-FL are highlighted. In Alpha, only TR-OL and Center-OL are highlighted. In Beta, Center-TL, Center-TR and TL-FR are highlighted. In Gamma, Center-TL, Center-TR, Center-FR, TR-FL and OR-OL are highlighted. In the combined frequency bands, Center-OL, OR-TL and TR-FL are highlighted.

Figure 8 presents the percentage of segmented data classified as MDD for each participant in our test dataset. On average, 45% ($\pm$7.6) of segments from MDD participants were classified as MDD, while 24% ($\pm$8.3) of segments from HBN participants were classified as MDD.

4.3 Power Spectra Density in Major Depressive Disorder Compared to Healthy Children and Adolescents

We further conducted an analysis to test whether PSD varies in MDD compared to healthy matched subjects using LMM. The use of PSD in this chapter was based on the assumption that depressive symptoms are linked to specific changes in brain oscillatory activity. PSD allows for a clear and structured analysis of how neural activity

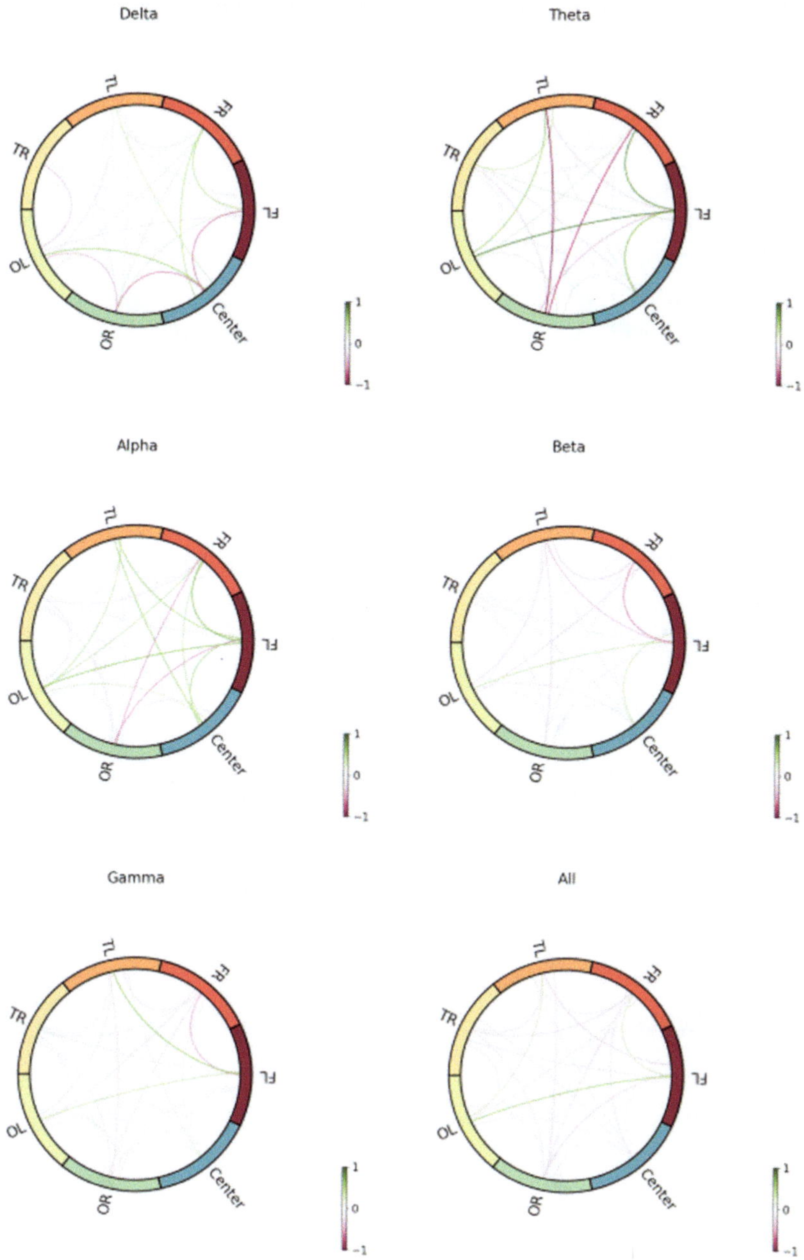

Fig. 6 Normalized Integrated Gradient bidirectional ROI importance factor, for each frequency band. Where positive value indicates positive bias of the ROI interaction toward the MDD class

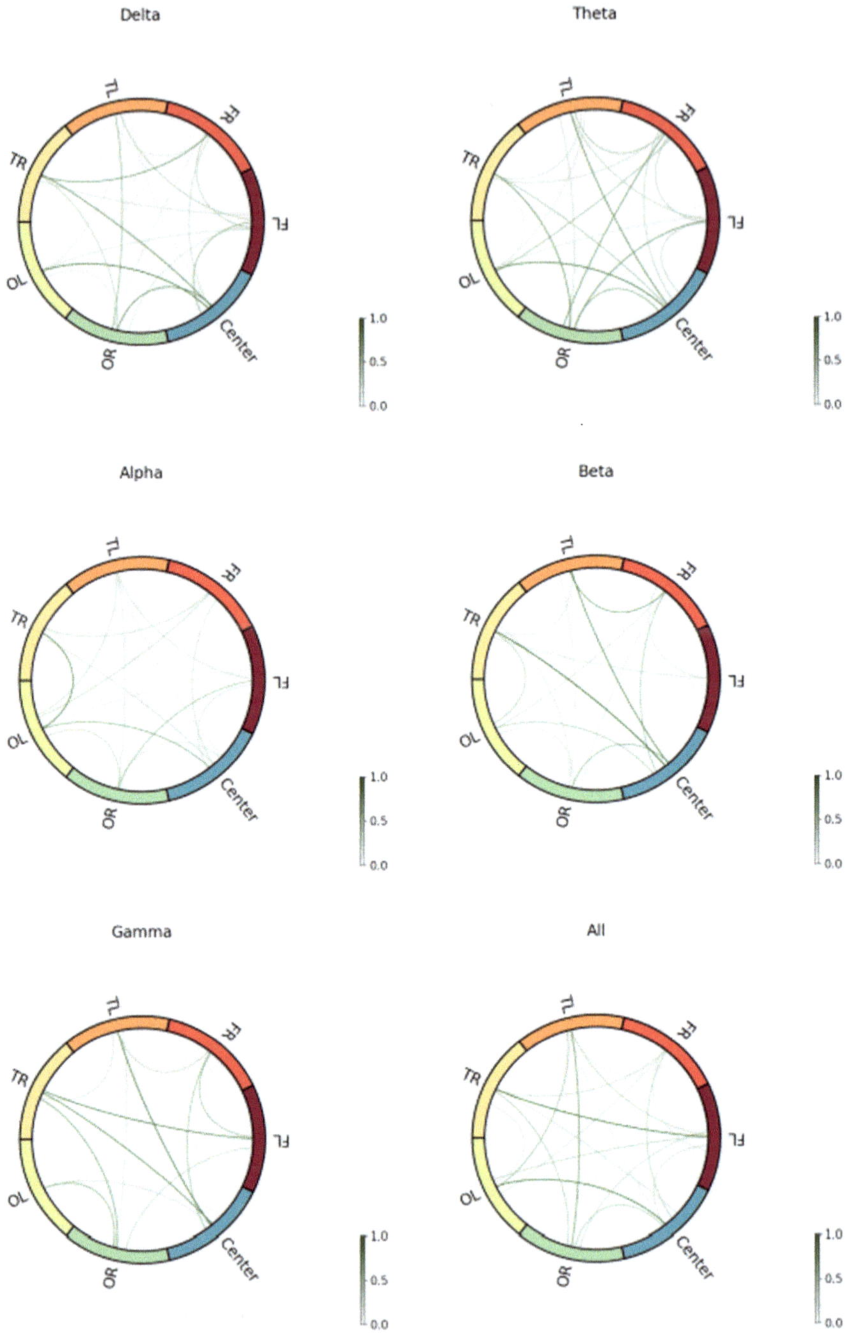

Fig. 7 Normalized PFI bidirectional ROI importance factor, for each frequency band

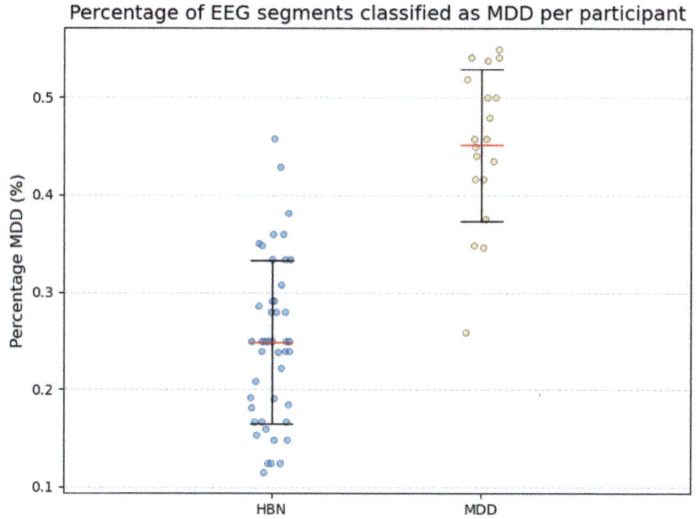

Fig. 8 Percentage of participants segments classified as MDD or HBN, for each participant

is distributed across these frequency bands. By quantifying the power within each band, we investigated potential neurophysiological differences between individuals with MDD and healthy controls. To address this question, we began by investigating whether relative PSD differed by gender (male vs. female). Figure 9 illustrates the gender-based Relative PSD per frequency band. The LMM analysis revealed no significant differences in PSD between the two gender groups.

4.3.1 Descriptive Results for Gender-Related Relative Power Spectra Density

In the HBN sample, relative EEG power values across frequency bands revealed consistent differences between males (n = 504) and females (n = 399). Males exhibited higher relative power in the delta band (M = 0.7614, SE = 0.00810) compared to females (M = 0.6993, SE = 0.00942), indicating stronger slow-wave activity at rest. In contrast, females showed greater relative power in the alpha band (M = 0.1118, SE = 0.00542) than males (M = 0.0937, SE = 0.00468), suggesting enhanced cortical idling or inhibitory processes in the female group. Similarly, beta band power was slightly elevated in females (M = 0.0599, SE = 0.00250) relative to males (M = 0.0546, SE = 0.00592), while gamma power was notably higher in females (M = 0.0384, SE = 0.00404) than in males (M = 0.0163, SE = 0.00081), pointing to possible sex-related differences in high-frequency oscillatory activity. Theta band power was also somewhat increased in females (M = 0.0906, SE = 0.00230) compared to males (M = 0.0741, SE = 0.00194), further supporting the pattern of relatively greater power in mid-range frequency bands among females.

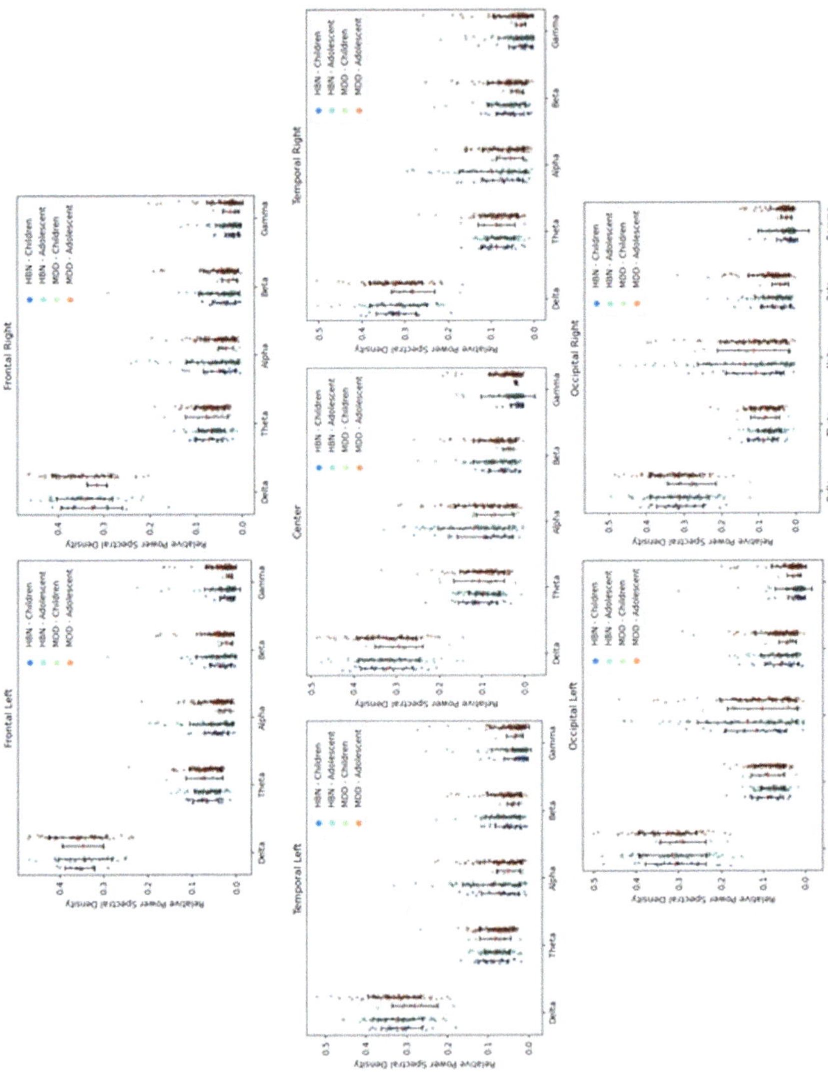

Fig. 9 Age-related relative power spectra density

In the MDD group, males (n = 224) and females (n = 385) displayed distinct spectral profiles. Delta power was again higher in males (M = 0.6509, SE = 0.01304) compared to females (M = 0.6144, SE = 0.00784), though the overall values were reduced relative to the HBN sample. Alpha power remained elevated in females (M = 0.1541, SE = 0.00624) compared to males (M = 0.1506, SE = 0.00905), consistent with the findings from the healthy group. Beta band power was comparable, with slightly higher values in males (M = 0.0860, SE = 0.00810) versus females (M = 0.0813, SE = 0.00248). Gamma power followed a similar pattern to the HBN group, with females exhibiting higher values (M = 0.0429, SE = 0.00203) than males (M = 0.0326, SE = 0.00188). Theta band activity was greater in females (M = 0.1074, SE = 0.00278) than in males (M = 0.0798, SE = 0.00320), suggesting that sex differences in oscillatory activity are preserved in clinical populations and may reflect underlying neurophysiological distinctions between male and female participants. Given the absence of significant gender-related effects, all participants were combined into a single dataset for subsequent analyses.

4.3.2 Descriptive Results for Relative Power Spectra Density by Age Group

In the HBN sample, age-related differences in relative EEG power were evident across all frequency bands. Children (n = 672) showed markedly higher delta power (M = 0.7619, SE = 0.00704) than adolescents (n = 231; M = 0.6528, SE = 0.01161), reflecting the expected dominance of slow-wave activity in younger individuals. In contrast, alpha band power was substantially greater in adolescents (M = 0.1732, SE = 0.00931) than in children (M = 0.0771, SE = 0.00301), consistent with neurodevelopmental shifts toward increased cortical inhibition and synchronization during adolescence. Similarly, beta power was slightly higher in adolescents (M = 0.0643, SE = 0.00273) than in children (M = 0.0544, SE = 0.00459), while gamma power was comparable across groups, with adolescents showing a marginally higher mean (M = 0.0267, SE = 0.00169) than children (M = 0.0258, SE = 0.00246). Theta power remained relatively stable between groups, with adolescents (M = 0.0831, SE = 0.00236) and children (M = 0.0808, SE = 0.00186) exhibiting similar levels.

In the MDD group, a comparable pattern was observed with some amplification of age-related contrasts. Delta power was elevated in children (n = 70; M = 0.7026, SE = 0.01985) compared to adolescents (n = 539; M = 0.6181, SE = 0.00729), mirroring the findings in the healthy cohort. Alpha power was again higher among adolescents (M = 0.1554, SE = 0.00535) than children (M = 0.1331, SE = 0.01781), reinforcing the developmental trend. The difference in beta activity was more pronounced in the clinical sample, with adolescents showing nearly double the beta power (M = 0.0878, SE = 0.00373) of children (M = 0.0461, SE = 0.00331), possibly reflecting increased cortical arousal or altered attentional processing associated with adolescence in the context of depression. Gamma band power followed a similar trajectory, with adolescents (M = 0.0401, SE = 0.00161) exhibiting higher levels than children (M = 0.0312, SE = 0.00301). Theta power was also elevated in adolescents (M = 0.0986, SE = 0.00236) relative to children (M = 0.0870, SE = 0.00526), aligning with typical maturational increases in mid-frequency activity. Together, these descriptive findings indicate clear developmental differences in EEG spectral power, with children characterized by higher delta activity and adolescents by increased alpha, beta, and gamma power. These patterns were consistent across both healthy and clinical populations, suggesting robust neurodevelopmental shifts in oscillatory dynamics that are preserved even in the presence of mood disorders such as MDD.

To examine differences in relative power between participants with MDD and HBN, we employed a LMM, with relative power as the dependent variable, Group (MDD vs. HBN), ROI, and frequency band were entered as fixed effects, while participant ID was included as a random effect to account for inter-individual variability. The model was estimated using restricted maximum likelihood and incorporated all main effects and interaction terms. Model fit was acceptable, with an Akaike Information Criterion (AIC) of −23320.48 and a Bayesian Information Criterion (BIC) of −22231.93. The marginal R^2, representing the variance explained by fixed effects, was 52.2%, while the conditional R^2, accounting for both fixed and random effects, was 57.7%. When we looked at how participants varied individually, we noticed some differences in relative power. The intraclass correlation coefficient was 0.052, meaning that only a small amount of the variation in relative power could be explained by differences between participants. The variance for the intercept was 0.0005, showing that there were small differences in baseline relative power between individuals. The residual variance, which captures the unexplained variation, was 0.009, suggesting that most of the differences in relative power are due to factors not related to individual differences. This analysis included 7783 data points from 362 participants (see Appendix for MDD vs HBN Relative Power PSD; cf., Fig. 10).

In the random effect analysis, we tested the variability due to participant differences (ID) using a likelihood ratio test. The model had 71 parameters, and the AIC for the model was −22,481. The likelihood ratio test statistic was 288 (df = 1; p < 0.001) indicating that there is a significant amount of variation in the data that can be attributed to individual differences between participants.

Omnibus F-tests (using the Satterthwaite approximation for degrees of freedom) revealed significant main effects of ROI ($F(6, 12{,}240) = 9.314$, $p < 0.001$), and frequency band ($F(4, 12{,}240) = 2648.025$, $p < 0.001$). No significant main effect was observed for group ($F(1, 360) = 0.413$, $p = 0.521$), suggesting that relative power alone may not differentiate MDD from HBN participants. Significant interactions were found between group and frequency band ($F(4, 12{,}240) = 7.070$, $p < 0.001$), and between ROI and frequency band ($F(24, 12{,}240) = 6.454$, $p < 0.001$). However, the group × ROI interaction ($F(6, 12{,}240) = 0.314$, $p = 0.930$), and the three-way interaction among group, ROI, and frequency band ($F(24, 12{,}240) = 0.103$, $p = 1$), were not statistically significant. These results suggest that while overall relative power is not significantly different between diagnostic groups, specific frequency-band interactions may carry relevant information regarding group differences.

The results showed that, on average, there was no significant difference between the MDD and HBN groups in terms of relative power, as the group variable itself did not produce any notable effects. However, significant differences were found in the ROI and frequency bands. For example, the regions FL (frontal left), FR (frontal right), TL (temporal left), and TR (temporal right) exhibited varying levels of relative power, with the frontal regions showing negative effects, meaning less power, while temporal regions had some negative effects as well but to a lesser degree. Importantly, the temporal right region (TR—Center) and temporal left region (TL—Center) showed significant differences, with power levels significantly lower compared to other regions. In contrast, the occipital regions did not show significant effects, suggesting these areas might not be as relevant in distinguishing between MDD and HBN in terms of relative power.

Frequency bands also played an important role in explaining the differences in brain activity. The frequency bands Theta, Alpha, Beta, and Gamma all exhibited significant effects. The Theta and Gamma bands showed particularly large differences, with both bands having a negative association with relative power, indicating lower brain activity in these frequencies. The Alpha and Beta bands also showed significant differences but with smaller effects compared to Theta and Gamma.

When looking at the interactions between group, brain region, and frequency band, no significant three-way interactions were observed. This suggests that the combination of group (MDD or HBN), ROI, and frequency band did not create any additional complex patterns of activity. However, there were significant two-way interactions between the regions and frequency bands. For example, the interaction between the frontal left region and the Theta frequency band showed a significant reduction in relative power, and similar patterns were observed for the other frequency bands in the frontal and frontal right regions. Overall, these findings suggest that although the MDD and HBN groups did not show major differences in overall brain power, there are specific regions and frequency bands that play a key role in distinguishing between the groups.

Table 7 shows a summary of the important detected FFPDC connectivity ROIs interaction across frequency bands by the CNN module, detected by the integrated gradient and PDI methods. It also shows the detected relative PSD patterns through LMM module.

5 Discussion

In this chapter, we explored how deep learning techniques—specifically CNNs—can be used to identify neurological markers associated with MDD in children and adolescents. Building on earlier research and recent advancements in the field, we focused on rsFC derived from EEG data to distinguish between participants diagnosed with MDD and healthy controls. We implemented a CNN model based on the VGG16 architecture, following a detailed pre-processing pipeline. Brain connectivity was estimated using the MVARICA model, which helped us identify meaningful biomarkers across different frequency bands and brain ROIs.

Table 7 Summary table of the LMM method on the relative PSD, as well as the important connectivity ROI interactions, detected by the integrated gradient and the PFI methods

Frequency band	Integrated gradient		PFI	Relative power	
	HBN-biased	MDD-biased		Higher in HBN	Higher in MDD
Delta	Center-FL; Center-OR	Center-FR; Center-TL	Center-TR; Center-OL Center-OR; Center-FL TL-OR; TR-FR	–	–
Theta	OR-TL; OR-FR	OL-FL FR-FL FL-Center	Center-TL; Center-OL Center-OR; OR-FR; OR-TR; OR-FL	All ROI	–
Alpha	OR-FR; OR-FL	FL-TL FL-OL FL-FR FL-Center Center-TL	TR-OL; Center-OL	All ROI	–
Beta	FL-FR	–	Center-TL; Center-TR; TL-FR	All ROI	–
Gamma	–	FL-TL	Center-TL; Center-TR Center-FR; TR-FL; OR-OL	All ROI	–
All	–	OL-FR	Center-OL; OR-TL; TR-FL	FL; FR TL; TR; TR-Center TL-Center	–

We also examined how the CNN made its decisions and compared these insights with results from traditional statistical methods based on PSD. Additionally, we assessed the model's potential as a diagnostic tool by evaluating its performance in classifying participants using short segments of resting-state EEG recordings. To improve performance, we applied advanced validation strategies, including robust cross-validation, early stopping, dropout layers, class weighting, and multiple performance metrics to guard against overfitting. This study placed particular emphasis on PDC and its related variants, with a focus on ffPDC, which demonstrated the highest compatibility with CNN-based analysis.

The CNN model achieved strong performance using ffPDC features, with an F1-Score of 0.815, a MCC of 0.753, and Cohen's Kappa of 0.746. Based on these results, subsequent analyses primarily focused on FFPDC. Importantly, the model's performance remained stable when tested on sex-specific subsets of the data. However, we were unable to analyze age-related effects due to the limited availability of MDD data for children (n = 10 datasets). To interpret the model's predictions, we used Integrated Gradients and PFI. By averaging pixel scores across all connections involving a given ROI, we identified the most influential brain regions independent of frequency band. The Center and OL ROIs showed stronger connectivity in MDD participants, while the FR ROI had higher values in the HBN group. PFI also highlighted the Center ROI as particularly important.

We then extracted the most important features per method across different frequency bands, as shown in Figs. 6 and 7. Connection strengths were normalized across bands. One might argue that the prominence of the Center ROI is partly due to its peripheral placement on frequency band maps. However, its consistent salience in the gamma and combined frequency bands where edge effects are minimal supports its significance. These findings are consistent with previous work [75], which also identified the Center ROI as a key feature across multiple frequency bands.

The differences in highlighted features between Integrated Gradients and PFI stem from their methodological principles. Integrated Gradients measure how sensitive the model is to changes in a specific feature, indicating how much predictions shift when a feature's value is modified. This can reveal model dependencies on features that may not be causally important but are strongly correlated with the outcome. In contrast, PFI assesses global feature importance by measuring how much performance drops when a feature is randomized. This can overlook redundant features if others compensate. As such, the two methods complement each other and Integrated Gradients show how the model uses features, while PFI reveals which features are statistically critical to performance.

On average, the model classified 45% ($\pm$7.6) of segments from MDD participants as MDD, while 24% ($\pm$8.3) of segments from HBN participants were incorrectly classified as MDD. While this distinction is clear, identifying an optimal classification threshold at the participant level (e.g., via F1-score optimization) could improve the model's practical application as a diagnostic tool. However, further refinement is needed.

Another key consideration is the temporal variability of MDD-related neurobiomarkers. Although MDD is generally associated with persistent changes in brain function, rsFC patterns can fluctuate over time. Given the 15-s segment duration, some MDD segments may not display distinguishable abnormalities, appearing instead as healthy-like patterns. This could explain why the HBN group whose connectivity is likely more consistent had a higher correct classification rate. This finding aligns with Kaiser et al. [81], who reported that rsFC abnormalities in MDD tend to fluctuate. As such, our CNN may only detect MDD in segments where those abnormalities are momentarily active.

A statistical analysis of relative PSD values revealed no significant differences by sex or age. However, differences were observed in terms of ROIs and frequency bands, particularly through two-way interactions (e.g., group differences by ROI or frequency band alone). These results align somewhat with the CNN connectivity analyses, though the machine learning explanations were more detailed. Interestingly, when comparing Integrated Gradient explanations with results from LMM, the two methods did not always agree on the direction of effect. For example, LMM found higher relative power in the FL and Center ROIs for HBN participants, while Integrated Gradients typically associated higher FFPDC connectivity values with MDD. These discrepancies are expected, as the underlying metrics capture different aspects of brain activity and PSD focuses on local activity, while PDC-based connectivity captures directional interactions.

In summary, this study demonstrates that deep learning applied to EEG-based rsFC can offer valuable insights into the neurobiology of MDD in young individuals. It also highlights the importance of feature interpretation and the complementary nature of different explanation and validation methods in model development. Of course, this is to be expected, as the measurements are different in nature. In our study, we found that multiple brain regions were involved in MDD, rather than a single localized area corroborating prior research [49, 62, 90, 151, 167]. This supports the understanding that depression impacts the brain as a complex, interconnected network. Notably, we observed alterations in both alpha and beta brainwave activity in the right and left frontal regions—areas associated with attention, memory, and emotional regulation. These disruptions may help explain common depressive symptoms such as difficulty concentrating, forgetfulness, and poor sleep quality.

Interestingly, supporting prior findings (e.g., [98]), we observed increased delta activity between the centro-frontal right and centro-temporal left regions, while theta activity was elevated in the left occipito-frontal, frontal left-to-right, and frontal left-to-center connections in individuals with MDD compared to those with HBN. Notably, the frontal left region plays a key role in theta band increases, whereas the central region is more involved in delta increases. Increased delta activity is often linked to cortical hypoactivation or dysfunction during rest, which may disrupt executive control and emotional regulation networks (e.g., [41, 134]). These changes can contribute to slower information processing, potentially impacting learning and development in children and adolescents. Over time, this may manifest as increased rumination, reduced motivation, and impaired decision-making hallmark symptoms of MDD. Elevated theta activity, particularly in centro-frontal regions, may reflect a dysregulation of top-down control, where the frontal cortex struggles to effectively manage emotional responses driven by limbic structures. This imbalance is also associated with a greater tendency toward negative thinking or bias, commonly observed in individuals with, e.g., depression [11, 133]

Alpha suppression in the frontal cortex has been linked to increased emotional reactivity and persistent negative thinking. In children and adolescents, this often manifests as heightened rumination and difficulty disengaging from negative thoughts, contributing to emotional distress [163, 164]. At the same time, abnormal beta activity in the frontal regions may reflect the brain's effort to compensate for deficits in attention and cognitive control. However, this can lead to overactivation in areas involved in executive function and mood regulation, which may hinder active thinking, motor planning, and behavioral response [99, 109, 125].

We also observed a general reduction in gamma activity in participants with MDD compared to healthy individuals. In our study, this decrease was most prominent in the left and frontal regions, suggesting impaired global brain communication. These findings align with previous research linking reduced gamma activity to decreased cognitive flexibility and increased anhedonia in individuals with depression [99, 103, 149]. Such impairments in information integration and higher-order cognitive function may disrupt typical developmental trajectories in children and adolescents, leading to lower motivation, emotional numbness, slowed thinking, and repetitive negative thought patterns (e.g., [70, 80]). Overall, our results show children and adolescents with diagnosed with MDD show stronger low-frequency brain connections, especially between frontal and temporal areas. This may point to difficulties in how the brain handles thoughts and emotions, slower mental processing, and a tendency to focus inward patterns that fit with common depression symptoms like overthinking, trouble concentrating, and emotional imbalance [63, 148, 159].

In line with findings from adult studies and reinforcing our previous large-scale research on classifying children and adolescents with MDD versus HBN [74, 75, 78], our analysis of larger datasets showed promising accuracy. These results suggest that FC patterns could be valuable diagnostic markers for MDD in younger populations. Given that MDD is widely recognized as a disorder involving disrupted brain network connectivity, it's not surprising that models based on rsFC perform well. This is especially relevant as coherence-based measures continue to gain traction in the development of classification models.

5.1 Limitations, Future Direction, and Clinical Applications

Although the dataset included large amount of healthy and MDD patients, one important limitation is the imbalance between groups, with patients diagnosed with MDD compared to healthy controls. This uneven distribution reduces the statistical power of the study and may introduce bias during model training, since deep learning approaches such as CNNs generally benefit from larger and more balanced datasets to capture stable and generalizable patterns. The relatively small number of MDD cases also raises the risk of overfitting, where the model may learn group-specific noise rather than true disorder-related features, limiting its performance on new populations. In addition, the age difference between groups introduces potential confounding effects, as developmental changes could influence EEG features independently of diagnosis. Therefore, the results should be interpreted with caution, and future work should aim to validate these findings in larger, more age-matched, and balanced cohorts to strengthen their clinical relevance.

For future work, we recommend replicating this study using a larger dataset to ensure the findings are robust and generalizable. Additionally, it would be beneficial to evaluate and compare this method with various machine learning models across different connectivity frameworks, as this could help identify the most effective approach for capturing the complex neural patterns associated with MDD. Another promising direction is to explore the relationship between the model's classification performance at both the segment level and the participant level. By doing so, we may gain valuable insights into the potential of these models to serve as reliable diagnostic tools in clinical settings. Moreover, further investigation into how machine learning models especially deep learning models handle, learn, and detect the temporal properties of MDD could enhance our understanding of the disorder's dynamic nature. Such studies could reveal how these models capture changes over time in brain activity, which may be crucial for developing more precise and individualized treatment strategies.

Based on our series of computational diagnostic modellings and clinical perspectives, if implemented in clinical settings, this CNN-based EEG diagnostic tool could serve as a complementary screening aid for child and adolescent psychiatrists. For instance:

- **Initial Assessment**: During an intake session, a child who exhibits symptoms of depression but whose clinical presentation is unclear could undergo a short, 5-min resting-state EEG. The CNN model could flag abnormal connectivity patterns in real time, helping the clinician decide whether to prioritize more comprehensive MDD assessments or explore alternative diagnoses.
- **Monitoring Treatment Response**: In another case, for example, a clinician treating a 14-year-old with pharmacotherapy and cognitive behavioral therapy might use the model to compare EEG recordings from different time points. Improvements in connectivity features, such as reduced low-frequency frontal–temporal coupling, could suggest that the patient is responding neurologically to treatment, even before behavioral symptoms fully improve. This itself is a valuable sign in clinical domain.
- **Risk Identification**: A school psychologist working together with child and adolescents' psychiatrist in early intervention department might use the model to screen children flagged for emotional difficulties. If the model identifies elevated theta or delta connectivity patterns consistent with MDD, the student could be referred for a full psychological evaluation, potentially accelerating access to care.

These applications do not replace clinical judgment but illustrate how such models could be integrated into multi-method diagnostic workflows, especially in under-resourced settings where psychiatric evaluation may be delayed and early care is needed until a concrete diagnostic and therapeutic plan has been made.

6 Conclusions

In conclusion, this chapter demonstrates that CNN models, when paired with functional connectivity measures, are capable of effectively identifying potential neurobiological markers of MDD in both children and adolescents. Notably, the model achieved its strongest performance using the FFPDC metric, highlighting its suitability for such classification tasks. While these results are promising and suggest meaningful applications in clinical diagnostics, additional research is essential. This should include larger and more diverse participant samples, along with a thorough consideration of the various subtypes of depression and the influence of ongoing medication use. Such efforts are essential to better contextualize the identified neurobiological markers, moving beyond technical validation toward meaningful clinical interpretation. Ultimately, this approach can contribute to bridging the gap between research findings and practical applications laying the groundwork for more personalized and precision-based approaches in neuropsychiatric diagnosis and treatment.

Acknowledgements This book chapter was prepared using limited access to the datasets obtained from the USA BioBank provided by Child Mind Institute (CMI), Healthy Brain Network. This manuscript reflects the views of the authors and does not necessarily reflect the opinions or views of the CMI. EEG Datasets and phenotypes were credited by CMI to Dr. Amir Jahanian Najafabadi under the official data usage agreement signed by Bielefeld University.

Conflict of Interest Authors declare no conflict of interest.

Ethic Declaration The study utilized data from the publicly available Healthy Brain Network (HBN) dataset, which was approved by institutional ethics committees in the United States. Written informed consent was obtained from all participants (and their legal guardians for minors) in accordance with ethical and institutional guidelines [5].

Declaration Statement Authors declare that ChatGPT v.4 was solely used for language proofreading, specifically for grammar checking and improving the clarity of the text. It was not used for any form of content generation or interpretation.

Appendix

Power Spectra Density for MDD Versus HBN

See Figs. 10 and 11.

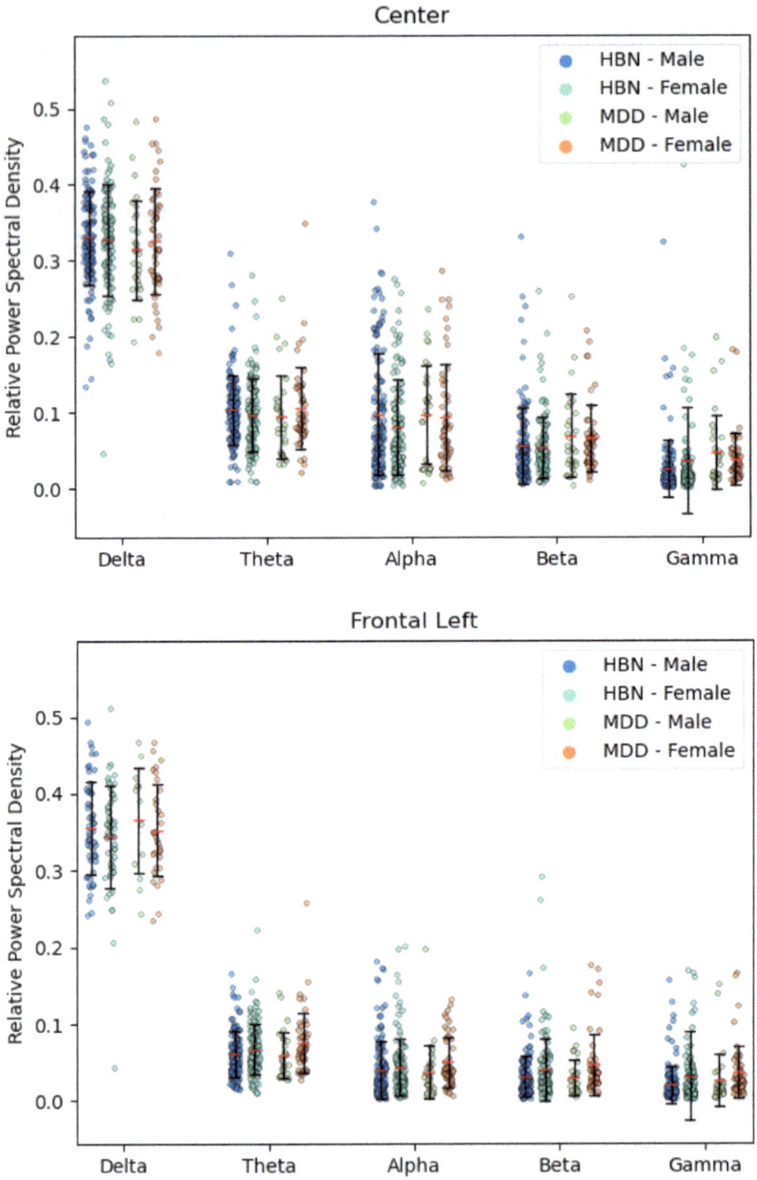

Fig. 10 Descriptive for gender-related power spectra density are illustrated for frequency bands (Male versus Female)

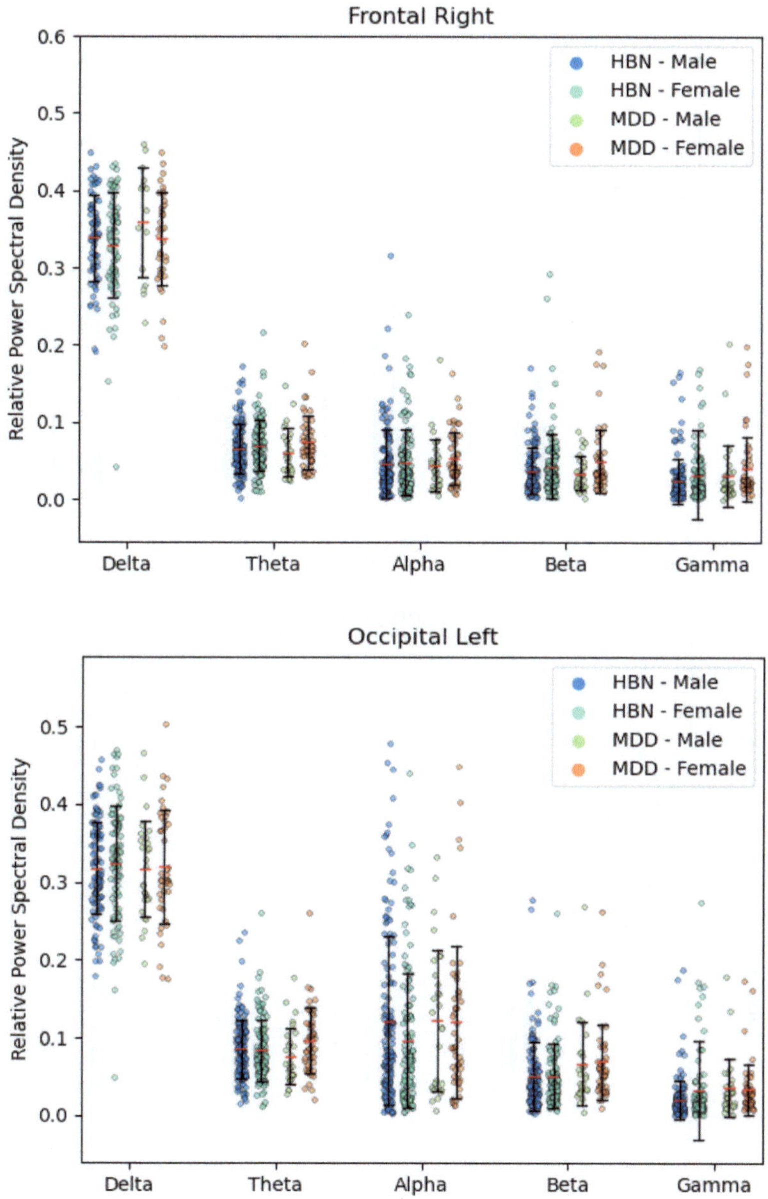

Fig. 10 (continued)

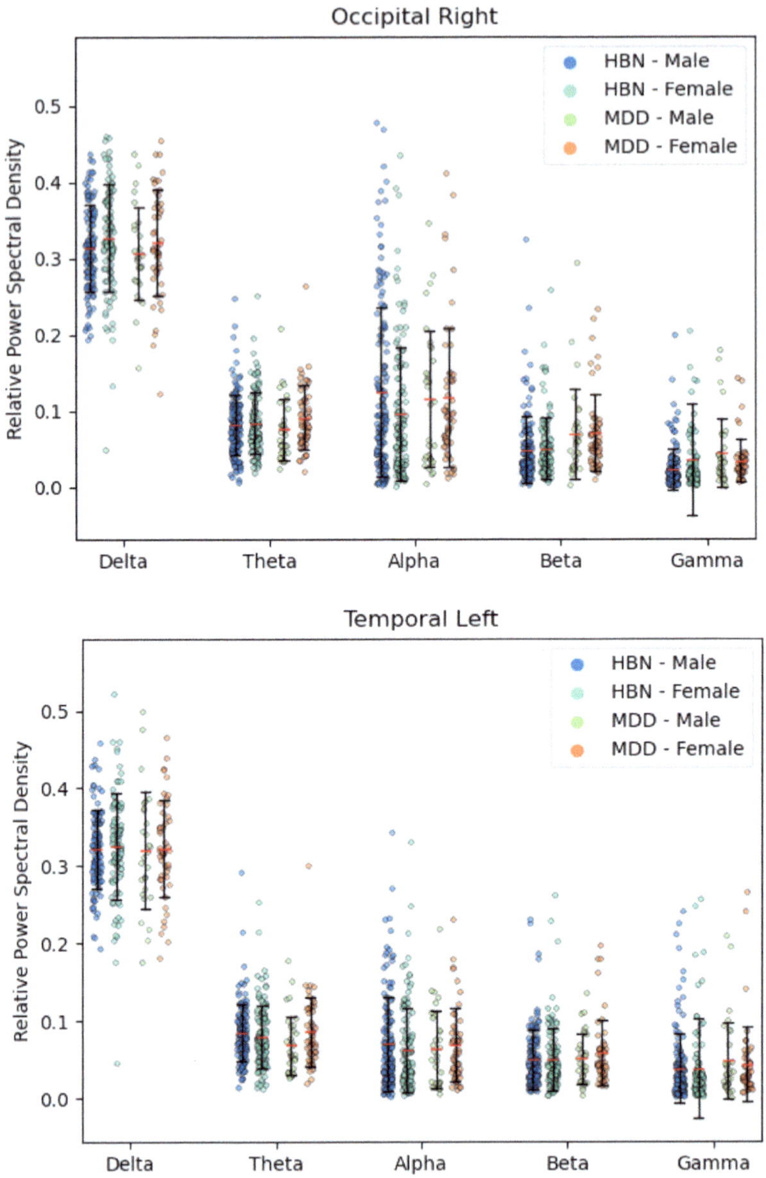

Fig. 10 (continued)

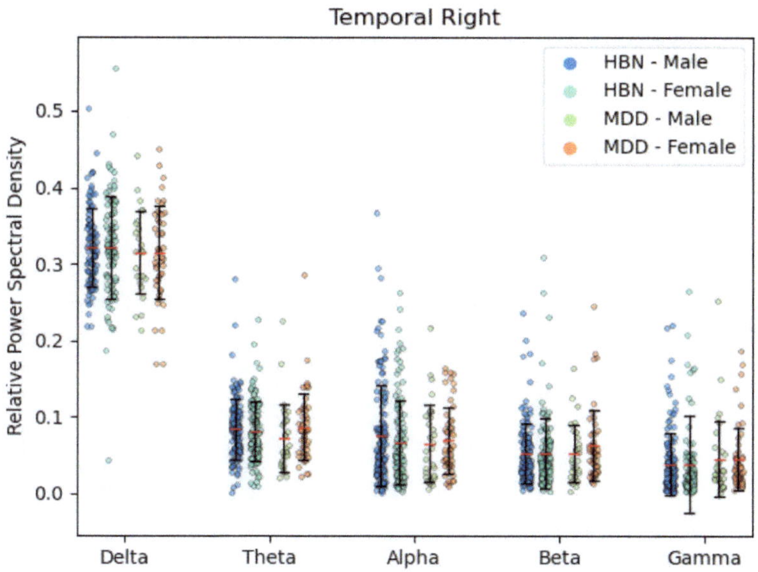

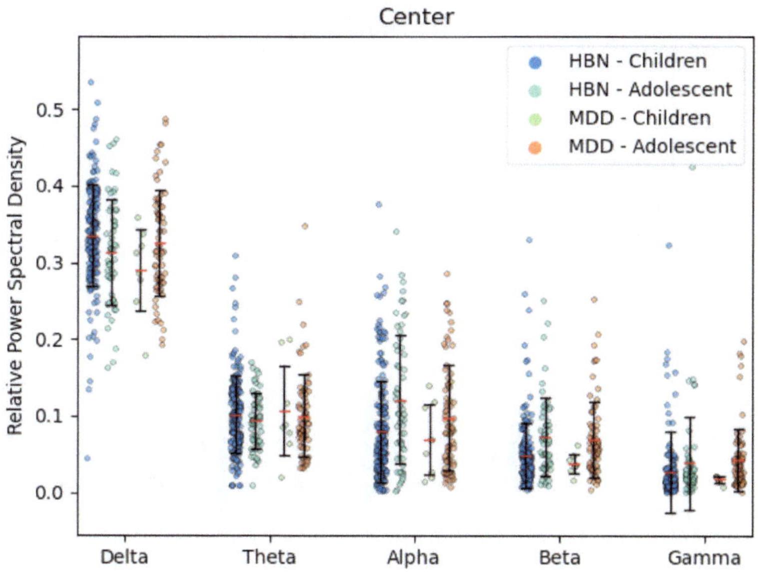

Fig. 10 (continued)

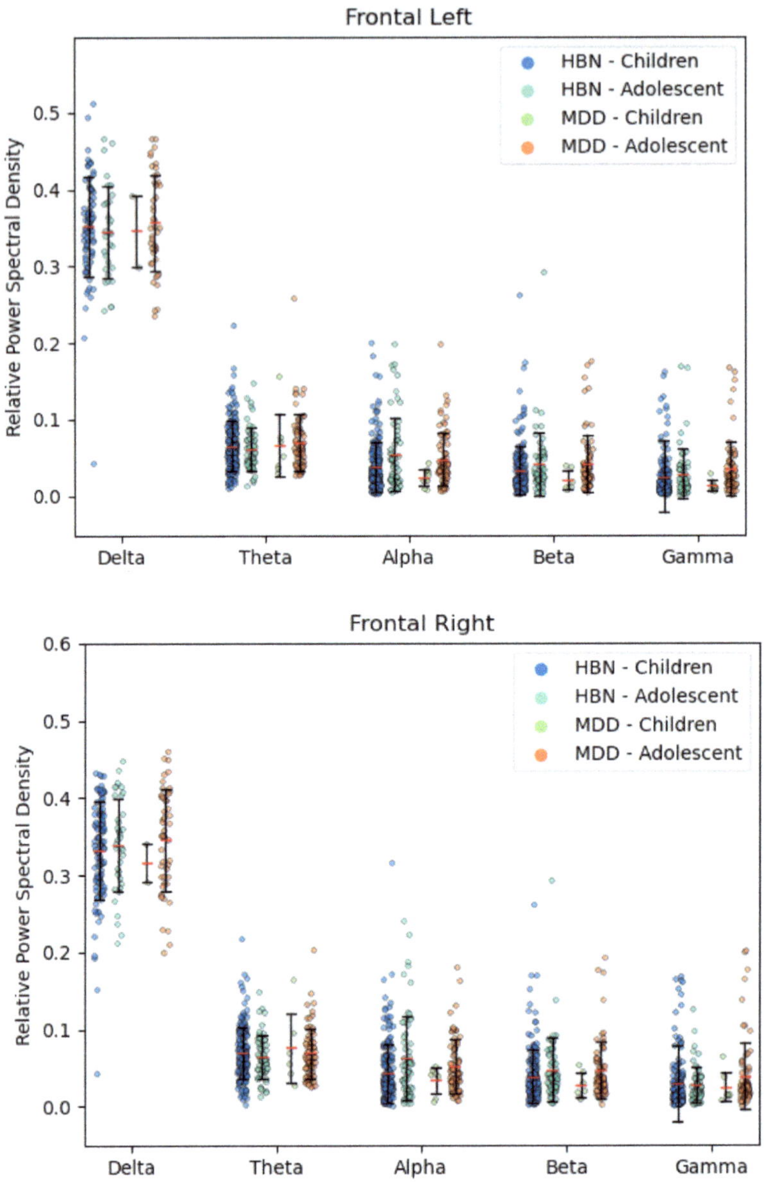

Fig. 10 (continued)

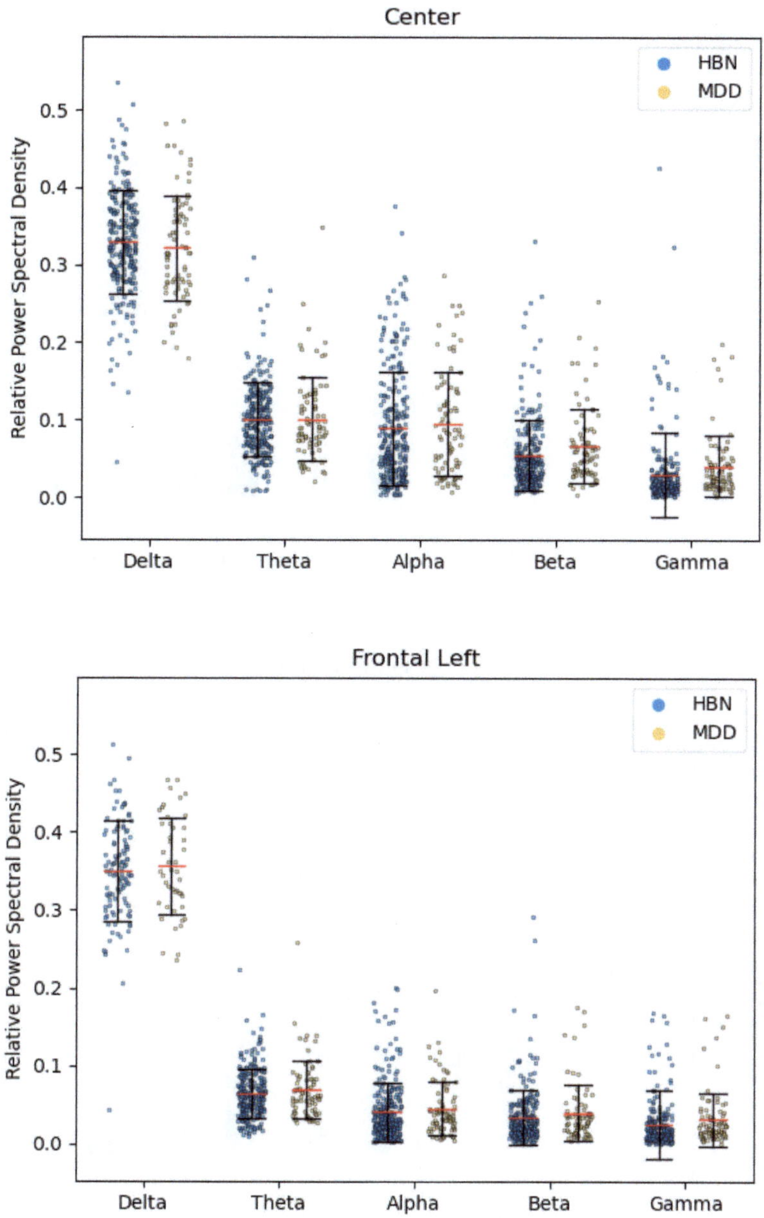

Fig. 11 Descriptive for HBN versus MDD are illustrated for frequency bands

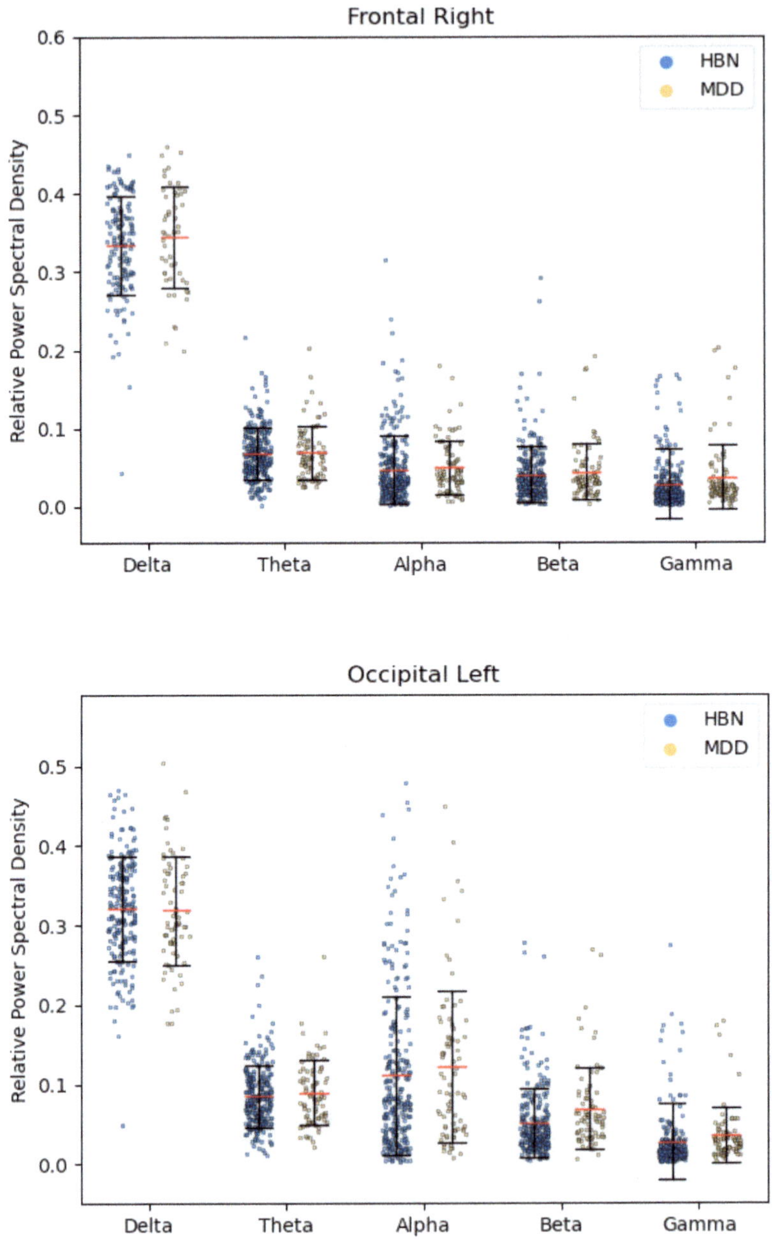

Fig. 11 (continued)

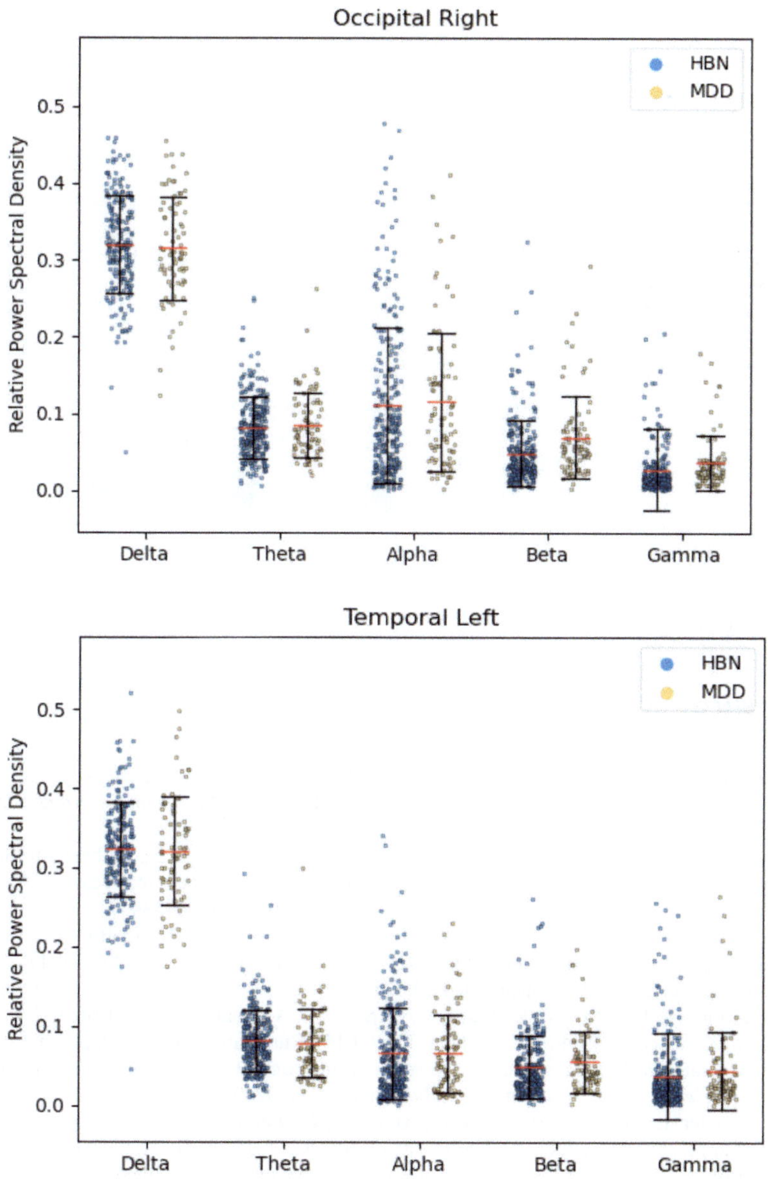

Fig. 11 (continued)

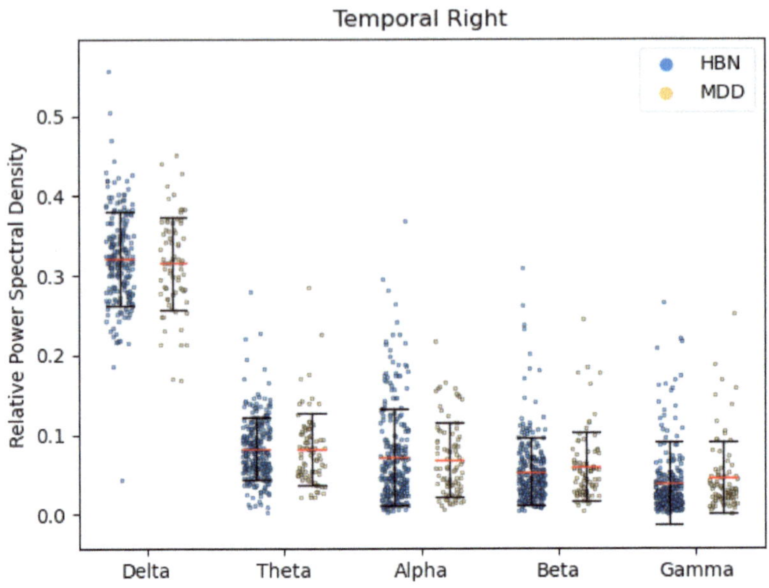

Fig. 11 (continued)

References

1. Acharya, U. R., Sudarshan, V. K., Adeli, H., Santhosh, J., Koh, J. E., & Adeli, A. (2015). Computer-Aided diagnosis of depression using EEG signals. *European Neurology*, *73*(5–6), 329–336. https://doi.org/10.1159/000381950
2. Acharya, U. R., Sudarshan, V. K., Adeli, H., Santhosh, J., Koh, J. E., Puthankatti, S. D., & Adeli, A. (2015). A novel depression diagnosis index using nonlinear features in EEG signals. *European Neurology*, *74*(1–2), 79–83. https://doi.org/10.1159/000438457
3. Ahmed Z, Wali A, Shahid S, Zikria S, Rasheed J, Asuroglu T. Psychiatric disorders from EEG signals through deep learning models. IBRO Neurosci Rep. 2024 Sep 24;17:300-310. https://doi.org/10.1016/j.ibneur.2024.09.003
4. Alderman, B. L., Olson, R. L., Bates, M. E., Selby, E. A., Buckman, J. F., Brush, C. J., Panza, E. A., Kranzler, A., Eddie, D., & Shors, T. J. (2015). Rumination in major depressive disorder is associated with impaired neural activation during conflict monitoring. *Frontiers in Human Neuroscience*, *9*. https://doi.org/10.3389/fnhum.2015.00269
5. Alexander, L. M., Escalera, J., Ai, L., Andreotti, C., Febre, K., Mangone, A., Vega-Potler, N., Langer, N., Alexander, A., Kovacs, M., Litke, S., O'Hagan, B., Andersen, J., Bronstein, B., Bui, A., Bushey, M., Butler, H., Castagna, V., Camacho, N.,... Milham, M. P. (2017). An open resource for transdiagnostic research in pediatric mental health and learning disorders. *Scientific Data*, *4*(1). https://doi.org/10.1038/sdata.2017.181
6. Alshaya, D. S. (2022). Genetic and epigenetic factors associated with depression: An updated overview. *Saudi Journal of Biological Sciences*, *29*(8), 103311. https://doi.org/10.1016/j.sjbs.2022.103311
7. Altmann, A., Toloşi, L., Sander, O., & Lengauer, T. (2010). Permutation importance: a corrected feature importance measure. *Bioinformatics*, *26*(10), 1340–1347. https://doi.org/10.1093/bioinformatics/btq134

8. Ambrosini, P. J. (2000). A review of pharmacotherapy of major depression in children and adolescents. *Psychiatric Services, 51*(5), 627–633. https://doi.org/10.1176/appi.ps.51.5.627
9. An, Z., Tang, K., Xie, Y., Tong, C., Liu, J., Tao, Q., Yan, C., Chen, X., Cao, L., Chen, W., Cheng, Y., Fang, Y., Gong, Q., Guo, W., Kuang, L., Li, B., Li, T., Liu, Y., Liu, Z.,... Feng, Y. (2024). Aberrant resting-state co-activation network dynamics in major depressive disorder. *Translational Psychiatry, 14*(1). https://doi.org/10.1038/s41398-023-02722-w
10. American Psychiatric Association. (2022). *Diagnostic and statistical manual of mental disorders* (5th ed., text rev.). https://doi.org/10.1176/appi.books.9780890425787
11. Babaeeghazvini, P., Rueda-Delgado, L. M., Gooijers, J., Swinnen, S. P., & Daffertshofer, A. (2021). Brain Structural and Functional Connectivity: A review of combined works of diffusion magnetic resonance Imaging and Electro-Encephalography. *Frontiers in Human Neuroscience, 15.* https://doi.org/10.3389/fnhum.2021.721206
12. Babiloni, C., Arakaki, X., Baez, S., Barry, R. J., Benussi, A., Blinowska, K., Bonanni, L., Borroni, B., Bayard, J. B., Bruno, G., Cacciotti, A., Carducci, F., Carino, J., Carpi, M., Conte, A., Cruzat, J., D'Antonio, F., Della Penna, S., Del Percio, C.,... Kamondi, A. (2025). Alpha Rhythm and Alzheimer's Disease: Has Hans Berger's dream come true? *Clinical Neurophysiology, 172,* 33–50. https://doi.org/10.1016/j.clinph.2025.02.256
13. Baccalá, L. A., & Sameshima, K. (2001). Partial directed coherence: a new concept in neural structure determination. *Biological Cybernetics, 84*(6), 463–474. https://doi.org/10.1007/pl0 0007990
14. Banaschewski, T., & Brandeis, D. (2006). Annotation: What electrical brain activity tells us about brain function that other techniques cannot tell us – a child psychiatric perspective. *Journal of Child Psychology and Psychiatry, 48*(5), 415–435. https://doi.org/10.1111/j.1469-7610.2006.01681.x
15. Bagh, K., and A. Jahanian Najafabadi. "Graph-Based Analysis of Brain Connectivity in Children and Adolescents Diagnosed with Major Depressive Disorder Compared to Aged-matched Healthy Control: An EEG Study." 2024 IEEE Signal Processing in Medicine and Biology Symposium (SPMB). IEEE, 2024.
16. Barone, J., & Rossiter, H. E. (2021). Understanding the role of sensorimotor beta oscillations. *Frontiers in Systems Neuroscience, 15.* https://doi.org/10.3389/fnsys.2021.655886
17. Barry, R. J., Clarke, A. R., Johnstone, S. J., & Brown, C. R. (2009). EEG differences in children between eyes-closed and eyes-open resting conditions. *Clinical Neurophysiology, 120*(10), 1806–1811. https://doi.org/10.1016/j.clinph.2009.08.006
18. Barry, R. J., Clarke, A. R., Johnstone, S. J., Magee, C. A., & Rushby, J. A. (2007). EEG differences between eyes-closed and eyes-open resting conditions. *Clinical Neurophysiology, 118*(12), 2765–2773. https://doi.org/10.1016/j.clinph.2007.07.028
19. Barry, R. J., & De Blasio, F. M. (2017a). EEG differences between eyes-closed and eyes-open resting remain in healthy ageing. *Biological Psychology, 129,* 293–304. https://doi.org/10.1016/j.biopsycho.2017.09.010
20. Beck, A. T., Steer, R. A., & Garbin, M. G. (1988) Psychometric properties of the Beck Depression Inventory: Twenty-five years of evaluation. Clinical Psychology Review, 8(1), 77–100.
21. Beck AT, Ward CH, Mendelson M, Mock J, Erbaugh J. An inventory for measuring depression. Arch Gen Psychiatry. 1961 Jun;4:561–571. https://doi.org/10.1001/archpsyc.1961.017 10120031004
22. Beirão, D., Monte, H., Amaral, M., Longras, A., Matos, C., & Villas-Boas, F. (2020). Depression in adolescence: a review. *Middle East Current Psychiatry, 27*(1). https://doi.org/10.1186/s43045-020-00050-z
23. Bhaskarachary C, Najafabadi AJ, Godde B. Machine learning supervised classification methodology for autism spectrum disorder based on resting-state electroencephalography (EEG) signals. In: 2020 IEEE Signal Processing in Medicine and Biology Symposium (SPMB) 2020 Dec 5 (pp. 1–4). IEEE.
24. Bienek, O., Allott, K., Antonucci, L., Bertolino, A., Bonivento, C., Borgwardt, S., Brambilla, P., Chisholm, K., Dannlowski, U., Lichtenstein, T. K., Kambeitz, J., Kambeitz-Ilankovic,

L., Koutsouleris, N., Lencer, R., Griffiths, S. L., Maggioni, E., Meisenzahl, E., Pantelis, C., Rosen, M.,... Piccin, S. (2024). Neurocognitive dysfunction in adolescents with recent onset major depressive disorder: a cross-sectional comparative study. *European Child & Adolescent Psychiatry*. https://doi.org/10.1007/s00787-024-02599-0

25. Bigdely-Shamlo, N., Mullen, T., Kothe, C., Su, K.-M., & Robbins, K. A. (2015). The PREP pipeline: standardized preprocessing for large-scale EEG analysis. Frontiers in Neuroinformatics, 9, 16. https://doi.org/10.3389/fninf.2015.00016
26. Brickwedde, M., Krüger, M.C. & Dinse, H.R. Somatosensory alpha oscillations gate perceptual learning efficiency. Nat Commun 10, 263 (2019). https://doi.org/10.1038/s41467-018-08012-0
27. Cai, H., Xie, X., Zhang, Q., Cui, X., Lin, J., Sim, K., Ungvari, G. S., Zhang, L., & Xiang, Y. (2021). Prevalence of Suicidality in Major Depressive Disorder: A Systematic Review and Meta-Analysis of Comparative Studies. *Frontiers in Psychiatry, 12*. https://doi.org/10.3389/fpsyt.2021.690130
28. Chao, Z. C., Dillon, D. G., Liu, Y., Barrick, E. M., & Wu, C. (2022a). Altered coordination between frontal delta and parietal alpha networks underlies anhedonia and depressive rumination in major depressive disorder. *Journal of Psychiatry and Neuroscience, 47*(6), E367–E378. https://doi.org/10.1503/jpn.220046
29. Chen ZS, Kulkarni PP, Galatzer-Levy IR, Bigio B, Nasca C, Zhang Y. Modern views of machine learning for precision psychiatry. Patterns (N Y). 2022 Nov 11;3(11):100602. https://doi.org/10.1016/j.patter.2022.100602. PMID: 36419447; PMCID: PMC9676543.
30. Chiarion G, Sparacino L, Antonacci Y, Faes L, Mesin L. Connectivity analysis in EEG data: a tutorial review of the state of the art and emerging trends. Bioengineering. 2023 Mar 17;10(3):372.
31. Ciodaro, G., Najafabadi, A. J., & Godde, B. (2020). Resting State EEG classification of children with ADHD. *2020 IEEE Signal Processing in Medicine and Biology Symposium (SPMB)*, 1–6. https://doi.org/10.1109/spmb50085.2020.9353628
32. Coleman, S. L., Sharpley, C. F., Vessey, K. A., Evans, I. D., Williams, R. J., & Bitsika, V. (2025). Gamma oscillations as correlates of depression: updating Fitzgerald and Watson (2018). Reviews in the Neurosciences. https://doi.org/10.1515/revneuro-2025-0023
33. Cui, L., Li, S., Wang, S., Wu, X., Liu, Y., Yu, W., Wang, Y., Tang, Y., Xia, M., & Li, B. (2024). Major depressive disorder: hypothesis, mechanism, prevention and treatment. *Signal Transduction and Targeted Therapy, 9*(1). https://doi.org/10.1038/s41392-024-01738-y
34. Cuijpers, P., Karyotaki, E., Ciharova, M., Miguel, C., Noma, H., & Furukawa, T. A. (2021). The effects of psychotherapies for depression on response, remission, reliable change, and deterioration: A meta-analysis. *Acta Psychiatrica Scandinavica, 144*(3), 288–299. https://doi.org/10.1111/acps.13335
35. Cuijpers, P., Karyotaki, E., Ciharova, M., Miguel, C., Noma, H., Stikkelbroek, Y., Weisz, J. R., & Furukawa, T. A. (2021). The effects of psychological treatments of depression in children and adolescents on response, reliable change, and deterioration: a systematic review and meta-analysis. *European Child & Adolescent Psychiatry, 32*(1), 177–192. https://doi.org/10.1007/s00787-021-01884-6
36. Cuijpers, P., Karyotaki, E., Weitz, E., Andersson, G., Hollon, S. D., & Van Straten, A. (2014). The effects of psychotherapies for major depression in adults on remission, recovery and improvement: A meta-analysis. *Journal of Affective Disorders, 159*, 118–126. https://doi.org/10.1016/j.jad.2014.02.026
37. Damborská, A., Honzírková, E., Barteček, R., Hořínková, J., Fedorová, S., Ondruš, Š., Michel, C. M., & Rubega, M. (2020). Altered directed functional connectivity of the right amygdala in depression: high-density EEG study. *Scientific Reports, 10*(1). https://doi.org/10.1038/s41598-020-61264-z

38. De Freitas, S. B., Marques, A. A., Bevilaqua, M. C., De Carvalho, M. R., Ribeiro, P., Palmer, S., Nardi, A. E., & Dias, G. P. (2016). Electroencephalographic findings in patients with major depressive disorder during cognitive or emotional tasks: a systematic review. Brazilian Journal of Psychiatry, 38(4), 338–346. https://doi.org/10.1590/1516-4446-2015-1834
39. Deboer, T. Technologies of sleep research. Cell. Mol. Life Sci. 64, 1227 (2007). https://doi.org/10.1007/s00018-007-6533-0
40. Deiber MP, Hasler R, Colin J, Dayer A, Aubry JM, Baggio S, Perroud N, Ros T. Linking alpha oscillations, attention and inhibitory control in adult ADHD with EEG neurofeedback. Neuroimage Clin. 2020;25:102145. https://doi.org/10.1016/j.nicl.2019.102145.
41. Ding, X., Yue, X., Zheng, R., Bi, C., Li, D., & Yao, G. (2019). Classifying major depression patients and healthy controls using EEG, eye tracking and galvanic skin response data. *Journal of Affective Disorders*, *251*, 156–161. https://doi.org/10.1016/j.jad.2019.03.058
42. Drevets, W. C., Price, J. L., & Furey, M. L. (2008). Brain structural and functional abnormalities in mood disorders: implications for neurocircuitry models of depression. *Brain Structure and Function*, *213*(1–2), 93–118. https://doi.org/10.1007/s00429-008-0189-x
43. Dupuy, J. M., Ostacher, M. J., Huffman, J., Perlis, R. H., & Nierenberg, A. A. (2011). A critical review of pharmacotherapy for major depressive disorder. *The International Journal of Neuropsychopharmacology*, *14*(10), 1417–1431. https://doi.org/10.1017/s1461145711000083
44. Engel, A. K., & Fries, P. (2010). Beta-band oscillations — signalling the status quo? *Current Opinion in Neurobiology*, *20*(2), 156–165. https://doi.org/10.1016/j.conb.2010.02.015
45. Faes, L., Erla, S., & Nollo, G. (2012). Measuring connectivity in linear multivariate processes: definitions, interpretation, and practical analysis. *Computational and Mathematical Methods in Medicine*, *2012*, 1–18. https://doi.org/10.1155/2012/140513
46. Feeney, A., Hock, R. S., Fava, M., Ortiz, J. M. H., Iovieno, N., & Papakostas, G. I. (2022). Antidepressants in children and adolescents with major depressive disorder and the influence of placebo response: A meta-analysis. *Journal of Affective Disorders*, *305*, 55–64. https://doi.org/10.1016/j.jad.2022.02.074
47. Fernández-Palleiro, P., Rivera-Baltanás, T., Rodrigues-Amorim, D., Fernández-Gil, S., Del Carmen Vallejo-Curto, M., Álvarez-Ariza, M., López, M., Rodriguez-Jamardo, C., Benavente, J. L., De Las Heras, E., Olivares, J. M., & Spuch, C. (2019). Brainwaves oscillations as a potential biomarker for major depression disorder risk. *Clinical EEG and Neuroscience*, *51*(1), 3–9. https://doi.org/10.1177/1550059419876807
48. Fernández-Ruiz, A., Oliva, A., Nagy, G. A., Maurer, A. P., Berényi, A., & Buzsáki, G. (2017). Entorhinal-CA3 Dual-Input Control of spike timing in the hippocampus by Theta-Gamma coupling. *Neuron*, *93*(5), 1213–1226.e5. https://doi.org/10.1016/j.neuron.2017.02.017
49. Filatova EV, Shadrina MI, Slominsky PA. Major Depression: One Brain, One Disease, One Set of Intertwined Processes. Cells. 2021 May 21;10(6):1283. https://doi.org/10.3390/cells10061283. PMID: 34064233; PMCID: PMC8224372.
50. Fingelkurts, A. A., Fingelkurts, A. A., Rytsälä, H., Suominen, K., Isometsä, E., & Kähkönen, S. (2006). Impaired functional connectivity at EEG alpha and theta frequency bands in major depression. *Human Brain Mapping*, *28*(3), 247–261. https://doi.org/10.1002/hbm.20275
51. Gallo, S., El-Gazzar, A., Zhutovsky, P., Thomas, R. M., Javaheripour, N., Li, M., Bartova, L., Bathula, D., Dannlowski, U., Davey, C., Frodl, T., Gotlib, I., Grimm, S., Grotegerd, D., Hahn, T., Hamilton, P. J., Harrison, B. J., Jansen, A., Kircher, T.,... Van Wingen, G. (2023). Functional connectivity signatures of major depressive disorder: machine learning analysis of two multicenter neuroimaging studies. *Molecular Psychiatry*, *28*(7), 3013–3022. https://doi.org/10.1038/s41380-023-01977-5

52. Ghiasi, S., Dell'Acqua, C., Benvenuti, S. M., Scilingo, E., Gentili, C., Valenza, G., & Greco, A. (2021). Classifying subclinical depression using EEG spectral and connectivity measures. *2021 43rd Annual International Conference of the IEEE Engineering in Medicine &Amp; Biology Society (EMBC)*, 2050–2053. https://doi.org/10.1109/embc46164.2021.9630044
53. Gold, C., Fachner, J., & Erkkilä, J. (2012). Validity and reliability of electroencephalographic frontal alpha asymmetry and frontal midline theta as biomarkers for depression. *Scandinavian Journal of Psychology, 54*(2), 118–126. https://doi.org/10.1111/sjop.12022
54. Gómez-Herrero, G., Atienza, M., Egiazarian, K., & Cantero, J. L. (2008). Measuring directional coupling between EEG sources. *NeuroImage, 43*(3), 497–508. https://doi.org/10.1016/j.neuroimage.2008.07.032
55. Gong, J., Wang, J., Qiu, S., Chen, P., Luo, Z., Wang, J., Huang, L., & Wang, Y. (2020). Common and distinct patterns of intrinsic brain activity alterations in major depression and bipolar disorder: voxel-based meta-analysis. *Translational Psychiatry, 10*(1). https://doi.org/10.1038/s41398-020-01036-5
56. Gramfort, A., Luessi, M., Larson, E., Engemann, D. A., Strohmeier, D., Brodbeck, C., Goj, R., Jas, M., Brooks, T., Parkkonen, L., & Hämäläinen, M. (2013). MEG and EEG data analysis with MNE-Python. *Frontiers in Neuroscience, 7*. https://doi.org/10.3389/fnins.2013.00267
57. Greenberg, P. E., Fournier, A., Sisitsky, T., Pike, C. T., & Kessler, R. C. (2015). The economic burden of adults with major depressive disorder in the United States (2005 and 2010). *The Journal of Clinical Psychiatry, 76*(02), 155–162. https://doi.org/10.4088/jcp.14m09298
58. Greenberg, P. E., Fournier, A., Sisitsky, T., Simes, M., Berman, R., Koenigsberg, S. H., & Kessler, R. C. (2021a). The Economic Burden of Adults with Major Depressive Disorder in the United States (2010 and 2018). *PharmacoEconomics, 39*(6), 653–665. https://doi.org/10.1007/s40273-021-01019-4
59. Greenberg, P. E., Kessler, R. C., Birnbaum, H. G., Leong, S. A., Lowe, S. W., Berglund, P. A., & Corey-Lisle, P. K. (2003). The economic burden of depression in the United States. *The Journal of Clinical Psychiatry, 64*(12), 1465–1475. https://doi.org/10.4088/jcp.v64n1211
60. Greicius, M. D., Flores, B. H., Menon, V., Glover, G. H., Solvason, H. B., Kenna, H., Reiss, A. L., & Schatzberg, A. F. (2007). Resting-State Functional Connectivity in Major Depression: Abnormally Increased Contributions from Subgenual Cingulate Cortex and Thalamus. *Biological Psychiatry, 62*(5), 429–437. https://doi.org/10.1016/j.biopsych.2006.09.020
61. Grunewald, M., Döhnert, M., Brandeis, D. et al. Attenuated LPP to Emotional Face Stimuli Associated with Parent- and Self-Reported Depression in Children and Adolescents. J Abnorm Child Psychol 47, 109–118 (2019). https://doi.org/10.1007/s10802-018-0429-3
62. Hamilton, J. P., Etkin, A., Furman, D. J., Lemus, M. G., Johnson, R. F., & Gotlib, I. H. (2012). Functional Neuroimaging of Major Depressive Disorder: A Meta-Analysis and new integration of baseline activation and neural response data. *American Journal of Psychiatry, 169*(7), 693–703. https://doi.org/10.1176/appi.ajp.2012.11071105
63. Han, Y., Dai, Z., Ridwan, M. C., Lin, P., Zhou, H., Wang, H., Yao, Z., & Lu, Q. (2020). Connectivity of the frontal cortical oscillatory dynamics underlying inhibitory control during a Go/No-Go task as a predictive biomarker in major depression. *Frontiers in Psychiatry, 11*. https://doi.org/10.3389/fpsyt.2020.00707
64. Harmony, T. (2013). The functional significance of delta oscillations in cognitive processing. *Frontiers in Integrative Neuroscience, 7*. https://doi.org/10.3389/fnint.2013.00083
65. Harmony, T., Fernández, T., Silva, J., Bernal, J., Díaz-Comas, L., Reyes, A., Marosi, E., Rodríguez, M., & Rodríguez, M. (1996). EEG delta activity: an indicator of attention to internal processing during performance of mental tasks. *International Journal of Psychophysiology, 24*(1–2), 161–171. https://doi.org/10.1016/s0167-8760(96)00053-0
66. Hasan, M. R., Fatemi, M. I., Khan, M. M., Kaur, M., & Zaguia, A. (2021). Comparative Analysis of Skin Cancer (Benign vs. Malignant) Detection Using Convolutional Neural Networks. *Journal of Healthcare Engineering, 2021*, 1–17. https://doi.org/10.1155/2021/5895156
67. He, X., Zhang, Y., Chen, J., Xie, C., Gan, R., Wang, L., & Wang, L. (2016). Changes in theta activities in the left posterior temporal region, left occipital region and right frontal region related to mild cognitive impairment in Parkinson's disease patients. *International Journal of Neuroscience, 127*(1), 66–72. https://doi.org/10.3109/00207454.2016.1143823

68. Henriques, J. B., & Davidson, R. J. (1991). Left frontal hypoactivation in depression. *Journal of Abnormal Psychology, 100*(4), 535–545. https://doi.org/10.1037/0021-843x.100.4.535
69. Ho, C., Peng, S., Yu, Y., Chu, Y., Huang, S., & Kuo, P. (2024). In perspective of specific symptoms of major depressive disorder: Functional connectivity analysis of electroencephalography and potential biomarkers of treatment response. Journal of Affective Disorders, 367, 944–950. https://doi.org/10.1016/j.jad.2024.08.139
70. Huberty, T. J. (2012). Anxiety and depression in children and adolescents. In *Springer eBooks*. https://doi.org/10.1007/978-1-4614-3110-7
71. Imani, H., & Godde, B. (2024a). Cortical resting-state activities may predict anodal TDCS-induced facilitation of dynamic balance training. *Psycharxiv*. https://doi.org/10.31234/osf.io/aufx3
72. Imani, H., & Godde, B. (2024b). In older adults resting-state alpha power is associated with stronger effects of anodal tDCS over prefrontal cortex on dynamic balance. *Psycharxiv*. https://doi.org/10.31234/osf.io/zsp28
73. Jaworska, N., Blier, P., Fusee, W., & Knott, V. (2012). Alpha power, alpha asymmetry and anterior cingulate cortex activity in depressed males and females. *Journal of Psychiatric Research, 46*(11), 1483–1491. https://doi.org/10.1016/j.jpsychires.2012.08.003
74. Jahanian Najafabadi, A., & Bagh, K. (2024a). Resting-state Functional Connectivity for Major Depressive Disorder Using Electroencephalography: A Deep Learning Approach. *Psycharxiv*. https://doi.org/10.31234/osf.io/y5asq
75. Jahanian Najafabadi, A., & Bagh, K. (2024b). Deep Learning Analysis Approach for major depressive Disorder in children and adolescents. *Psycharxiv*. https://doi.org/10.31234/osf.io/zy3ak
76. Jahanian Najafabadi, A., and Khaled Bagh. "Resting-State Functional Connectivity in Children and Adolescents with Major Depressive Disorder: A Deep Learning Approach Using High-density EEG." *2024 IEEE Signal Processing in Medicine and Biology Symposium (SPMB)*. IEEE, 2024.
77. Jahanian Najafabadi, A. J., Khosravi, M., Khosravi, R., Liepelt, R., & Imani, H. (2025). Resting-state and Task-related Cortical Activities Predict Sense of Ownership and Agency: An Explorative Tool-use Study. *Psycharxive*. https://doi.org/10.31234/osf.io/s4yd2_v1
78. Jahanian Najafabadi, A. J., Küster, D., Putze, F., & Godde, B. (2023). Association between resting-state and task-related EEG and practice effect in young and older adults. *Psycharxiv*. https://doi.org/10.31234/osf.io/nquym
79. Jones, N. P., Siegle, G. J., Muelly, E. R., Haggerty, A., & Ghinassi, F. (2010). Poor performance on cognitive tasks in depression: Doing too much or not enough? *Cognitive Affective & Behavioral Neuroscience, 10*(1), 129–140. https://doi.org/10.3758/cabn.10.1.129
80. Jones, J. D., Fraley, R. C., Stern, J. A., Lejuez, C. W., & Cassidy, J. (2024). Developmental trajectories of adolescent internalizing symptoms and parental responses to distress. *Development and Psychopathology*, 1–12. https://doi.org/10.1017/s0954579424000361
81. Kaiser, R. H., Andrews-Hanna, J. R., Wager, T. D., & Pizzagalli, D. A. (2015). Large-Scale network dysfunction in major depressive disorder. *JAMA Psychiatry, 72*(6), 603. https://doi.org/10.1001/jamapsychiatry.2015.0071
82. Kamenov, K., Twomey, C., Cabello, M., Prina, A. M., & Ayuso-Mateos, J. L. (2016). The efficacy of psychotherapy, pharmacotherapy and their combination on functioning and quality of life in depression: a meta-analysis. *Psychological Medicine, 47*(3), 414–425. https://doi.org/10.1017/s0033291716002774
83. Karamacoska, D., Barry, R. J., & Steiner, G. Z. (2017). Resting state intrinsic EEG impacts on go stimulus-response processes. *Psychophysiology, 54*(6), 894–903. https://doi.org/10.1111/psyp.12851
84. Karamacoska, D., Barry, R. J., & Steiner, G. Z. (2019). Using principal components analysis to examine resting state EEG in relation to task performance. *Psychophysiology, 56*(5). https://doi.org/10.1111/psyp.13327

85. Karamacoska, D., Barry, R. J., Steiner, G. Z., Coleman, E. P., & Wilson, E. J. (2018). Intrinsic EEG and task-related changes in EEG affect Go/NoGo task performance. *International Journal of Psychophysiology, 125*, 17–28. https://doi.org/10.1016/j.ijpsycho.2018.01.015
86. Karrouri, R., Hammani, Z., Benjelloun, R., & Otheman, Y. (2021). Major depressive disorder: Validated treatments and future challenges. *World Journal of Clinical Cases, 9*(31), 9350–9367. https://doi.org/10.12998/wjcc.v9.i31.9350
87. Karyotaki, E., Smit, Y., Henningsen, K. H., Huibers, M., Robays, J., De Beurs, D., & Cuijpers, P. (2016). Combining pharmacotherapy and psychotherapy or monotherapy for major depression? A meta-analysis on the long-term effects. *Journal of Affective Disorders, 194*, 144–152. https://doi.org/10.1016/j.jad.2016.01.036
88. Keefe, R. S. E., McClintock, S. M., Roth, R. M., Doraiswamy, P. M., Tiger, S., & Madhoo, M. (2014). Cognitive effects of pharmacotherapy for major depressive Disorder. *The Journal of Clinical Psychiatry, 75*(08), 864–876. https://doi.org/10.4088/jcp.13r08609
89. Kennedy, S. H. (2008). Core symptoms of major depressive disorder: relevance to diagnosis and treatment. *Dialogues in Clinical Neuroscience, 10*(3), 271–277. https://doi.org/10.31887/dcns.2008.10.3/shkennedy
90. Kerestes, R., Davey, C. G., Stephanou, K., Whittle, S., & Harrison, B. J. (2013). Functional brain imaging studies of youth depression: A systematic review. *NeuroImage Clinical, 4*, 209–231. https://doi.org/10.1016/j.nicl.2013.11.009
91. Kim, E. Y., Kim, S. H., Lee, H. J., Lee, N. Y., Kim, H. Y., Park, C. H. K., & Ahn, Y. M. (2019). A randomized, double-blind, 6-week prospective pilot study on the efficacy and safety of dose escalation in non-remitters in comparison to those of the standard dose of escitalopram for major depressive disorder. *Journal of Affective Disorders, 259*, 91–97. https://doi.org/10.1016/j.jad.2019.08.057
92. Kim, H., Ährlund-Richter, S., Wang, X., Deisseroth, K., & Carlén, M. (2016). Prefrontal parvalbumin neurons in control of attention. *Cell, 164*(1–2), 208–218. https://doi.org/10.1016/j.cell.2015.11.038
93. Kim, M., & Lee, U. (2019). Alpha oscillation, criticality, and responsiveness in complex brain networks. *Network Neuroscience, 4*(1), 155–173. https://doi.org/10.1162/netn_a_00113
94. Lam, R. W., Kennedy, S. H., McIntyre, R. S., & Khullar, A. (2014). Cognitive dysfunction in major depressive Disorder: Effects on psychosocial functioning and implications for treatment. *The Canadian Journal of Psychiatry, 59*(12), 649–654. https://doi.org/10.1177/070674371405901206
95. Leuchter, A. F., Cook, I. A., Hunter, A. M., & Korb, A. S. (2009). A new paradigm for the prediction of antidepressant treatment response. Dialogues in Clinical Neuroscience, 11(4), 435–446. https://doi.org/10.31887/dcns.2009.11.4/afleuchter
96. Li, S., Zhang, X., Cai, Y., Zheng, L., Pang, H., & Lou, L. (2023). Sex difference in incidence of major depressive disorder: an analysis from the Global Burden of Disease Study 2019. *Annals of General Psychiatry, 22*(1). https://doi.org/10.1186/s12991-023-00486-7
97. Li, Y., Cao, D., Wei, L., Tang, Y., & Wang, J. (2015). Abnormal functional connectivity of EEG gamma band in patients with depression during emotional face processing. *Clinical Neurophysiology, 126*(11), 2078–2089. https://doi.org/10.1016/j.clinph.2014.12.026
98. Li, Y., Kang, C., Qu, X., Zhou, Y., Wang, W., & Hu, Y. (2016). Depression-Related brain connectivity analyzed by EEG Event-Related Phase Synchrony Measure. *Frontiers in Human Neuroscience, 10*. https://doi.org/10.3389/fnhum.2016.00477
99. Li, Y., Kang, C., Wei, Z., Qu, X., Liu, T., Zhou, Y., & Hu, Y. (2017). Beta oscillations in major depression – signalling a new cortical circuit for central executive function. *Scientific Reports, 7*(1). https://doi.org/10.1038/s41598-017-18306-w
100. Li, P., Yokoyama, M., Okamoto, D., Nakatani, H., & Yagi, T. (2023). Depressive states in healthy subjects lead to biased processing in frontal-parietal ERPs during emotional stimuli. *Scientific Reports, 13*(1). https://doi.org/10.1038/s41598-023-44368-0
101. Lin, Y., Huang, S., Mao, J., Li, M., Haihambo, N., Wang, F., Liang, Y., Chen, W., & Han, C. (2024). The neural oscillatory mechanism underlying human brain fingerprint recognition

using a portable EEG acquisition device. *NeuroImage, 294*, 120637. https://doi.org/10.1016/j.neuroimage.2024.120637

102. Lipari, R. N., & Park-Lee, E. (2019). Key substance use and mental health indicators in the United States: results from the 2018 National Survey on Drug Use and Health. Substance Abuse and Mental Health Services Administration, 1–67.
103. Liu, X., Liu, S., Li, M., Su, F., Chen, S., Ke, Y., & Ming, D. (2022). Altered gamma oscillations and beta–gamma coupling in drug-naive first-episode major depressive disorder: Association with sleep and cognitive disturbance. *Journal of Affective Disorders, 316*, 99–108. https://doi.org/10.1016/j.jad.2022.08.022
104. Liu, X., Zhang, H., Cui, Y., Zhao, T., Wang, B., Xie, X., Liang, S., Sha, S., Yan, Y., Zhao, X., & Zhang, L. (2024). EEG-based major depressive disorder recognition by neural oscillation and asymmetry. Frontiers in Neuroscience, 18. https://doi.org/10.3389/fnins.2024.1362111
105. Luck, S. J. (2005). *An introduction to the Event-Related Potential technique.* http://library.mpib-berlin.mpg.de/toc/z2008_2218.pdf
106. Lyons, S., Beck, I., & Depue, B. (2025). Depression is marked by differences in structural covariance between deep-brain nuclei and sensorimotor cortex. *NeuroImage*, 121127.https://doi.org/10.1016/j.neuroimage.2025.121127
107. Lütkepohl, H. (2005). New introduction to multiple time series analysis. In *Springer eBooks.* https://doi.org/10.1007/978-3-540-27752-1
108. Malhi, G. S., & Mann, J. J. (2018). Depression. *The Lancet, 392*(10161), 2299–2312. https://doi.org/10.1016/s0140-6736(18)31948-2
109. Margoni, M., Preziosa, P., Rocca, M. A., & Filippi, M. (2023). Depressive symptoms, anxiety and cognitive impairment: emerging evidence in multiple sclerosis. *Translational Psychiatry, 13*(1). https://doi.org/10.1038/s41398-023-02555-7
110. Marshall, P. S., Forstot, M., Callies, A., Peterson, P. K., & Schenck, C. H. (1997). Cognitive slowing and working memory difficulties in chronic fatigue syndrome. *Psychosomatic Medicine, 59*(1), 58–66. https://doi.org/10.1097/00006842-199701000-00008
111. Masi, G., Liboni, F., & Brovedani, P. (2010). Pharmacotherapy of major depressive disorder in adolescents. *Expert Opinion on Pharmacotherapy, 11*(3), 375–386. https://doi.org/10.1517/14656560903527226
112. Mathew, A. R., Pettit, J. W., Lewinsohn, P. M., Seeley, J. R., & Roberts, R. E. (2011). Comorbidity between major depressive disorder and anxiety disorders: shared etiology or direct causation? *Psychological Medicine, 41*(10), 2023–2034. https://doi.org/10.1017/s0033291711000407
113. Mattar MG, Talmi D. Patterns of Neural Oscillations in Emotional Memory Discrimination. Neuron. 2019 May 22;102(4):715–717. https://doi.org/10.1016/j.neuron.2019.05.006.
114. McCulloch, A., Kroll, L., Glass, J., & Dubicka, B. (2021). A systematic review of the characteristics of adolescents with major depressive disorder in randomised controlled treatment trials. *The European Journal of Psychiatry, 36*(1), 1–10. https://doi.org/10.1016/j.ejpsy.2021.07.001
115. McVoy, M., Aebi, M. E., Loparo, K., Lytle, S., Morris, A., Woods, N., Deyling, E., Tatsuoka, C., Kaffashi, F., Lhatoo, S., & Sajatovic, M. (2019). Resting-State Quantitative Electroencephalography Demonstrates Differential Connectivity in Adolescents with Major Depressive Disorder. *Journal of Child and Adolescent Psychopharmacology, 29*(5), 370–377. https://doi.org/10.1089/cap.2018.0166
116. Mendoza, J. (2019). Circadian insights into the biology of depression: Symptoms, treatments and animal models. *Behavioural Brain Research, 376*, 112186. https://doi.org/10.1016/j.bbr.2019.112186
117. Michel, C. M., & Brunet, D. (2019). EEG Source Imaging: A practical review of the analysis steps. *Frontiers in Neurology, 10.* https://doi.org/10.3389/fneur.2019.00325
118. Miljevic, A., Bailey, N. W., Vila-Rodriguez, F., Herring, S. E., & Fitzgerald, P. B. (2021). Electroencephalographic Connectivity: A fundamental guide and checklist for optimal study design and evaluation. *Biological Psychiatry Cognitive Neuroscience and Neuroimaging, 7*(6), 546–554. https://doi.org/10.1016/j.bpsc.2021.10.017

119. Mitiureva, D., Sysoeva, O., Proshina, E., Portnova, G., Khayrullina, G., & Martynova, O. (2024). Comparative analysis of resting-state EEG functional connectivity in depression and obsessive-compulsive disorder. *Psychiatry Research Neuroimaging*, *342*, 111828. https://doi.org/10.1016/j.pscychresns.2024.111828
120. Monni A, Collison KL, Hill KE, Oumeziane BA, Foti D. The novel frontal alpha asymmetry factor and its association with depression, anxiety, and personality traits. Psychophysiology. 2022 Nov;59(11):e14109. https://doi.org/10.1111/psyp.14109. Epub 2022 May 26. PMID: 35616309; PMCID: PMC9532346.
121. Moretta, T., & Benvenuti, S. M. (2023). Familial risk for depression is associated with reduced P300 and late positive potential to affective stimuli and prolonged cardiac deceleration to unpleasant stimuli. *Scientific Reports*, *13*(1). https://doi.org/10.1038/s41598-023-33534-z
122. Mouradian, C., Naboulsi, D., Yangui, S., Glitho, R. H., Morrow, M. J., & Polakos, P. A. (2017). A comprehensive survey on FoG Computing: State-of-the-Art and Research challenges. *IEEE Communications Surveys & Tutorials*, *20*(1), 416–464. https://doi.org/10.1109/comst.2017.2771153
123. Mullen, S. (2018). Major depressive disorder in children and adolescents. *Mental Health Clinician*, *8*(6), 275–283. https://doi.org/10.9740/mhc.2018.11.275
124. Murphy, M. J., & Peterson, M. J. (2014). Sleep disturbances in depression. *Sleep Medicine Clinics*, *10*(1), 17–23. https://doi.org/10.1016/j.jsmc.2014.11.009
125. Newson, J. J., & Thiagarajan, T. C. (2019). EEG Frequency Bands in Psychiatric Disorders: A review of resting State studies. *Frontiers in Human Neuroscience*, *12*. https://doi.org/10.3389/fnhum.2018.00521
126. Nutt, D., Wilson, S., & Paterson, L. (2008). Sleep disorders as core symptoms of depression. *Dialogues in Clinical Neuroscience*, *10*(3), 329–336. https://doi.org/10.31887/dcns.2008.10.3/dnutt
127. Obuobi-Donkor, G., Nkire, N., & Agyapong, V. I. O. (2021). Prevalence of major depressive disorder and correlates of thoughts of death, suicidal behaviour, and death by suicide in the Geriatric Population—A General Review of Literature. *Behavioral Sciences*, *11*(11), 142. https://doi.org/10.3390/bs11110142
128. Olatunji BO, Williams NL, Tolin DF, Abramowitz JS, Sawchuk CN, Lohr JM, Elwood LS. The Disgust Scale: item analysis, factor structure, and suggestions for refinement. Psychological assessment. 2007 Sep;19(3):281.
129. Oliver, A., Pile, V., Elm, D., & Lau, J. Y. F. (2019). The Cognitive Neuropsychology of Depression in Adolescents. *Current Behavioral Neuroscience Reports*, *6*(4), 227–235. https://doi.org/10.1007/s40473-019-00187-0
130. Orsolini, L., Latini, R., Pompili, M., Serafini, G., Volpe, U., Vellante, F., Fornaro, M., Valchera, A., Tomasetti, C., Fraticelli, S., Alessandrini, M., La Rovere, R., Trotta, S., Martinotti, G., Di Giannantonio, M., & De Berardis, D. (2020). Understanding the Complex of Suicide in Depression: from Research to Clinics. *Psychiatry Investigation*, *17*(3), 207–221. https://doi.org/10.30773/pi.2019.0171
131. Özçoban, M. A., & Tan, O. (2025). Electroencephalographic markers in Major Depressive Disorder: insights from absolute, relative power, and asymmetry analyses. Frontiers in Psychiatry, 15. https://doi.org/10.3389/fpsyt.2024.1480228
132. Park K, Yoon S, Cho S, Choi Y, Lee S-H and Choi K-H (2022) Final validation of the mental health screening tool for depressive disorders: A brief online and offline screening tool for major depressive disorder. Front. Psychol. 13:992068. https://doi.org/10.3389/fpsyg.2022.992068
133. Park, Y., Cha, Y., Kim, H., Kim, Y., Woo, J. H., Cho, H., Mashour, G. A., Xu, T., Lee, U., Hong, S., Honey, C. J., & Moon, J. (2025). Sub-second Fluctuation between Top-Down and Bottom-Up Modes Distinguishes Diverse Human Brain States. *bioRxiv (Cold Spring Harbor Laboratory)*. https://doi.org/10.1101/2025.03.12.642768
134. Passynkova, N. R., & Volf, N. V. (2001). Seasonal affective disorder: spatial organization of EEG power and coherence in the depressive state and in light-induced and summer remission. *Psychiatry Research Neuroimaging*, *108*(3), 169–185. https://doi.org/10.1016/s0925-4927(01)00122-6

135. Patil, P., Du, J., & Tibshirani, R. J. (2024). Optimal ridge regularization for Out-of-Distribution prediction. *arXiv (Cornell University)*. https://doi.org/10.48550/arxiv.2404.01233
136. Peng, H., Xia, C., Wang, Z., Zhu, J., Zhang, X., Sun, S., Li, J., Huo, X., & Li, X. (2019). Multivariate pattern analysis of EEG-Based Functional Connectivity: A study on the identification of depression. *IEEE Access, 7*, 92630–92641. https://doi.org/10.1109/access.2019.2927121
137. Pirani S, Kulhanek C, Wainwright K, Osman A. The reasons for living inventory for young adults (RFL-YA-II). Assessment. 2021 Apr;28(3):942–954.
138. Pizzagalli, D. A., Peccoralo, L. A., Davidson, R. J., & Cohen, J. D. (2005). Resting anterior cingulate activity and abnormal responses to errors in subjects with elevated depressive symptoms: A 128-channel EEG study. *Human Brain Mapping, 27*(3), 185–201. https://doi.org/10.1002/hbm.20172
139. Racine, N., McArthur, B. A., Cooke, J. E., Eirich, R., Zhu, J., & Madigan, S. (2021). Global prevalence of depressive and anxiety symptoms in children and adolescents during COVID-19. *JAMA Pediatrics, 175*(11), 1142. https://doi.org/10.1001/jamapediatrics.2021.2482
140. Ray, K. L., Griffin, N. R., Shumake, J., Alario, A., Allen, J. J., Beevers, C. G., & Schnyer, D. M. (2023). Altered electroencephalography resting state network coherence in remitted MDD. Brain Research, 1806, 148282. https://doi.org/10.1016/j.brainres.2023.148282
141. Roh, S., Park, E., Shim, M., & Lee, S. (2016). EEG beta and low gamma power correlates with inattention in patients with major depressive disorder. *Journal of Affective Disorders, 204*, 124–130. https://doi.org/10.1016/j.jad.2016.06.033
142. Rolls, E. T., Cheng, W., & Feng, J. (2020). The orbitofrontal cortex: reward, emotion and depression. *Brain Communications, 2*(2). https://doi.org/10.1093/braincomms/fcaa196
143. Salcedo S, Chen YL, Youngstrom EA, Fristad MA, Gadow KD, Horwitz SM, Frazier TW, Arnold LE, Phillips ML, Birmaher B, Kowatch RA, Findling RL. Diagnostic Efficiency of the Child and Adolescent Symptom Inventory (CASI-4R) Depression Subscale for Identifying Youth Mood Disorders. J Clin Child Adolesc Psychol. 2018 Sep- Oct;47(5):832–846. https://doi.org/10.1080/15374416.2017.1280807.
144. Schumacher, A., Campisi, S. C., Khalfan, A. F., Merriman, K., Williams, T. S., & Korczak, D. J. (2023). Cognitive functioning in children and adolescents with depression: A systematic review and meta-analysis. *European Neuropsychopharmacology, 79*, 49–58. https://doi.org/10.1016/j.euroneuro.2023.11.005
145. Sghirripa, S., Graetz, L., Merkin, A., Rogasch, N. C., Semmler, J. G., & Goldsworthy, M. R. (2020). Load-dependent modulation of alpha oscillations during working memory encoding and retention in young and older adults. *Psychophysiology, 58*(2). https://doi.org/10.1111/psyp.13719
146. Shorey, S., Ng, E. D., & Wong, C. H. J. (2021). Global prevalence of depression and elevated depressive symptoms among adolescents: A systematic review and meta-analysis. *British Journal of Clinical Psychology, 61*(2), 287–305. https://doi.org/10.1111/bjc.12333
147. Stockings E, Degenhardt L, Lee YY, Mihalopoulos C, Liu A, Hobbs M, Patton G. Symptom screening scales for detecting major depressive disorder in children and adolescents: a systematic review and meta-analysis of reliability, validity and diagnostic utility. Journal of affective disorders. 2015 Mar 15;174:447–463.
148. Straub, J., Brown, R., Malejko, K., Bonenberger, M., Grön, G., Plener, P. L., & Abler, B. (2019). Adolescent depression and brain development: evidence from voxel-based morphometry. *Journal of Psychiatry and Neuroscience, 44*(4), 237–245. https://doi.org/10.1503/jpn.170233
149. Strelets, V. B., Garakh, Z. V., & Novototskii-Vlasov, V. Y. (2007). Comparative study of the gamma rhythm in normal conditions, during examination stress, and in patients with first depressive episode. *Neuroscience and Behavioral Physiology, 37*(4), 387–394. https://doi.org/10.1007/s11055-007-0025-4
150. Su, Z., Zhang, H., Wang, Y., Chen, B., Zhang, Z., Wang, B., Liu, J., Shi, Y., & Zhao, X. (2024). Neural oscillation in bipolar disorder: a systematic review of resting-state electroencephalography studies. *Frontiers in Neuroscience, 18*. https://doi.org/10.3389/fnins.2024.1424666

151. Su, C., Tang, Y., Tang, N., Liu, N., Li, J., Qi, S., Wang, H., & Huang, Z. (2025). Unveiling the dynamic effects of major depressive disorder and its rTMS interventions through energy landscape analysis. *Frontiers in Neuroscience, 19.* https://doi.org/10.3389/fnins.2025.1444999
152. Sundararajan, M., Taly, A., & Yan, Q. (2017). Axiomatic Attribution for Deep Networks. Proceedings of the 34th International Conference on Machine Learning (ICML), 70, 3319–3328.
153. Tei, S., Faber, P. L., Lehmann, D., Tsujiuchi, T., Kumano, H., Pascual-Marqui, R. D., Gianotti, L. R. R., & Kochi, K. (2009). Meditators and Non-Meditators: EEG source imaging during resting. *Brain Topography, 22*(3), 158–165. https://doi.org/10.1007/s10548-009-0107-4
154. Thapar, A., Collishaw, S., Pine, D. S., & Thapar, A. K. (2012). Depression in adolescence. *The Lancet, 379*(9820), 1056–1067. https://doi.org/10.1016/s0140-6736(11)60871-4
155. Thibodeau, R., Jorgensen, R. S., & Kim, S. (2006). Depression, anxiety, and resting frontal EEG asymmetry: A meta-analytic review. *Journal of Abnormal Psychology, 115*(4), 715–729. https://doi.org/10.1037/0021-843x.115.4.715
156. Tian, H., Wang, Z., Meng, Y., Geng, L., Lian, H., Shi, Z., Zhuang, Z., Cai, W., & He, M. (2024). Neural mechanisms underlying cognitive impairment in depression and cognitive benefits of exercise intervention. *Behavioural Brain Research, 476*, 115218. https://doi.org/10.1016/j.bbr.2024.115218
157. Vairavan, S., Rashidisabet, H., Li, Q. S., Ness, S., Morrison, R. L., Soares, C. N., Uher, R., Frey, B. N., Lam, R. W., Kennedy, S. H., Trivedi, M., Drevets, W. C., & Narayan, V. A. (2023). Personalized relapse prediction in patients with major depressive disorder using digital biomarkers. *Scientific Reports, 13*(1). https://doi.org/10.1038/s41598-023-44592-8
158. Van De Wiel, M. A., Van Nee, M. M., & Rauschenberger, A. (2021). Fast cross-validation for multi-penalty high-dimensional ridge regression. *Journal of Computational and Graphical Statistics, 30*(4), 835–847. https://doi.org/10.1080/10618600.2021.1904962
159. Villalobos, D., Pacios, J., & Vázquez, C. (2021). Cognitive Control, Cognitive Biases and Emotion Regulation in Depression: A new proposal for an Integrative interplay model. *Frontiers in Psychology, 12.* https://doi.org/10.3389/fpsyg.2021.628416
160. Wagner, S., Müller, C., Helmreich, I., Huss, M., & Tadić, A. (2014). A meta-analysis of cognitive functions in children and adolescents with major depressive disorder. *European Child & Adolescent Psychiatry, 24*(1), 5–19. https://doi.org/10.1007/s00787-014-0559-2
161. Wang, X., Chen, H., Liu, Y., Zhao, Z., & Zang, S. (2023). Association between depression status in adolescents and cognitive performance over the subsequent six years: A longitudinal study. *Journal of Affective Disorders, 329*, 105–112. https://doi.org/10.1016/j.jad.2023.02.051
162. Watson, B. O., Ding, M., & Buzsáki, G. (2017). Temporal coupling of field potentials and action potentials in the neocortex. *European Journal of Neuroscience, 48*(7), 2482–2497. https://doi.org/10.1111/ejn.13807
163. Wenzlaff, R.M., Luxton, D.D. (2003). The Role of Thought Suppression in Depressive Rumination. *Cognitive Therapy and Research* **27**, 293–308. https://doi.org/10.1023/A:1023966400540
164. Weissman, D. G., Bitran, D., Miller, A. B., Schaefer, J. D., Sheridan, M. A., & McLaughlin, K. A. (2019). Difficulties with emotion regulation as a transdiagnostic mechanism linking child maltreatment with the emergence of psychopathology. *Development and Psychopathology, 31*(3), 899–915. https://doi.org/10.1017/s0954579419000348
165. Welch, P. (1967). The use of fast Fourier transform for the estimation of power spectra: A method based on time averaging over short, modified periodograms. *IEEE Transactions on Audio and Electroacoustics, 15*(2), 70–73. https://doi.org/10.1109/tau.1967.1161901
166. Whitton, A. E., Deccy, S., Ironside, M. L., Kumar, P., Beltzer, M., & Pizzagalli, D. A. (2017). Electroencephalography source functional connectivity reveals abnormal High-Frequency communication among Large-Scale functional networks in depression. *Biological Psychiatry Cognitive Neuroscience and Neuroimaging, 3*(1), 50–58. https://doi.org/10.1016/j.bpsc.2017.07.001

167. Young, I. M., Dadario, N. B., Tanglay, O., Chen, E., Cook, B., Taylor, H. M., Crawford, L., Yeung, J. T., Nicholas, P. J., Doyen, S., & Sughrue, M. E. (2023). Connectivity model of the anatomic substrates and network abnormalities in major depressive disorder: A coordinate meta-analysis of resting-state functional connectivity. *Journal of Affective Disorders Reports*, *11*, 100478. https://doi.org/10.1016/j.jadr.2023.100478
168. Zbozinek, T. D., Rose, R. D., Wolitzky-Taylor, K. B., Sherbourne, C., Sullivan, G., Stein, M. B., Roy-Byrne, P. P., & Craske, M. G. (2012). DIAGNOSTIC OVERLAP OF GENERALIZED ANXIETY DISORDER AND MAJOR DEPRESSIVE DISORDER IN a PRIMARY CARE SAMPLE. *Depression and Anxiety*, *29*(12), 1065–1071. https://doi.org/10.1002/da.22026
169. Zhang, A., Zhang, Q., Zhao, Z., Li, Q., Li, F., Hu, Y., Huang, X., Kuang, W., Kemp, G. J., Zhao, Y., & Gong, Q. (2025). The neural association between symptom and cognition in major depressive Disorder: A Network Control Theory study. *Human Brain Mapping*, *46*(5). https://doi.org/10.1002/hbm.70198
170. Zhang, H., Liu, X., Su, Z., Wang, Y., Chen, B., Zhang, Z., Wang, B., Zhou, J., Zhang, L., & Zhao, X. (2024). Alpha oscillation mediates the interaction between suicide risk and symptom severity in Major Depressive Disorder. *Frontiers in Neuroscience*, *18*. https://doi.org/10.3389/fnins.2024.1429019
171. Zhang, Y., Zhang, Z., Du, F., Song, J., Huang, S., Mao, J., Xiang, W., Wang, F., Liang, Y., Chen, W., Lin, Y., & Han, C. (2024). Shared oscillatory mechanisms of alpha-band activity in prefrontal regions in eyes open and closed state using a portable EEG acquisition device. *Scientific Reports*, *14*(1). https://doi.org/10.1038/s41598-024-78173-0
172. Zikereya, T., Lin, Y., Zhang, Z., Taguas, I., Shi, K., & Han, C. (2024). Different oscillatory mechanisms of dementia-related diseases with cognitive impairment in closed-eye state. *NeuroImage*, 120945. https://doi.org/10.1016/j.neuroimage.2024.120945

Effect of the Initial Condition on SCG Clustering Using Unsupervised Machine Learning

Sherif A. Farahat, Richard H. Sandler, and Hansen A. Mansy

Abstract Seismocardiographic (SCG) signal represents the vibrations of the chest surface associated with the cardiac cycle. It is clinically valuable however its variability limits its applications. Clustering SCG beats into groups of similar morphologies could reduce SCG variability and leverage its clinical utility. The K-Medoid algorithm is used extensively in clustering SCG however it requires an initial condition (i.e., two SCG beats that serve as initial medoids), and the final clustering solution might be sensitive to the choice of these initial conditions. This study aimed to investigate the effect of the initial condition on SCG final clustering solution. SCG and electrocardiography (ECG) were simultaneously acquired from 19 healthy young adult subjects for one minute. Signals were filtered and segmented. SCG baseline wander was used to calculate signals that correlated with the respiratory flow rate and lung volume. Then K-Medoid was applied along with Dynamic Time Warping (DTW) as a distance measure. All possible initial conditions were tested. All subjects had multiple clustering solutions and initiating the clustering algorithm with only one random initial condition is not recommended since it can result in sparse clustering and other non-optimal solutions. The number of final unique clustering solutions is much smaller than the number of possible initial conditions. The solution that had the maximum difference between the inter-cluster and intra-cluster variability did not always have the highest purity (i.e., separability between the clusters based on the respiratory flow rate and lung volume). Initiating the clustering algorithm with lung volume or respiratory flow rate based initial conditions does not always guarantee that maximum variability difference solution will be obtained. The breathing phase

S. A. Farahat (✉) · R. H. Sandler · H. A. Mansy
Biomedical Acoustic Research Lab, University of Central Florida, Orlando, FL, USA
e-mail: sherif.farahat@ucf.edu

H. A. Mansy
e-mail: Hansen.Mansy@ucf.edu

S. A. Farahat
Mechanical Power Engineering Department, Zagazig University, Zagazig, Egypt

R. H. Sandler · H. A. Mansy
Biomedical Acoustisc Research Company, Orlando, FL, USA

A. Ahmed and J. Picone (eds.), *Signal Processing and Biomedical Engineering Research*, https://doi.org/10.1007/978-3-032-18509-9_2

angle of maximum variability difference solution had two ranges: 3–46° and 106–148°. The study suggested a possible time-efficient procedure to obtain maximum variability difference and highest purity solutions based on considering a subset of the segmented SCG beats (i.e., 10% that are not chosen randomly) for initial conditions.

Keywords SCG · Clustering · K-Medoid · Initial condition · Variability · Purity

1 Introduction

Seismocardiographic (SCG) signal represents the vibrations of the chest surface associated with cardiac muscle contraction-relaxation cycle, blood momentum changes and opening and closure of cardiac valves [1–4]. Accelerometers are often used to collect SCG signals [4], however, non-contact methods (e.g., ultrasound [5] and Doppler [6]) were also used. SCG is clinically valuable due to the potential correlation between its fiducial points and the different mechanical events in the heart [1, 4, 7]. Therefore, it was used to estimate cardiac contractility [8] and cardiac time intervals (CTI) [9] and monitor sleep apnea [10] and heart failure [11, 12]. Unlike standard stethoscopes that acquire audible sound, SCG acquires both audible and subaudible sounds [3, 4] which can extend its clinical utility [12].

SCG variability might reduce the likelihood of extracting clinically valuable features [4, 13, 14]. Different factors were reported to induce variability including posture [13, 15], motion [16], exercise [17, 18], sensor location [7, 19], and respiration [20–22].

Respiration is a source of SCG variability that may be driven by cardiopulmonary interactions that are associated with periodic changes in lung volume and intrathoracic pressure [23]. During inspiration, the right ventricle preload (i.e., cardiac myocytes stretching prior to contraction [24]) increases, the left ventricle preload decreases [23], the left ventricle afterload (i.e., the force generated by the heart to allow blood ejection [24]) increases [23], the intensity of the first heart sound decreases [25], the intensity of the of the second heart sound increases [25], and the time interval between the two heart sounds decreases [25]. In a recent study [20], it was shown that, compared to normal breathing, SCG variability decreased during breath holding (BH) at zero airway gauge pressure and increased as the pressure deviated from zero. The airway pressure in that study [20] was used as a correlate to the intrathoracic pressure. The same study [20] addressed BH at two different lung volumes (i.e., end inspiration and end expiration) and it was found that the average SCG variability of these two lung volumes was lower than the variability when combining them as one cluster (i.e., not accounting for the lung volume difference). Therefore, it was concluded that lung volume and heart location (which change during respiration) were significant sources of SCG variability.

Supervised machine learning was used in SCG analysis to predict coronary artery disease [26, 27], aortic stenosis [12], atrial fibrillation [28], and respiration phases [29–31]. Unlike supervised machine learning, unsupervised machine learning (i.e., clustering) does not require prior knowledge of data clusters. Moreover, clustering SCG beats into groups of similar morphologies could reduce SCG variability which can help extract more accurate SCG features [14]. SCG beats are often obtained by segmenting SCG using the timing of the R peak of the electrocardiography (ECG) signal [14, 19, 31]. Three methods were previously used for performing SCG clustering: (1) respiratory-flow-rate-based grouping (i.e., inspiration vs expiration) [22], (2) lung-volume-based grouping (i.e., high vs low lung volume) [22], and (3) unsupervised machine learning [14, 32]. All these methods were found to reduce SCG variability, but unsupervised machine learning was more effective [14].

Among the unsupervised machine learning methods, that are used in SCG clustering, is K-Medoid algorithm [14, 20, 32]. K-Medoid algorithm starts by choosing an initial condition (i.e., initial medoids) then assigning SCG beats to the nearest "medoid." Once all beats are assigned to a cluster, new medoids are determined and clustering is repeated until cluster assignments stop changing [14]. The initial condition (choice of initial medoids) is either chosen randomly [14] or based on physiology (e.g., high/low lung volume or inspiration/expiration) [32]. However, these choices might not guarantee that a good clustering solution is obtained as demonstrated by the current study results. The literature has not addressed, in detail, the effect of the initial condition on SCG clustering as different initial conditions might result in different clustering solutions. Therefore, the objectives of the current study are to

1. investigate the effect of the initial condition on SCG final clustering solution
2. investigate the characteristics of different clustering solutions
3. recommend a procedure for choosing the initial condition such that certain clustering solutions can be obtained

2 Methods

2.1 Data Acquisition

After IRB approval and subjects' consent, SCG and electrocardiography (ECG) were simultaneously acquired from 19 healthy subjects for one minute. Subjects were instructed to breathe at a steady comfortable rate and depth of respiration (i.e., tidal volume). Subjects were not given feedback about their breathing pattern. Demographics of the subjects are shown in Table 1. SCG was acquired using a uniaxial accelerometer (dorsoventral direction) attached at the 4th intercoastal space near the left sternal border by a double-sided medical tape. The chosen accelerometer location was reported to have high separability between SCG clusters and high signal-to-noise ratio [33]. The data was acquired at sampling rate of 10 kHz and later down sampled. Spirometry was not acquired. Instead, signals correlated to breathing flow rate and lung volume were extracted from raw SCG signal as described later in sect. 2.4.

Table 1 Demographics of the study's subjects

Number of subjects	19 (9 females, 10 males)
Age (years)	20.58 ± 1.43
Weight (kg)	65.85 ± 9.53
Height (cm)	171.63 ± 9.93
Body mass index (BMI)	22.39 ± 3.11 (Normal range:18.5–24.9)

2.2 Filtering

ECG was bandpass filtered (0.5–50 Hz), and SCG was bandpass filtered (0.5–250 Hz). The signals were filtered forward and backward to avoid phase distortion. The filter pass bands lower limits were selected to remove baseline wondering (e.g., breathing, motion artifacts, etc.). The higher limits were selected to remove high frequency noise like breathing sounds and speech. Filtered signals were downsampled to 500 Hz [14, 20] which provided enough time resolution and helped reduce the computation time of clustering compared to higher sampling rates [14, 20].

2.3 Segmentation

ECG R peaks were detected using Pan-Tompkins algorithm [34] as shown in Fig. 1. Pan-Tompkins algorithm is based on analyzing different slopes, amplitudes, and widths and applying certain thresholds on the ECG signal to accurately detect the R peaks [34]. Filtered SCG was segmented into beats such that each beat started 250 ms before the corresponding ECG R peak timing and ended 250 ms before the following ECG R peak. This approach was previously used [14, 19, 31] but with a time interval shorter than 250 ms (i.e., usually 100 ms). All the segmented beats, used in the clustering, were visually checked to ensure that there was no noise.

In the current study, the 250 ms interval ensured that each segmented SCG beat corresponded to a complete sequential ECG cycle (i.e., P-QRS-T waves). Figure 2 shows two segmented SCG and ECG beats that started 250 and 100 ms before the timing of the R wave and ended 250 and 100 ms before the timing of the following R wave, respectively. According to Fig. 2, using 100 ms excluded the P wave of the corresponding ECG beat and included the P wave of the following ECG beat (i.e., QRS-T-P). In Fig. 2, SCG and ECG beats were normalized by removing the mean value and dividing by the standard deviation according to Eq. 1. where X, μ_X, and σ_X are the signal, signal mean, and signal standard deviation respectively

$$X_{normalized} = (X - \mu_X)/\sigma_X \tag{1}$$

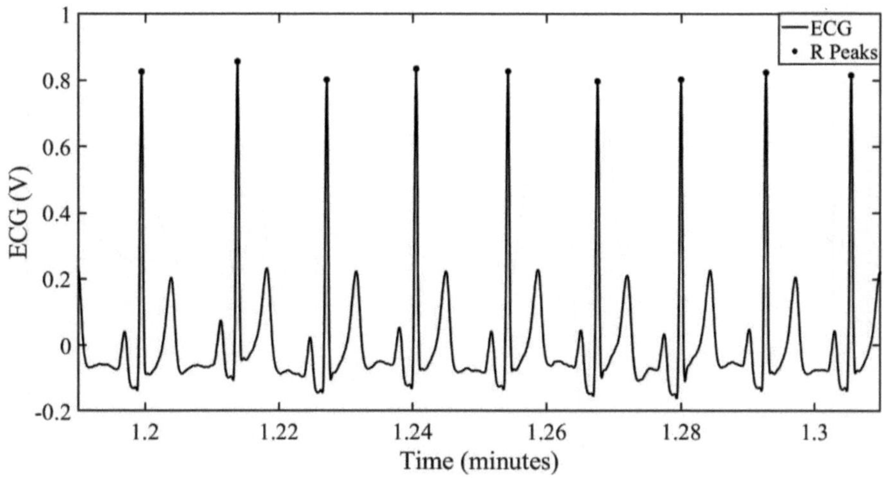

Fig. 1 Detecting R peaks using Pan-Tompkins algorithm after filtering and down-sampling ECG

Figure 3 shows two examples of the filtered segmented SCG beats (from the same subject); they are different not only in their durations (due to heart rate differences) but also in their morphologies.

2.4 *Breathing Signals Extraction*

In the current study, spirometry was not acquired. Rather, chest wall movement was extracted by applying a bandpass filter to raw SCG (passband: 0.1–0.4 Hz). It was then integrated once and twice to get signals that previously have been shown to correlate well with the respiratory flow rate (FL) and lung volume (LV), respectively [35]. Figure 4 shows an example of the extracted respiratory flow rate and lung volume signals. Here, each signal was normalized according to Eq. 1.

To track the timing of segmented SCG beats during the respiratory cycle, each SCG beat was assigned the FL and LV values at the timing of the corresponding ECG R wave [14, 32]. Figure 5 shows an example of the distribution of SCG beats in the normalized FL-LV space where each beat was represented by a point.

2.5 *SCG Clustering*

There are three categories of time series clustering algorithms: partitional, hierarchical, and model-based [36]. Among the partitional methods mostly used are K-medoid [32] and K-means algorithms [37]. In partitional methods, each cluster has a

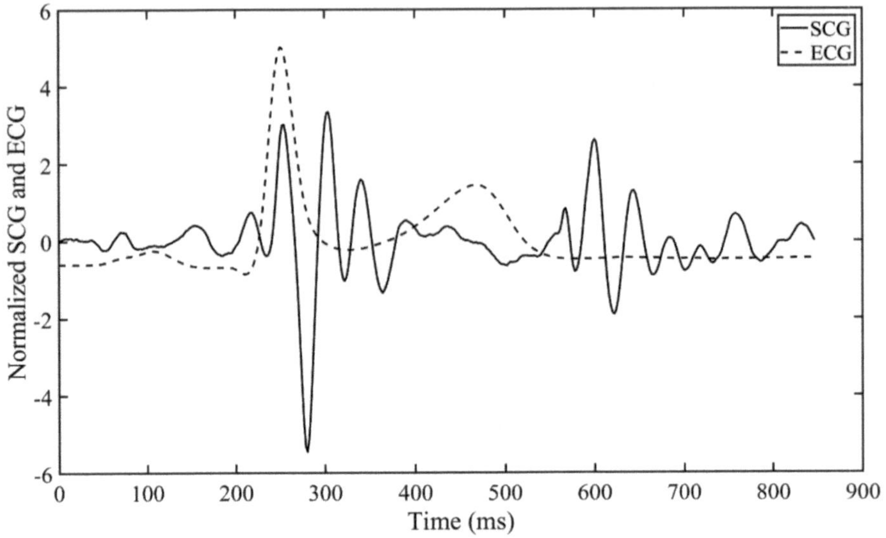

(a)

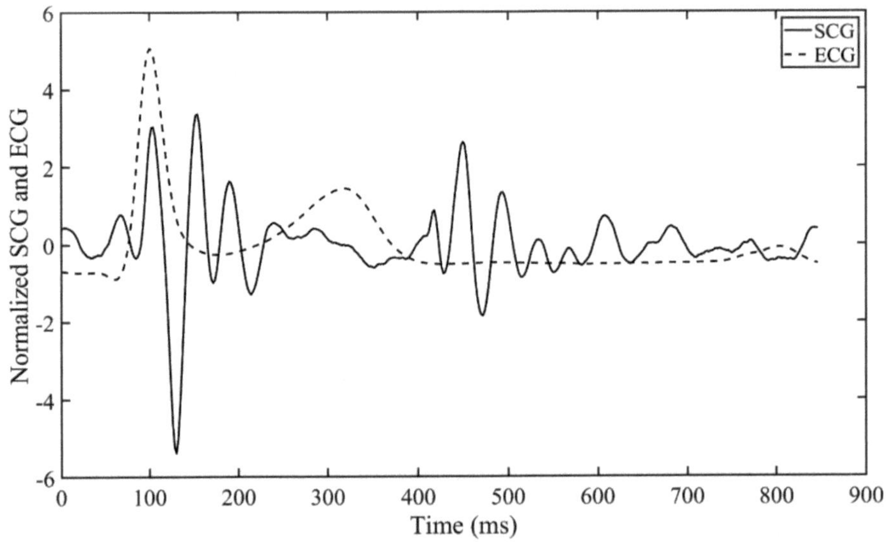

(b)

Fig. 2 Two segmented SCG and ECG beats that started **a** 250 and **b** 100 ms before the timing of the R wave and ended 250 and 100 ms before the timing of the following R wave respectively. Unlike100 ms, 250 ms case included the P wave of the corresponding ECG beat. In this figure, SCG and ECG beats were normalized by removing the mean value and dividing by the standard deviation according to Eq. 1

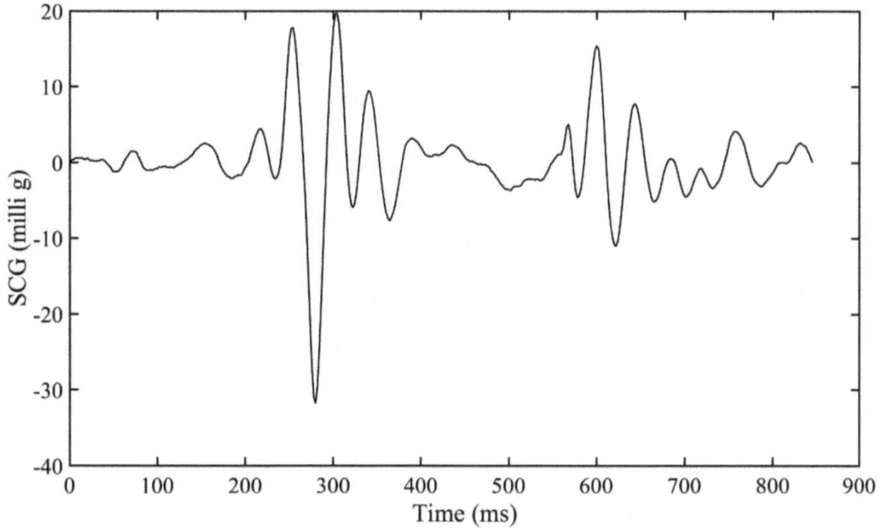

(a)

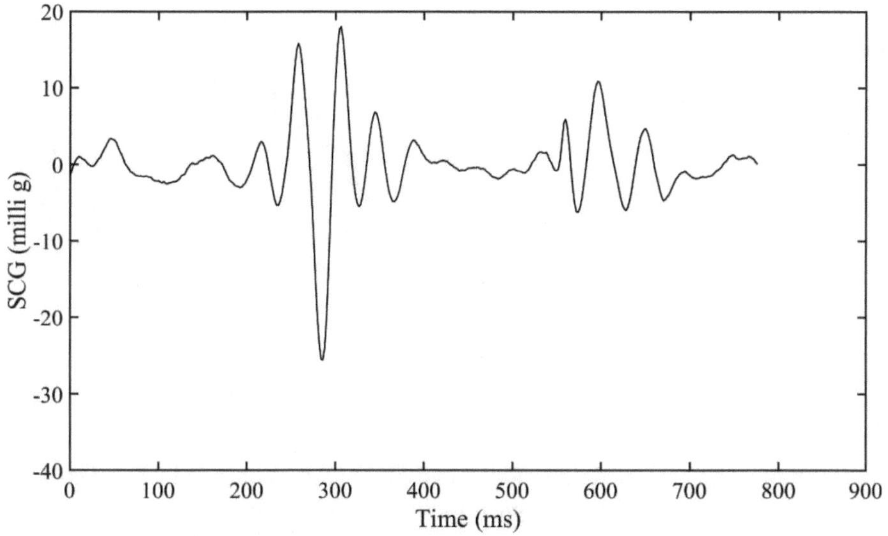

(b)

Fig. 3 Examples of the filtered segmented SCG beats. They are different not only in their durations (due to heart rate differences) but also in their morphologies. The "g" in the unit of SCG stands for the acceleration of gravity (9.81 m/s^2)

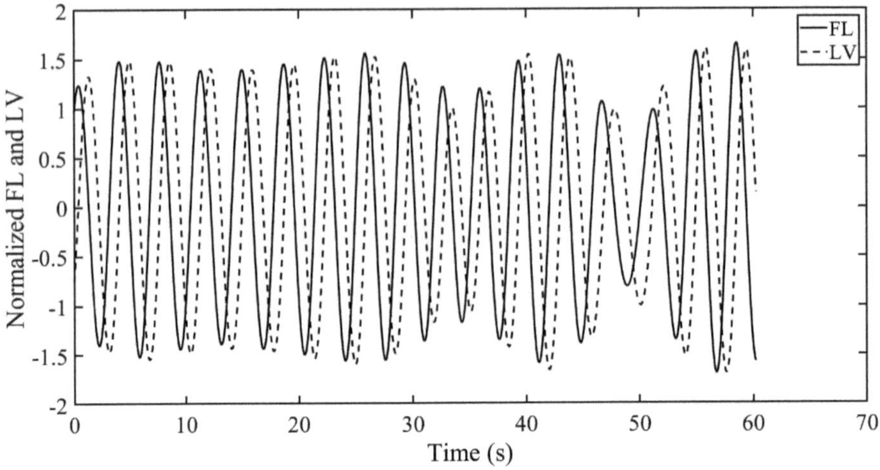

Fig. 4 Example of the signals extracted from SCG that correlated with respiratory flow rate and lung volume signals. These signals were normalized by removing the mean value and dividing by the standard deviation according to Eq. 1

Fig. 5 Example of the distribution of the segmented SCG beats in the normalized FL-LV space

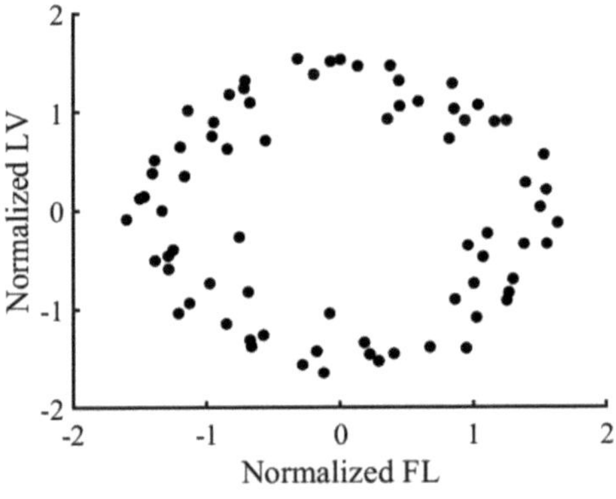

representative, i.e., the medoid in K-medoid [32] and the centroid in K-means [37]. Objects (e.g., SCG beats) are assigned to the cluster whose representative is closer in morphology to the assigned object. The medoid, in the K-medoid method, is the cluster member that is collectively closest to the rest of the members in the cluster [32]. The centroid in K-means is the average of the members in the cluster [37]; which requires SCG beats to be of the same length. If the beats were trimmed to satisfy this condition, this might lead to information loss [37]. Partitional methods iterate until cluster assignment stops changing [37–39]. Hierarchical methods are either agglomerative or divisive. Agglomerative algorithms assign each object to one cluster and merges clusters, according to a cost function that depends on the similarity between the data points in the clusters [40], while divisive algorithms do the opposite [36]. These methods are sensitive to outliers [14, 36]. In model-based clustering methods, a model is assumed for each cluster and each object is assigned to the cluster that fits its model [36]. In the current study, K-medoid was used to avoid the shortcomings of the other methods.

2.5.1 Distance Measure: Dynamic Time Warping (DTW)

Dynamic time warping (DTW) algorithm was used to measure the dissimilarity between SCG beats; this is required for cluster assignment. It was first introduced in speech pattern matching [41] and later it was used extensively in SCG clustering [14, 19, 37]. Ahdy et al. [32] showed how DTW was efficient in introducing local non-linear alignment prior to measuring the dissimilarity (also known as DTW distance) between SCG beats. However, DTW requires more computational cost than Euclidian distance [37]. The reader is referred to [32] for details about the implementation of the DTW algorithm in SCG distance measurements.

2.5.2 K-Medoid Algorithm

The number of clusters considered in the current study was two following recommendations from the literature that were based on the elbow method and average Silhouette value [42]. The steps of implementing the K-medoid algorithm are as follows:

1. Choose initial condition (i.e., two SCG beats as initial medoids).
2. Assign each SCG beat to the cluster whose medoid is closer (i.e., lower DTW distance). The two clusters at this step represent the initial clustering solution.
3. Update the medoids based on Eq. 2 where C_j, X_{i_j}, n_j, and DTW are the medoid of cluster "j," the ith beat in cluster "j," the number of beats in cluster "j," and the DTW distance. The medoid of a cluster is the beat which has the minimum sum of DTW distances from the rest of the beats in that cluster; that's why the "*argmin*" function was used in Eq. 2.

$$C_j = argmin_{y\epsilon\left\{X_{1_j},\ X_{2_j},\ \ldots,\ X_{i_j},\ldots,\ X_{n_j}\ \right\}}\left(\sum_{i=1}^{n_j} DTW\left(y, X_{i_j}\right)\right) \tag{2}$$

4. Repeat steps 2 and 3 until cluster assignment stops changing. The final clusters represent the final clustering solution.
5. Repeat steps 1 to 4 for all possible initial conditions (i.e., all SCG beat pairs without duplications).

2.5.3 Intra-cluster and Inter-cluster Variabilities

The intra-cluster and inter-cluster variabilities were calculated using Eqs. 3 and 4 [13], respectively. It is preferred that clustering solutions have a high inter-cluster variability to ensure the two clusters are dissimilar from each other and a low intra-cluster variability to ensure similar morphology within the same cluster [38].

$$Intra-cluster\ variability = \frac{1}{n_1 + n_2}\left(\sum_{i=1}^{n_1} DTW(C_1, x_{i1}) + \sum_{j=1}^{n_2} DTW\left(C_2, x_{j2}\right)\right) \quad (3)$$

$$Inter-cluster\ variability = \frac{1}{n_1 + n_2}\left(\sum_{i=1}^{n_1} DTW(C_2, x_{i1}) + \sum_{j=1}^{n_2} DTW\left(C_1, x_{j2}\right)\right) \quad (4)$$

2.5.4 Relative Weight of Clusters

The relative weight of each cluster was calculated by dividing the number of the beats in the cluster by the total number of beats. Figure 6 shows SCG over two breathing cycles in the same subject. Breathing was represented by the estimated FL signal. Both SCG and FL were normalized according to Eq. 1. In Fig. 6, SCG_1 (i.e., the wave between each two dashed lines) and SCG_2 (i.e., the wave between each two center lines) are the SCG wave structures corresponding to the first and second heart sounds respectively. Changes in SCG morphology were similar among these two breathing cycles. For example, during inspiration (i.e., positive FL), the peak-to-peak amplitude of SCG_1 (i.e., the vertical distance between the two circles) was lower than during expiration. During inspiration, the amplitude of SCG_2 at the "square" timing was lower than the amplitude at the "diamond" timing; the opposite happened during expiration. These consistent changes in SCG morphology suggested that breathing may be a consistent source of SCG variability. Therefore, the relative weights of the two clusters obtained from K-medoid algorithm might be similar to the inspiration-expiration ratio. In the current study, if one of the final clusters obtained had a relative weight less than 20%, it would be considered sparse, and the corresponding final clustering solution would be called "sparse solution."

2.5.5 Purity of Clusters

Linear support vector machines (SVMs) [43] were employed to find the decision boundary between the two clusters of all final unique solutions based on previous studies [14, 32]. Linear SVM was implemented in MATLAB R2024b using the

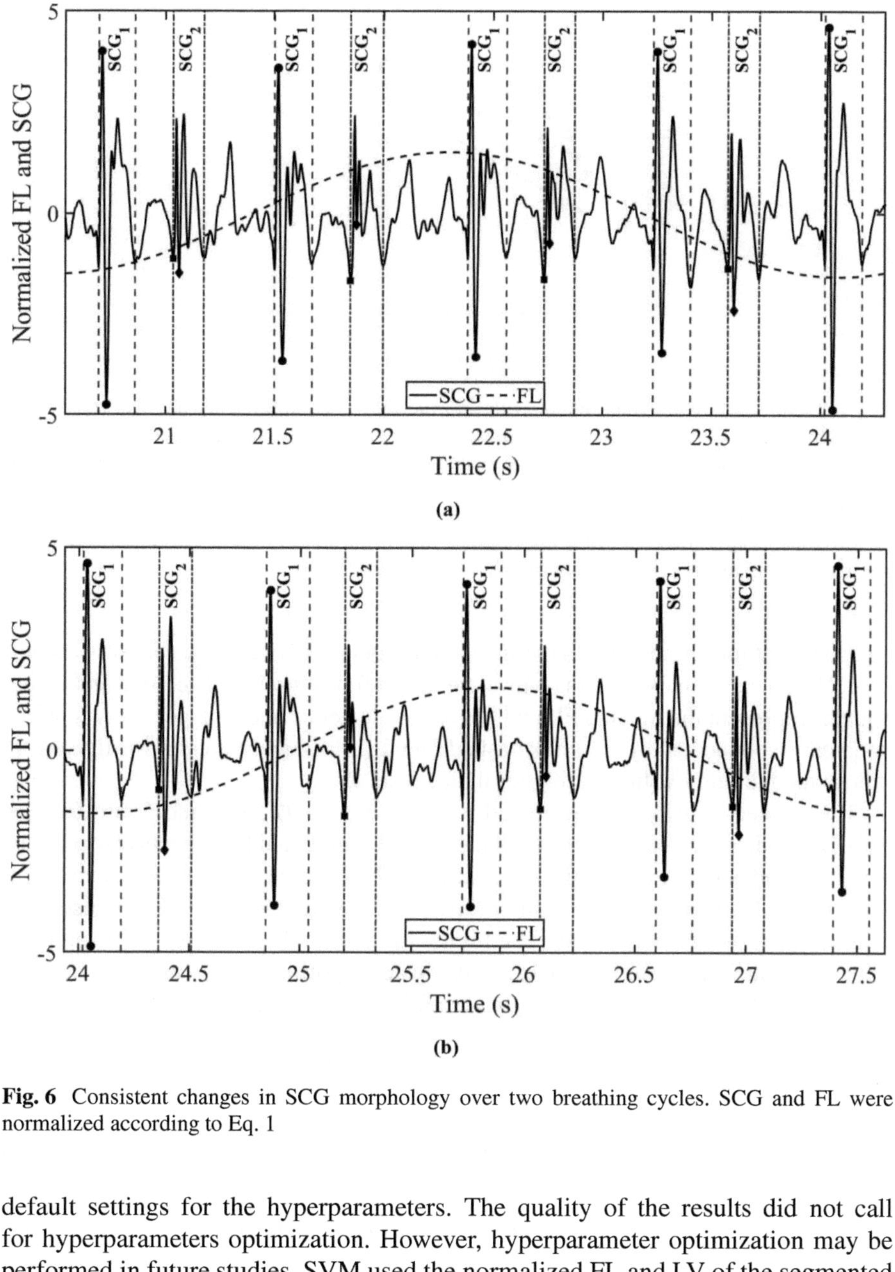

Fig. 6 Consistent changes in SCG morphology over two breathing cycles. SCG and FL were normalized according to Eq. 1

default settings for the hyperparameters. The quality of the results did not call for hyperparameters optimization. However, hyperparameter optimization may be performed in future studies. SVM used the normalized FL and LV of the segmented SCG beats as features. Figure 7 shows an example of a clustering solution in the FL-LV space. Each beat was represented by a point. Filled and empty circles represent the two clusters (i.e., clusters 1 and 2 respectively). The solid line is the decision boundary calculated by SVM. The distribution of the clusters in the FL-LV space

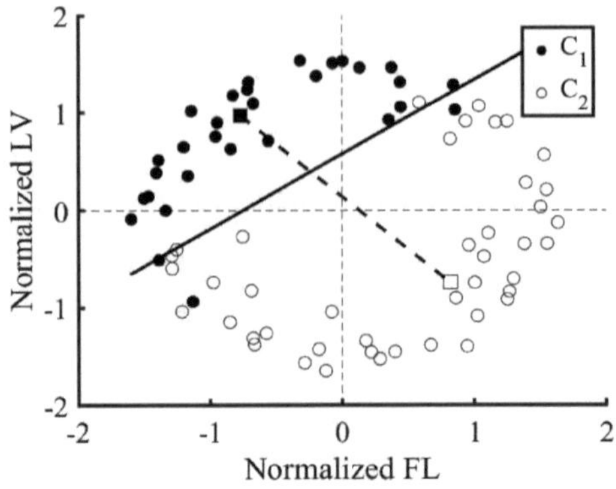

Fig. 7 Example of a clustering solution in the normalized FL-LV space. Each beat was represented by a point. Filled and empty circles represent the two clusters (i.e., clusters 1 and 2 respectively). The solid line is the decision boundary calculated by SVM. The distribution of the clusters in the FL-LV space suggests that the choice of a linear decision boundary between clusters was acceptable. The filled and empty squares represent the median FL and LV of the clusters 1 and 2 respectively; the two squares are connected by the thick dashed line. In all subjects, cluster 1 is the cluster whose median lung volume is higher than cluster 2.

suggests that the choice of a linear decision boundary between clusters was acceptable. Nonlinear boundaries can be investigated in future studies. The filled and empty squares represent the median FL and LV of the clusters 1 and 2 respectively; the two squares are connected by the thick dashed line. In all subjects, cluster 1 is the cluster whose median lung volume is higher than cluster 2.

The clusters' purity was estimated by Eq. 5 where *TP*, *TN*, and *N* are the true positives, true negatives, and the total number of segmented SCG beats respectively. Positives and negatives represent clusters 1 and 2 respectively. Purity was also calculated for each cluster separately using Eqs. 6 and 7, where *FP*, *FN* are the false positives and negatives respectively. The higher the purity, the more separable and less heterogeneous the clusters. Higher purity clustering solutions are preferred because they are likely to produce more distinctive features from each cluster.

$$Purity = (TP + TN)/N \tag{5}$$

$$Purity\ of\ cluster\ 1 = TP/(TP + FP) \tag{6}$$

$$Purity\ of\ cluster\ 2 = TN/(TN + FN) \tag{7}$$

2.5.6 Breathing Phase Angle

The breathing phase angle is the angle of the line connecting the two points representing the median FL and LV of each cluster (i.e., the angle of the thick dashed line in Fig. 7) according to Eq. 8. It was calculated for final clustering solutions as one of solution characteristics since it reflects how the clusters are distributed in the FL-LV space.

$$\textit{Breathing phase angle} = \tan^{-1}(\Delta LV/\Delta FL) \tag{8}$$

2.5.7 Similarity of Final Clustering Solutions

Similarity of the final clustering solutions to each other was calculated by comparing the cluster assignments (i.e., how many beats are common between solutions). Suppose two solutions need to be compared, namely solutions #1 and #2, and let C_{ij} represent cluster "j" in solution "i". Both "i" and "j" are either 1 or 2 since we have two solutions and two clusters per solution. Two values of the similarity can be calculated: one by comparing C_{11} and C_{12} to C_{21} and C_{22} respectively and the other by comparing C_{11} and C_{12} to C_{22} and C_{21} respectively. The higher of the two similarities was only considered and was represented as a percentage (i.e., #common beats/total #beats).

3 Results and Discussion

3.1 *Effect of the Initial Condition*

All possible initial conditions (i.e., initial choice of medoid pairs without duplications) were considered. Several different initial conditions were found to lead to the same final clustering solutions. Table 2 shows the number of all possible initial conditions, number of final unique solutions, and number of sparse solutions for the study subjects. All subjects had multiple unique solutions, which suggests that clustering using one initial condition is not recommended. In Table 2, the number of initial conditions was different among subjects, although the acquired data segment was the same duration (i.e., one minute). This is because of the differences in the heart rate among subjects where the higher the heart rate, the higher the number of SCG beats and the number of total possible initial conditions. To elaborate more, assume the heart rate is 75 BPM, then one minute is equivalent to 75 beats. In this case, the total number of all possible initial conditions (i.e., SCG beat pairs without duplications) would be in the order of 2700. This number would increase if the heart rate increased due to the increase in the number of beats.

Table 2 Number of all possible initial conditions, final unique solutions, and sparse solutions for the study subjects

Subject	Gender	# All possible initial conditions	#Final unique solutions (%)	#Sparse solutions (%)
1	F	2628	23 (0.88)	4 (17.39)
2	F	2926	35 (1.20)	3 (8.57)
3	F	4278	36 (0.84)	15 (41.67)
4	F	3916	20 (0.51)	1 (5.00)
5	F	2775	48 (1.73)	9 (18.75)
6	F	3403	5 (0.15)	1 (20.00)
7	F	4005	7 (0.17)	0 (0)
8	F	2775	6 (0.22)	1 (16.67)
9	F	2278	32 (1.40)	6 (18.75)
10	M	2628	14 (0.53)	1 (7.14)
11	M	3916	14 (0.36)	1 (7.14)
12	M	2080	17 (0.82)	12 (70.59)
13	M	2211	10 (0.45)	0 (0)
14	M	2701	3 (0.11)	0 (0)
15	M	3240	25 (0.77)	1 (4.00)
16	M	3916	7 (0.18)	0 (0)
17	M	2485	6 (0.24)	0 (0)
18	M	3828	44 (1.15)	2 (4.55)
19	M	3486	54 (1.55)	8 (14.81)

The number of final unique clustering solutions is less than the number of initial clustering solutions as shown in Table 2 and was 0.11–1.73% of the number of possible initial conditions. The final unique clustering solutions were identified, and their characteristics were studied. Some of the final unique clustering solutions were sparse and were reported in Table 2 as a percentage of the number of final unique solutions (range = 0 to 71%). The existence of sparse solutions suggests that initiating the clustering algorithm with a random initial condition is not recommended since it could result in a sparse final solution. Figure 8 shows two final unique clustering solutions from the same subject. One solution has a more balanced cluster weight while the other is sparse. Sparsity of clusters renders sparse solutions unreliable for accurate feature extraction due to the small number of data points in the sparse cluster.

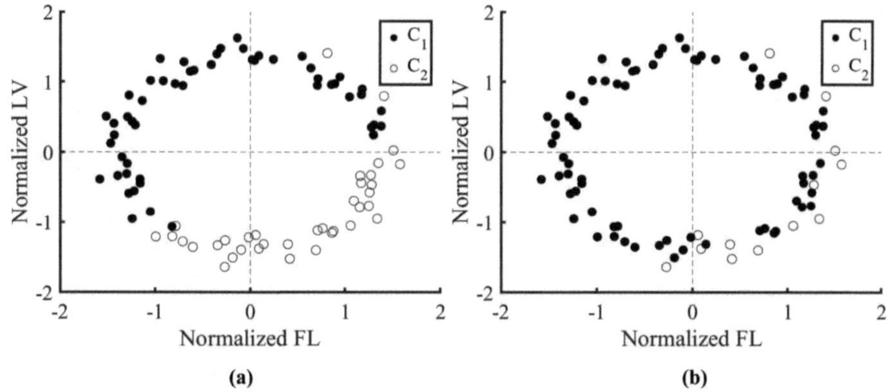

Fig. 8 Two final unique clustering solutions from the same subject; **a** The solution has a more balanced cluster weight, **b** The solution is sparse

3.2 Characteristics of Relevant Final Clustering Solutions

The objective of clustering in the current study is to find clusters that are distinct from each other. Therefore, the non-sparse clustering solution that had the maximum difference between inter- and intra-cluster variabilities was considered as the reference solution to which other solutions were compared. This solution will be called the maximum variability difference solution (MVDS). If two solutions had the same maximum variability difference, the MVDS with the higher cluster purity (estimated by Eq. 5) was chosen to represent MVDS. The clustering solutions that were compared to MVDS included:

1. The Highest Purity Solution (HPS): This is the non-sparse solution whose two clusters are better separated in the FL-LV space than other final unique solutions. The degree of separation was quantified by the clusters' purity, estimated by Eq. 5. If two or more final unique solutions had the same highest cluster purity value as in Fig. 9, the one that has a higher average of the individual cluster purities (i.e., average of the purities calculated by Eqs. 6 and 7) was chosen to represent HPS. The average of the individual cluster purities was 88.5% for the solution in Fig. 9a and 82.5% for the solution in Fig. 9b. Therefore, Fig. 9a was selected to be HPS.
2. The clustering solution that started with LV based initial condition [32]. This means that, initially, SCG beats were divided into two groups representing high and low lung volumes. Then the medoids of these groups served as the initial condition for the K-medoid algorithm.
3. The clustering solution that started with FL based initial condition [32]. This means that, initially, SCG beats were divided into two groups representing inspiration and expiration. Then the medoids of these groups served as the initial condition for the K-mcdoid algorithm.

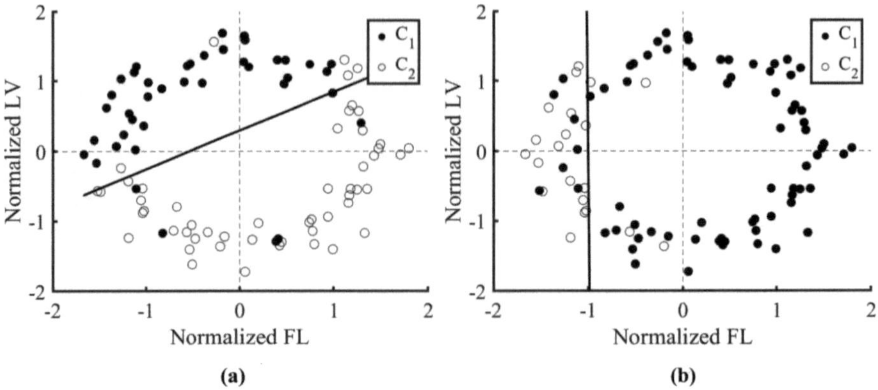

Fig. 9 Two clustering solutions, from the same subject, with the same purity as estimated by Eq. 5 but solution in **a.** has a higher average of individual clusters purities than solution in **b**. Therefore, the solution in **(a)** was chosen to represent the HPS

Figure 10 shows the purity of MVDS and HPS. The vertical black dashed line separates female from male subjects. The horizontal black dashed line at 85% purity defines a threshold of high purity in the current study. Figure 10 suggests that MVDS had relatively high purity (i.e., > 85%) except in 4 subjects (2, 5, 7, and 11). Figure 11 shows the similarity of HPS to MVDS. The vertical black dashed line separates female from male subjects. The horizontal black dashed line at 85% similarity defines a threshold of high similarity in the current study. Fig. 10 shows that HPS had a purity higher than or equal to MVDS, however the similarity between MVDS and HPS was not always high (i.e., ≥ 85%) as in subjects 2, 3, 4, 5, 7, 8, 11, and 19 as shown in Fig. 11. Despite the low similarity, the purity of MVDS in subjects 3, 4, 8, and 19 was relatively high (i.e., ≥ 85%).

Figure 12 shows the variability difference of HPS as a percentage of that of MVDS. This was calculated by dividing the variability difference (Inter-cluster variability—Intra-cluster variability) of HPS by the variability difference of MVDS and representing it as a percentage. The vertical black dashed line separates female from male subjects. The horizontal black dashed line at 85% defines a threshold of high variability difference in the current study. HPS had noticeably lower variability difference in some subjects suggesting that HPS could not accurately substitute for MVDS

Figure 13 shows the distribution of the clusters, in the FL-LV space, of MVDS for subjects 13 and 7 as two examples of high and low purity, respectively. High purity of MVDS suggests that respiration was the main source of SCG variability. When the purity of the MVDS was relatively low, there might be other sources of SCG variability that were more dominant than respiration (e.g., hormonal changes, physical or emotional stress, etc.), or it may be a sign that the effect of respiration on SCG variability might not be as consistent as expected.

Figure 14 shows the distribution of clusters in the FL-LV space for MVDS and HPS for subject 8; the two solutions had comparable purities, but low similarity, and HPS had a lower percent variability difference. In general, as the similarity

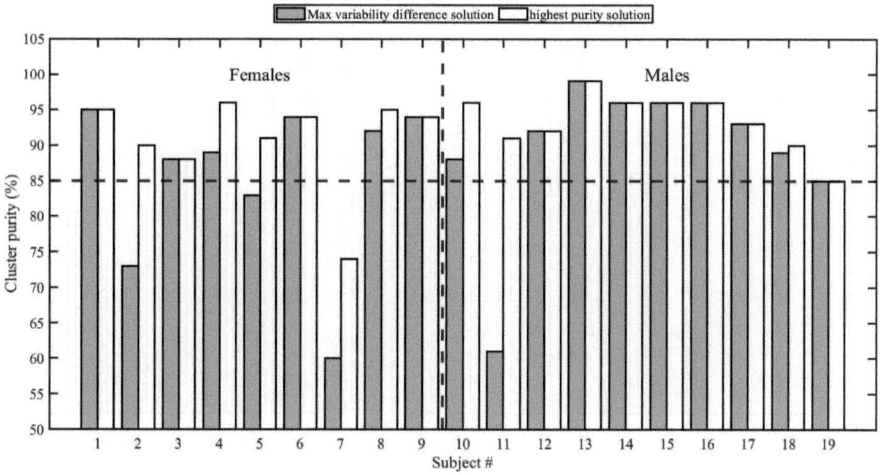

Fig. 10 The cluster purity (calculated based on Eq. 5) of the maximum variability difference solution (MVDS) and the highest purity solution (HPS). The vertical black dashed line separates female from male subjects. The horizontal black dashed line at 85% purity defines a threshold of high purity. In most subjects the purity of MVDS was comparable to HPS. For 3 subjects (2, 7, 11) the purity of MVDS appeared noticeably lower

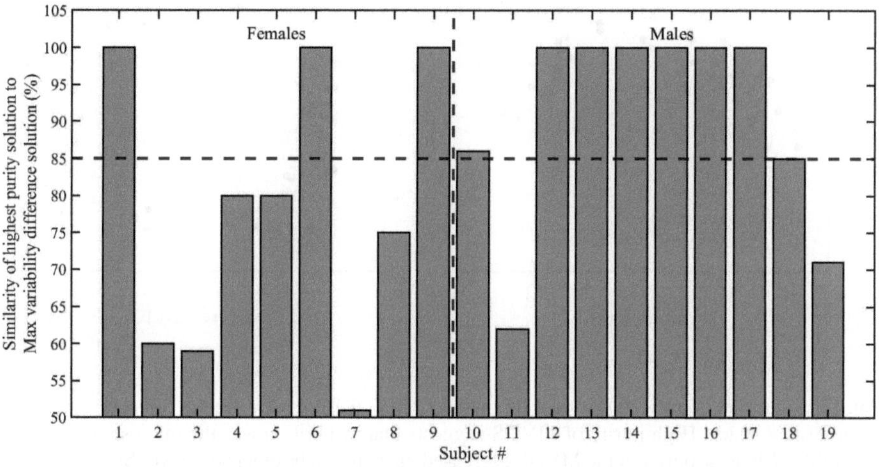

Fig. 11 Similarity between HPS and MVDS. The vertical black dashed line separates female from male subjects. The horizontal black dashed line at 85% similarity defines a threshold of high similarity. Similarity appeared high in more than half of the study subjects. Notable low similarity (e.g., subject 2,3, 4, 5, 7, 8, 11, 19) suggests that these solutions are significantly different

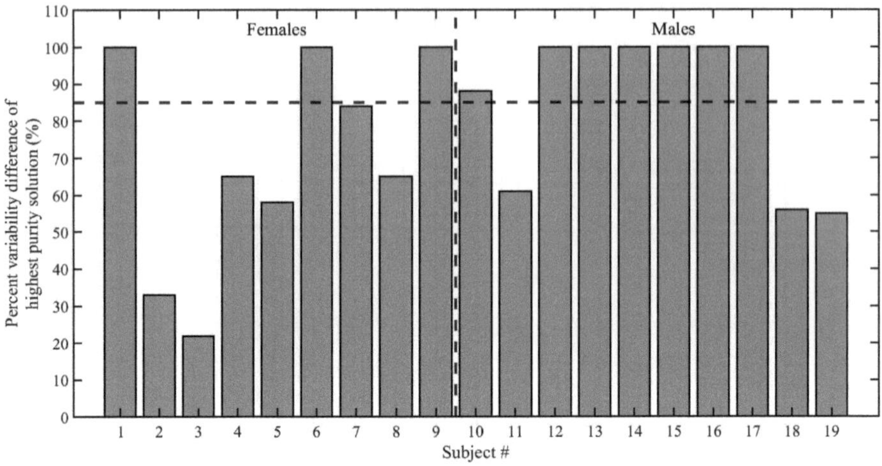

Fig. 12 Percent variability difference of the highest purity solution (HPS). The vertical black dashed line separates female from male subjects. The horizontal black dashed line at 85% defines a threshold of high variability difference. The HPS had noticeably low variability difference in some subjects suggesting that HPS could not accurately substitute for the MVDS

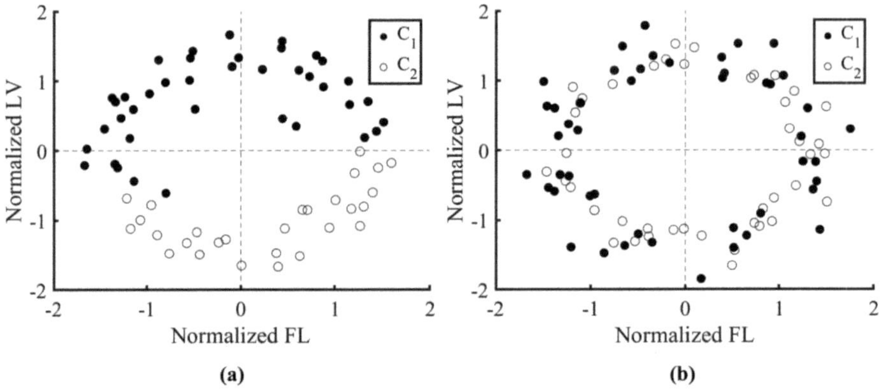

Fig. 13 Examples of **a** High purity and **b** Low purity MVDS from two different subjects represented in the FL-LV space. High purity of MVDS suggests that respiration was the main source of SCG variability. When the purity of the MVDS decreased, there might be other sources of SCG variability that were more dominant than respiration (e.g., hormonal changes, physical or emotional stress, etc.), or the effect of respiration might not be as consistent as it expected

decreased, the percent variability difference of HPS decreased as shown in Fig. 12. So, collectively, Figs. 10 to 14 showed that HPS is not always close to MVDS even if both have comparable purities because HPS might have low similarity to MVDS and accordingly low variability difference. This suggests the HPS may not be an accurate substitute for MVDS as mentioned above.

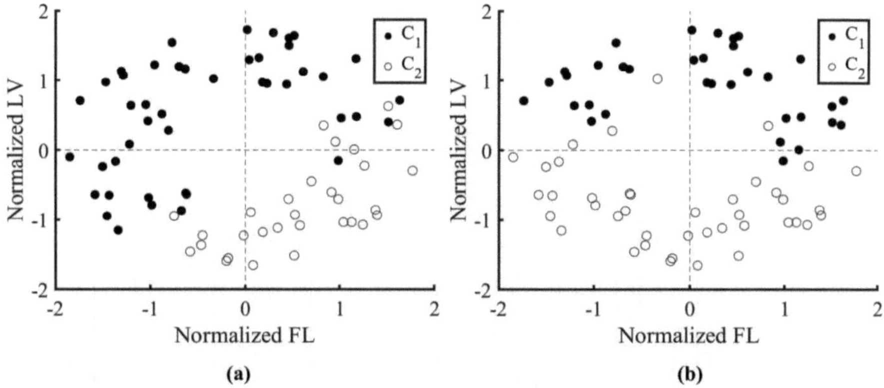

Fig. 14 The distribution of clusters in the FL-LV space for **a** MVDS and **b** HPS for subject 8. The two solutions had comparable purities, but low similarity, and HPS had a lower percent variability difference

Figure 15 shows the similarity of the final solutions with LV or FL based initial conditions compared with MVDS. Figure 16 shows the percent variability difference of these two solutions. Collectively, the two solutions had high similarity to MVDS only in 13/19 subjects. As the similarity decreased, the percent variability difference also decreased (where only 8/19 subjects had high variability difference values in Fig. 16). Therefore, it can be concluded that initiating the clustering algorithm with LV or FL based initial conditions does not always guarantee that MVDS will be obtained.

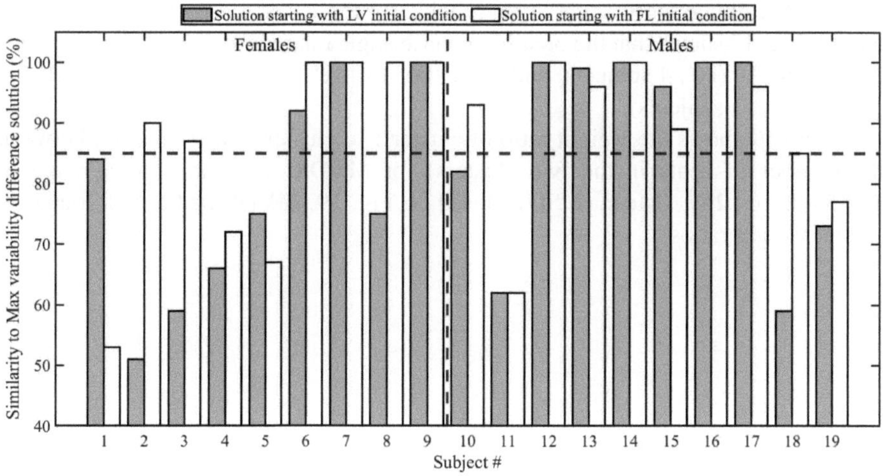

Fig. 15 Similarity of the final solutions that started with LV or FL based initial conditions to MVDS. Collectively, the two solutions had high similarity to MVDS in 13/19 subjects

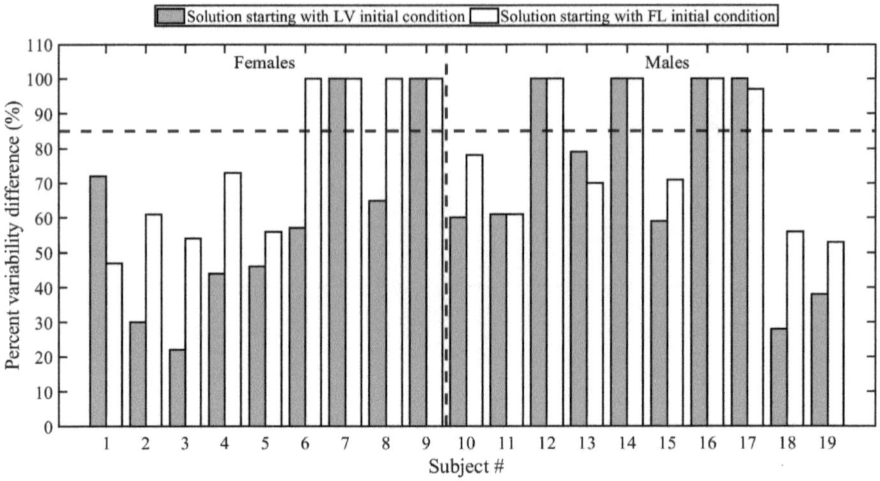

Fig. 16 Percent variability difference of the solutions that started with LV or FL based initial conditions. Collectively, the two solutions had high similarity to MVDS only in 8/19 subjects. This suggests that these initial conditions may not be effective in finding the MVDS

Figure 17 shows the breathing phase angle of MVDS for all subjects. The breathing phase angle helps quantify the position of the clusters in the FL-LV space as discussed and shown in Fig. 7. As the cluster purity increases, the breathing phase angle becomes more meaningful. MVDS of subjects 7 and 11 had very low purities (60 and 61%) as shown in Fig. 10. Figure 18 shows the clusters distribution, in the FL-LV space, for these two subjects; it shows that clusters were highly overlapping. Therefore, the breathing phase angle is not reliable for these cases. Apart from subjects 7 and 11, Fig. 17 shows that the breathing phase angle ranged from 3 to 46° in subjects 2, 3, 6, and 19 (i.e., 4 subjects) and from 106° to 148° in subjects 1, 4, 5, 8–10, and 12–18 (i.e., 13 subjects).

Figure 19 shows the weight ratio of cluster 1 to cluster 2 (i.e., the ratio between the number of beats in the two clusters) for MVDS. The ratio has a range of 142.53 ± 104.26% (Mean ± SD). Based on Fig. 19, MVDS had a wide range of cluster weight ratios.

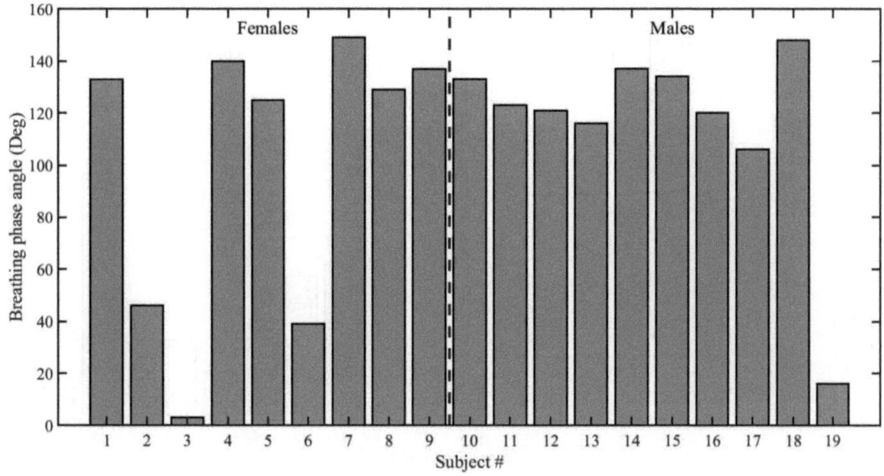

Fig. 17 The breathing phase angle of MVDS. The angle was 106–148 degrees in 13 of the 19 subjects. Note that the angles in subjects 7 and 11 are not meaningful since clustering was not consistent with respiration as shown in Fig. 18

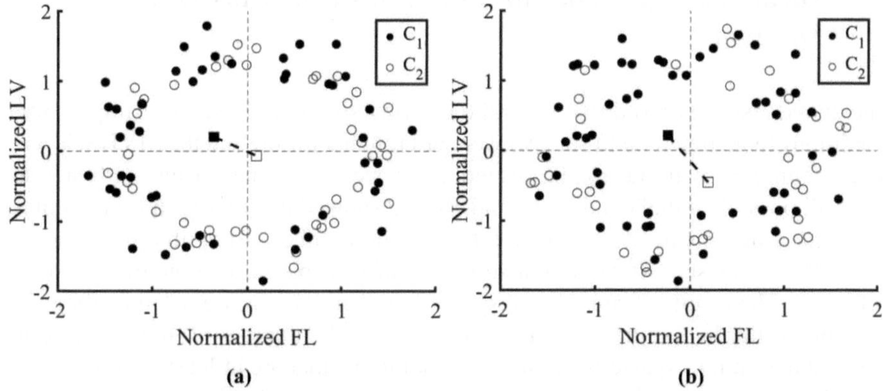

Fig. 18 Clusters distribution of MVDS in the FL-LV space of subjects **a** 7 and **b** 11. The clusters were not well separated (i.e., due to the very low purity) thus the breathing phase angle was not meaningful in these cases

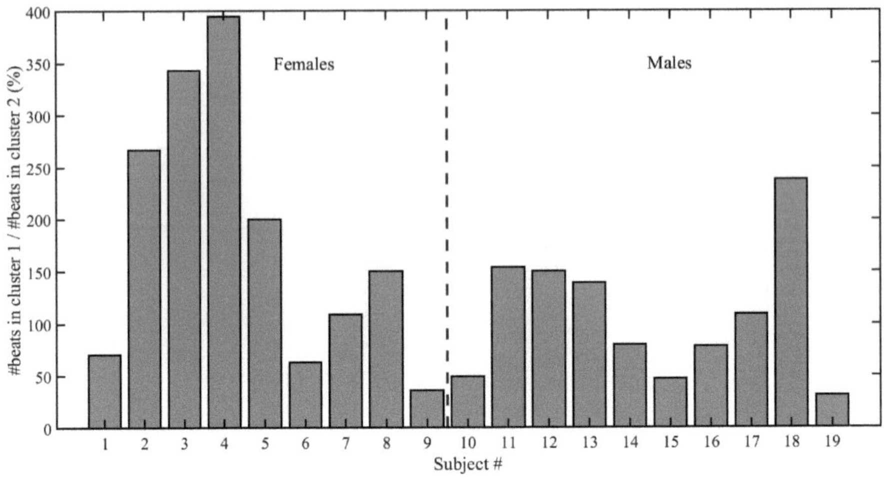

Fig. 19 The clusters weight ratio of MVDS. The ratio had a wide range (about 25–400%)

3.3 Efficient Procedures for Finding the Maximum Variability Difference and Highest Purity Clustering Solutions

The above results showed that several clustering solutions exist, and simple initial conditions (e.g., based on FL or LV) were not reliable for finding the MVDS. Therefore, running many initial conditions would be necessary. The current study investigated all possible initial conditions, and Table 2 showed that there is a small number of unique solutions (< 2%) compared to the total number of possible initial conditions. This suggests that running a subset of all possible initial conditions may be sufficient. The results of running subsets of initial conditions are listed in Table 3.

Table 3 (third column) shows the percentage of the number of initial conditions that resulted in non-sparse final clustering solutions that are at least 85% similar to MVDS. It can be seen that many initial conditions (range: 9–98%) led to MVDS or near-MVDS. This suggests that it may not be necessary to consider all possible initial conditions to reach MVDS or near-MVDS. Therefore, the current study investigated the utility of using a subset of all initial conditions (range: 5–100%) for reaching a near-MVDS. Here, the SCG beats chosen for the subset were not selected randomly. Rather, when a beat was included in the subset, nearby beats were not included (as these may have similar morphologies). For example, when the subset contains 10% of the total SCG beats, then beats #1, #11, and #21, etc., were used to build the subset. Table 3 (fourth column) shows that the minimum percentage of beats needed to find MVDS or near-MVDS is 5–10%.

Similarly, Table 4 has the same information as Table 3 but for HPS. Table 4 shows that the minimum percentage of beats needed to find the HPS or non-sparse near-HPS

Table 3 The percentage of the number of initial conditions that resulted in non-sparse final clustering solutions that are at least 85% similar to MVDS and the minimum percentage of beats needed to find a non-sparse clustering solution that is at least 85% similar to MVDS. These results suggest that considering 5–10% of initial conditions may be sufficient, which can reduce the computational burden by at least 10 times

Subject	Gender	Number of initial conditions which resulted in non-sparse final clustering solutions that are at least 85% similar to MVDS (%)	The minimum percentage of beats needed to find a non-sparse clustering solution that is at least 85% similar to MVDS (%)
1	F	1149 (43.72)	5
2	F	726 (24.81)	5
3	F	2467 (57.67)	5
4	F	783 (19.99)	5
5	F	398 (14.34)	10
6	F	3277 (96.29)	5
7	F	2702 (67.47)	5
8	F	2005 (72.25)	5
9	F	1921 (84.33)	5
10	M	1621 (61.68)	5
11	M	3175 (81.08)	5
12	M	1618 (77.79)	5
13	M	2163 (97.83)	5
14	M	2665 (98.67)	5
15	M	2429 (74.97)	5
16	M	3768 (96.22)	5
17	M	2404 (96.74)	5
18	M	2753 (71.92)	5
19	M	323 (9.27)	5

is also 5–10%. This may lead to reducing the computational burden by at least 10 times, while reaching the desired results (i.e., near-MVDS and near-HPS).

Since clustering is a calculation intensive task, using a subset of initial conditions can significantly decrease the calculation burden. It is to be noted that running only one initial condition for one-minute SCG recording took about 15 seconds in the current study (corresponding to about 12 hours for considering all initial conditions in a typical subject), while clustering a 5-minute recording would take an estimated 6.5 minutes per initial condition. This suggests that reducing calculation burden is particularly important when clustering long SCG data recordings. Several approaches may be attempted for long recordings. For example, clustering may start by choosing a one-minute representative part of the recording and a 10% subset of initial conditions. The resulting final solution may then be used as an initial condition for other one-minute segments of the recording or for the whole recording. Given the results of

Table 4 Percentage of the number of initial conditions which resulted in non-sparse final clustering solutions that are at least 85% similar to HPS and the minimum percentage of beats needed to find a non-sparse clustering solution that is at least 85% similar to HPS

Subject	Gender	Number of initial conditions which resulted in non-sparse final clustering solutions that are at least 85% similar to HPS (%)	The minimum percentage of beats needed to find a non-sparse clustering solution that is at least 85% similar to HPS (%)
1	F	1149 (43.72)	5
2	F	191 (6.53)	10
3	F	20 (0.47)	10
4	F	1126 (28.75)	5
5	F	584 (21.05)	10
6	F	3277 (96.29)	5
7	F	222 (5.54)	10
8	F	1328 (47.86)	5
9	F	1921 (84.33)	5
10	M	2437 (92.73)	5
11	M	539 (13.76)	10
12	M	1618 (77.79)	5
13	M	2163 (97.83)	5
14	M	2665 (98.67)	5
15	M	2429 (74.97)	5
16	M	3768 (96.22)	5
17	M	2404 (96.74)	5
18	M	2753 (71.92)	5
19	M	229 (6.5691)	10

Fig. 17, another complementary or independent approach for choosing an effective initial condition may be to split the data by breathing phase angles around 0–45 degrees and 105–150 degrees. These approaches will be investigated in future studies.

4 Conclusions

The effect of the initial condition on SCG clustering using unsupervised machine learning was investigated. All subjects had multiple clustering solutions and initiating the clustering algorithm with only one random initial condition is not recommended since it can result in sparse clustering and other non-optimal solutions. The number of final unique clustering solutions is much smaller than the number of possible initial conditions. Four final unique clustering solutions were investigated in detail: maximum variability difference solution (MVDS), highest purity solution (HPS), and the final solutions that started with either LV or FL based initial conditions. MVDS was considered the reference case to which other solutions were compared. This is because the primary purpose of clustering in the current study is to increase inter-cluster variability relative to intra-cluster variability. High purity of MVDS suggests that respiration was the main source of SCG variability. When the purity of the MVDS decreased, there might be other sources of SCG variability that were more dominant than respiration (e.g., hormonal changes, physical or emotional stress, etc.), or the effect of respiration might not be as consistent. HPS is not always identical to MVDS even if both have comparable purities because HPS might have low similarity to MVDS and accordingly low variability difference. Initiating the clustering algorithm with LV or FL based initial conditions does not always guarantee that MVDS will be obtained. The breathing phase angle of MVDS had two ranges: 3-46° and 106-148°. MVDS had a wide range of cluster weight ratio. The study suggested a possible time-efficient procedure to obtain MVDS and HPS based on considering a subset of the segmented SCG beats (i.e., 10% that are not chosen randomly) for initial conditions. Future work will aim to address the study limitations and include testing more subjects, longer recordings, different ethnicities, pathological data, and different breathing maneuvers. The future work will also investigate other ways to determine the initial conditions that result in MVDS and HPS.

Acknowledgments This material is based upon work supported by the National Science Foundation under Grant No. FW-HTF-P 2026516. The research reported in this publication was also supported by the National Heart and Lung Institute of the National Institutes of Health under award number R44 HL099053. Any opinions, findings, and conclusions or recommendations expressed in this material are those of the author(s) and do not necessarily reflect the views of the National Science Foundation or National Institute of Health.

COI Statement Richard H. Sandler and Hansen A. Mansy are part owners of Biomedical Acoustics Research Company, which is the primary recipient of the NIH grant R44HL099053, as such they may benefit financially as a result of the outcomes of the research work reported in this publication.

References

1. Crow, R. S., Hannan, P., Jacobs, D., Hedquist, L., & Salerno, D. M. (1994). Relationship between Seismocardiogram and Echocardiogram for Events in the Cardiac Cycle. *American Journal of Noninvasive Cardiology*, *8*, 39–46.
2. Salerno, D. M., & Zanetti, J. (1991). Seismocardiography for monitoring changes in left ventricular function during ischemia. *CHEST*, *100*(**4**), 991–993.
3. Salerno, D. M., Zanetti, J. M., Green, L. A., Mooney, M. R., Madison, J. D., & Van Tassel, R. A. (1991). Seismocardiographic changes associated with obstruction of coronary blood flow during balloon angioplasty. *The American Journal of Cardiology*, *68*(2), 201–207.
4. Taebi, A., Solar, B. E., Bomar, A. J., Sandler, R. H., & Mansy, H. A. (2019). Recent Advances in Seismocardiography. *Vibration*, *2*(1), 64–86.
5. Shirkovskiy, P., Laurin, A., Jeger-Madiot, N., Chapelle, D., Fink, M., & Ing, R. (2018). Airborne ultrasound surface motion camera: Application to seismocardiography. *Applied Physics Letters*, *112*(21), 213702.

6. Morbiducci, U., Scalise, L., De Melis, M., & Grigioni, M. (2007). Optical Vibrocardiography: A Novel Tool for the Optical Monitoring of Cardiac Activity. *Annals of Biomedical Engineering, 35*(1), 45–58. https://doi.org/10.1007/s10439-006-9202-9
7. Lin, W., Chou, W., Chang, P., Chou, C., Wen, M., Ho, M., Lee, W., Hsieh, M., Lin, C., Tsai, T., & Lee, M. (2018). Identification of Location Specific Feature Points in a Cardiac Cycle Using a Novel Seismocardiogram Spectrum System. *IEEE Journal of Biomedical and Health Informatics, 22*(2), 442–449. https://doi.org/10.1109/JBHI.2016.2620496
8. Tavakolian, K., Portacio, G., Tamddondoust, N. R., Jahns, G., Ngai, B., Dumont, G. A., & Blaber, A. P. (2012, September). Myocardial contractility: A seismocardiography approach. 2012 Annual International Conference of the IEEE Engineering in Medicine and Biology Society.
9. Tavakolian, K., Blaber, A. P., Ngai, B., & Kaminska, B. (2010). Estimation of hemodynamic parameters from seismocardiogram. 2010 Computing in Cardiology.
10. Morillo, D. S., Ojeda, J. L. R., Foix, L. F. C., & Jiménez, A. L. (2010). An Accelerometer-Based Device for Sleep Apnea Screening. *IEEE Transactions on Information Technology in Biomedicine, 14*(2), 491–499. https://doi.org/10.1109/TITB.2009.2027231
11. Dhar, R., Hossen, R., Gamage, P. T., Sandler, R. H., Raval, N. Y., Mentz, R. J., & Mansy, H. A. (2024). Heart Failure Readmission Prediction Using Seismocardiogram Signal. *Authorea Preprints.*
12. Yang, C., Aranoff, N. D., Green, P., & Tavassolian, N. (2020). Classification of Aortic Stenosis Using Time–Frequency Features From Chest Cardio-Mechanical Signals. *IEEE Transactions on Biomedical Engineering, 67*(6), 1672–1683.
13. Azad, M. K. (2020). *Seismocardiographic Signal Variability and Pulmonary Phase Detection in Adults* University of Central Florida].
14. Gamage, P. T., Azad, M. K., Taebi, A., Sandler, R. H., & Mansy, H. A. (2020). Clustering of SCG Events Using Unsupervised Machine Learning. In I. Obeid, I. Selesnick, & J. Picone (Eds.), *Signal Processing in Medicine and Biology: Emerging Trends in Research and Applications* (pp. 205–233). Springer International Publishing. https://doi.org/10.1007/978-3-030-36844-9_7
15. Javaid, A. Q., Dowling, S., Etemadi, M., Heller, J. A., Roy, S., Klein, L., & Inan, O. T. (2016, 11-14 Sept. 2016). Quantification of posture induced changes in wearable seismocardiogram signals for heart failure patients. 2016 Computing in Cardiology Conference (CinC).
16. Jain, P. K., & Tiwari, A. K. (2016, 16-18 Oct. 2016). A novel method for suppression of motion artifacts from the seismocardiogram signal. 2016 IEEE International Conference on Digital Signal Processing (DSP).
17. Dhar, R., Darwish, S. E., Darwish, S. A., Sandler, R. H., & Mansy, H. A. (2025). Effect of respiration and exercise on seismocardiographic signals. *Computers in Biology and Medicine, 185*, 109600. https://doi.org/10.1016/j.compbiomed.2024.109600
18. Libonati, J. R., Colby, A. M., Caldwell, T. M., Kasparian, R., & Glassberg, H. L. (1999). Systolic and diastolic cardiac function time intervals and exercise capacity in women. *Medicine & Science in Sports & Exercise, 31*(2).
19. Azad, M. K., D'Angelo, J., Gamage, P. T., Ismail, S., Sandler, R. H., & Mansy, H. A. (2021). Spatial Distribution of Seismocardiographic Signals. In I. Obeid, I. Selesnick, & J. Picone (Eds.), *Biomedical Signal Processing: Innovation and Applications* (pp. 129–159). Springer International Publishing. https://doi.org/10.1007/978-3-030-67494-6_5
20. Ahdy, S., Hassan, T., Rahman, B., Sandler, R. H., & Mansy, H. A. (2024). SCG variability and spectral energy distribution during normal breathing and breath hold at different lung volumes and airway pressures. *Scientific Reports, 14*(1), 17904. https://doi.org/10.1038/s41598-024-68590-6
21. Pandia, K., Inan, O. T., Kovacs, G. T. A., & Giovangrandi, L. (2012). Extracting respiratory information from seismocardiogram signals acquired on the chest using a miniature accelerometer. *Physiological measurement, 33*(10), 1643-1660. https://doi.org/10.1088/0967-3334/33/10/1643

22. Taebi, A., & Mansy, H. A. (2017, 2-2 Dec. 2017). Grouping similar seismocardiographic signals using respiratory information. 2017 IEEE Signal Processing in Medicine and Biology Symposium (SPMB).
23. Lara, S., & Desmond, B. (1999). Cardiovascular Effects of Mechanical Ventilation. *Archives of Disease in Childhood*, *80*(5), 475.
24. Klabunde, R. (2011). *Cardiovascular Physiology Concepts*. Lippincott Williams & Wilkins.
25. Pandia, K. (2017). *Seismocardiographic Assessment of Cardiopulmonary Health* (Publication Number 28115540) [Ph.D., Stanford University]. ProQuest Dissertations & Theses A&I; ProQuest Dissertations & Theses Global. Ann Arbor.
26. Iftikhar, Z., Lahdenoja, O., Jafari Tadi, M., Hurnanen, T., Vasankari, T., Kiviniemi, T., Airaksinen, J., Koivisto, T., & Pänkäälä, M. (2018). Multiclass Classifier based Cardiovascular Condition Detection Using Smartphone Mechanocardiography. *Scientific Reports*, *8*(1), 9344. https://doi.org/10.1038/s41598-018-27683-9
27. Poliac, M. O., Zanetty, J. M., Salerno, D., & Wilcox, G. L. (1991). Seismocardiogram (SCG) interpretation using neural networks. Computer-Based Medical Systems-Proceedings of the Fourth Annual IEEE Symposium.
28. Lahdenoja, O., Hurnanen, T., Iftikhar, Z., Nieminen, S., Knuutila, T., Saraste, A., Kiviniemi, T., Vasankari, T., Airaksinen, J., Pänkäälä, M., & Koivisto, T. (2018). Atrial Fibrillation Detection via Accelerometer and Gyroscope of a Smartphone. *IEEE Journal of Biomedical and Health Informatics*, *22*(1), 108-118. https://doi.org/10.1109/JBHI.2017.2688473
29. Solar, B. E., Taebi, A., & Mansy, H. A. (2017, 2-2 Dec. 2017). Classification of seismocardiographic cycles into lung volume phases. 2017 IEEE Signal Processing in Medicine and Biology Symposium (SPMB).
30. Taebi, A., Solar, B. E., & Mansy, H. A. (2018, 19-22 April 2018). An Adaptive Feature Extraction Algorithm for Classification of Seismocardiographic Signals. SoutheastCon 2018.
31. Zakeri, V., Akhbardeh, A., Alamdari, N., Fazel-Rezai, R., Paukkunen, M., & Tavakolian, K. (2017). Analyzing Seismocardiogram Cycles to Identify the Respiratory Phases. *IEEE Transactions on Biomedical Engineering*, *64*(8), 1786–1792. https://doi.org/10.1109/TBME.2016.2621037
32. Ahdy, S., Azad, M. K., Sandler, R. H., Raval, N., & Mansy, H. A. (2023). Spatial Distribution of Seismocardiographic Signal Clustering. In I. Obeid, J. Picone, & I. Selesnick (Eds.), *Signal Processing in Medicine and Biology: Innovations in Big Data Processing* (pp. 87-119). Springer International Publishing. https://doi.org/10.1007/978-3-031-21236-9_4
33. Ahdy, S., Azad, M., Sandler, R., Raval, N., & Mansy, H. (2021, 4-4 Dec. 2021). Spatial Distribution of Seismocardiographic Signal Clustering. 2021 IEEE Signal Processing in Medicine and Biology Symposium (SPMB).
34. Pan, J., & Tompkins, W. J. (1985). A Real-Time QRS Detection Algorithm. *IEEE Transactions on Biomedical Engineering*, *BME-32*(3), 230–236. https://doi.org/10.1109/TBME.1985.325532
35. Azad, M., Gamage, P., Sandler, R., Raval, N., & Mansy, H. (2018). Detection of respiratory phase and rate from chest surface measurements. *J. Appl. Biotechnol. Bioeng*, *5*(6), 10.15406.
36. Warren Liao, T. (2005). Clustering of time series data—a survey. *Pattern Recognition*, *38*(11), 1857–1874. https://doi.org/10.1016/j.patcog.2005.01.025
37. Paparrizos, J., & Gravano, L. (2015). k-shape: Efficient and accurate clustering of time series. Proceedings of the 2015 ACM SIGMOD international conference on management of data.
38. Azad, M. K., Gamage, P. T., Sandler, R. H., Raval, N., & Mansy, H. A. (2019, 7-7 Dec. 2019). Seismocardiographic Signal Variability During Regular Breathing and Breath Hold in Healthy Adults. 2019 IEEE Signal Processing in Medicine and Biology Symposium (SPMB)
39. Gamage, P. T., Khurshidul Azad, M., Taebi, A., Sandler, R. H., & Mansy, H. A. (2018, 1-1 Dec. 2018). Clustering Seismocardiographic Events using Unsupervised Machine Learning. 2018 IEEE Signal Processing in Medicine and Biology Symposium (SPMB).
40. Dasgupta, S. (2016). A cost function for similarity-based hierarchical clustering. Proceedings of the forty-eighth annual ACM symposium on Theory of Computing

41. Sakoe, H., & Chiba, S. (1978). Dynamic programming algorithm optimization for spoken word recognition. *IEEE Transactions on Acoustics, Speech, and Signal Processing*, *26*(1), 43–49. https://doi.org/10.1109/TASSP.1978.1163055
42. Gamage, P. T. (2020). *Seismocardiography–Genesis, And Utilization of Machine Learning for Variability Reduction and Improved Cardiac Health Monitoring* University of Central Florida].
43. Cortes, C., & Vapnik, V. (1995). Support-vector networks. *Machine Learning*, *20*(3), 273–297. https://doi.org/10.1007/BF00994018

Evaluating Genomic Language Models: Experimental Validation, Multi-Omics Integration, and Ethical-Legal Challenges

Mohammad Saleh Refahi, Hyunwoo Yoo, Gavin Hearne, Chaz Allegra, James R. Brown, Bahrad A. Sokhansanj, Stephen Woloszynek, Keyush Ramjattun, Shanmukha Rao Allipilli, Robi Polikar, and Gail L. Rosen

Abstract The rapid expansion of genomic data, driven by advancements in sequencing and multi-omics technologies, calls for analytical approaches that go beyond traditional methods to extract deeper biological insights. Genomic Language Mod-

M. S. Refahi · H. Yoo · G. Hearne · J. R. Brown · B. A. Sokhansanj · K. Ramjattun · S. R. Allipilli · G. L. Rosen (✉)
Electrical and Computer Engineering, Drexel University, Philadelphia, PA, United States
e-mail: glr26@drexel.edu

M. S. Refahi
e-mail: sr3622@drexel.edu

H. Yoo
e-mail: hty23@drexel.edu

G. Hearne
e-mail: glh52@drexel.edu

J. R. Brown
e-mail: jb4633@drexel.edu

B. A. Sokhansanj
e-mail: bahrad@molhealtheng.com

K. Ramjattun
e-mail: kr3288@drexel.edu

S. R. Allipilli
e-mail: sa3935@drexel.edu

C. Allegra · R. Polikar
Electrical and Computer Engineering, Rowan University, Glassboro, NJ, United States
e-mail: allegr32@students.rowan.edu

R. Polikar
e-mail: polikar@rowan.edu

S. Woloszynek
Beth Israel Deaconess Medical Center, Boston, MA, United States
e-mail: swolosz1@bidmc.harvard.edu

A. Ahmed and J. Picone (eds.), *Signal Processing and Biomedical Engineering Research*, https://doi.org/10.1007/978-3-032-18509-9_3

els (GLMs) have emerged as powerful tools capable of integrating diverse omics data—including genomic, transcriptomic, epigenomic, proteomic, and metabolomic information—providing a comprehensive view of biological systems. By leveraging high-dimensional data across multiple biological scales, these models not only process large-scale sequence data but also uncover meaningful biological patterns, functional relationships, and novel regulatory mechanisms. In this work, we present a comprehensive review of GLMs, examining their core methodologies, tokenization strategies, and computational frameworks. We also conduct an experimental comparison of multiple GLMs to evaluate their efficiency, scalability, and generalizability across diverse genomic datasets. As GLMs integrate with multi-omics data, the need for models that not only process vast datasets but also extract meaningful biological insights becomes increasingly critical. The growing scale and fidelity of multi-omics data demand approaches capable of capturing complex relationships across genes, proteins, metabolites, and disease mechanisms. Deep learning models, including large language models (LLMs) and GLMs, hold promise in bridging molecular data with clinical insights, unlocking new opportunities in personalized medicine and biomedical discovery. While GLMs offer transformative potential, their application raises significant ethical and legal concerns, including data privacy, informed consent, proprietary data access, and biases in training datasets that may impact fairness and representation. Addressing these challenges is crucial to ensure the responsible and equitable use of genomic AI technologies.

Keywords Genomic language models · Metagenomics and microbiome · Host–microbe interactions · Ethics and policy · Data privacy and governance

1 Introduction

Genomes serve as repositories of genetic instructions, governing cellular replication, genetic recombination, and the transmission of heritable information. They are the product of billions of years of evolution, shaped by natural selection, genetic drift, and environmental pressures. The information genomes encode reflects an organism's current and past phenotypes, which in turn sheds light on past and present environmental pressures affecting these organisms as well as the underlying genetic basis of specific disease states and environmental adaptations [1–5].

Traditional approaches to examine genomic complexity have relied on sequence alignment and motif discovery methods. Sequence alignment algorithmically identifies regions of nucleotide (or protein) sequence similarity, which can be performed locally (Smith-Waterman) or globally (Needleman-Wunsch), with recent approaches capable of aligning entire genomes against reference genomes to identify ultraconserved segments of DNA [6, 7]. Motif discovery algorithms, on the other hand, use computational and statistical means to identify recurring short motifs of DNA, RNA, or proteins that occur more often than chance.

Despite their successes, however, these approaches often fail to capture the full multi-dimensional complexity of the non-linear and downstream interactions that specific regions of DNA exploit [8–10]. These approaches typically focus on linearly organized patterns and local subsequence similarity; however, information encoded in genomes extends far beyond a subsequence's local environment. Genomic DNA is folded, organized into loops and structural domains, allowing different regions to come into contact with sequences far downstream. For example, specific regions of DNA that regulate transcription, called enhancers, are often vast genomic distances from their associated gene, and while evolutionarily conserved, the relative position of an enhancer to the gene it regulates can vary considerably among organisms [9]. Thus, predicting enhancer-promoter pairs (the on/off and regulatory switches for gene expression, respectively) based on their proximity within a linear genome is not optimal [9]. Moreover, a given gene can have multiple enhancers working synergistically to control expression [10]. These complex relationships are often ignored by conventional motif discovery approaches, which tend to opt for more simplified models relying on independence between motifs [10].

Previous research often neglected noncoding regions of DNA, now considered essential for elucidating genome function [11]. Work has shown that noncoding regions often house promoters and enhancers, demonstrating their role in the regulation of gene expression [12]. This in turn has implications in, for example, human disease phenotypes given that the vast majority of disease-associated DNA variants are found in noncoding regions [13]. Accordingly, examples of noncoding DNA mutations having functional manifestations have been shown [14].

In recent years, the development of complex large language models has provided a means to analyze long spans of raw text from exceedingly large corpora of documents. These models leverage deep neural network architectures with upwards of billions of tunable parameters (weights) to extract patterns and associations from language. They are trained to generate coherent responses and accurate predictions. The input for these models is raw text from documents, tokenized into a vocabulary of preprocessed terms or subsequences. These tokens are then transformed into numeric representations called embeddings that capture semantic and contextual nuances [15]. The embeddings are progressively refined using a self-attention mechanism [16], which dynamically weighs the interactions among tokens across the entire sequence. This process enables each token's numerical representation to seamlessly integrate both immediate and distant contextual cues from its surrounding environment. Additionally, Large Language Models (LLMs) contain multiple layers of weights; hence, deeper layers build richer, more complex representations (such as semantic meaning) and longer-range dependencies between tokens.

Put simply, genomic sequences are collections of sequential symbols arranged in a meaningful order with a specific context. It should be no surprise then that this genomic "language" can be learned by LLMs in a manner not dissimilar to English or Spanish. This has paved the way for Genomic Language Models (GLMs), which have emerged as transformative solutions by integrating state-of-the-art natural language processing (NLP) techniques with genomic data analysis. GLMs extract rich, context-aware representations directly from raw, unlabeled DNA sequences with slight modifications. Given that there is no natural split between words or tokens, continuous overlapping fixed-length subsequences of nucleotides (termed k-mers) or commonly identified dynamically determined subsequences (via byte-pair encoding) are used as inputs [15, 17]. From here, tokens can be numerically embedded and the GLM trained accordingly. Like their human language counterparts, the deep layers of the GLM architecture represent complex combinations of DNA patterns. Shallow layers may capture gross differences between protein-coding regions of DNA; a deeper layer may capture promoter-specific elements such as CpG islands [18, 19] and the presence of a TATA box [20, 21], while even deeper layers may identify specific splice sites [22]. Together, over iterations of training, the embedding for a TATA box may be linked to other promoter-specific embeddings as well as those representing gene expression modulators such as enhancer regions and termination signals. When mapped together, a numeric representation of a gene expression signal may emerge.

Thus, the embeddings resulting from LLMs enable the identification of genomic regions with specific nucleotide compositions and environments (context), potentially representing different gene types and the prediction of essential housekeeping genes [23, 24]. GLMs can learn both coding and noncoding regions, including regulatory elements and repetitive sequences, providing insights into gene regulation and phenotypic traits [24, 25]. Models like DNABERT [26], HyenaDNA [24], and Evo [27, 28] have made significant contributions to genomic sequence analysis. DNABERT introduced k-mer tokenization for sequence classification, HyenaDNA leveraged convolution-based architectures to improve efficiency in processing long genomic sequences, and Evo utilized taxonomically diverse datasets to enhance microbial genome representation. These GLMs have been instrumental in tasks such as promoter identification, taxonomy classification, and functional annotation. Despite their success, challenges remain in refining tokenization approaches and curating comprehensive training datasets to enhance performance in large-scale metagenomic applications [23, 29].

While GLMs have revolutionized genomic analysis by capturing complex sequence patterns, they still face fundamental limitations. One major issue is their reliance on sequence-based representations, often leading to repetitive patterns in embeddings. This repetition fails to fully capture the dynamic interactions within the genome, such as regulatory networks, chromatin accessibility, and evolutionary pressures [30]. A promising solution involves integrating multi-omics data–encompassing layers such as gene expression (transcriptomics), chemical modifications to DNA (epigenomics), and small-molecule profiles (metabolomics)–into genomic language models (GLMs), allowing these models to capture biological function beyond the linear DNA sequence. Contrastive learning could further refine embeddings by distinguishing functionally relevant sequences from redundancies, while a mixture-of-experts approach could allow subnetworks to specialize in distinct genomic contexts [31, 32].

The increasing adoption of GLMs raises significant ethical and legal concerns, particularly regarding data privacy, consent, and potential misuse. Unlike conventional AI models, GLMs analyze inherently sensitive genomic data, which can reveal personal health risks, ancestry, and familial relationships [33]. While legal frameworks like the Genetic Information Nondiscrimination Act (GINA) [34] in the U.S. and the Genetic Non-Discrimination Act in Canada offer some protection, they may not fully address the risks associated with large-scale genomic AI. Moreover, integrating multi-omics data heightens privacy concerns, as combining multiple biological layers increases the potential for re-identification. To ensure responsible AI-driven genomics, future policies must focus on transparent data stewardship, robust consent mechanisms, and regulatory safeguards that balance scientific innovation with individual rights [35]

In this chapter, we first compare the core components of Genomic Language Models (GLMs), analyzing key aspects from tokenization to training and inference pipelines. Next, we examine widely used GLMs, highlighting their advantages and limitations. We then evaluate several common genomic language models on the same dataset, comparing their effectiveness in taxonomy and gene classification. Following this, we explore the integration of multi-omics data, discussing the challenges and opportunities it presents for enhancing genomic analysis. Finally, we address the ethical and legal considerations surrounding GLMs, focusing on privacy concerns, data governance, and responsible AI development in genomics.

2 Core Components of Genomic Language Models

2.1 Tokenization Methods

Tokenization is the process of converting a raw biological sequence into discrete computational units. In GLMs, tokenization is essential for transforming continuous nucleotide or protein sequences into a structured representation suitable for machine learning. This enables models to process biological sequences similarly to how natural language processing systems handle text. By breaking sequences into tokens, we can apply language modeling techniques, such as Transformers or recurrent neural networks, to learn meaningful representations of biological sequences. Due to the nature of genomic data, which lacks the simple demarcations of natural text, tokenization in a GLM context then presents a unique challenge and opportunity to improve model performance [36]. Existing models most frequently rely on three popular tokenization schemes: **character-based, k-mer, and byte-pair encoding**.

Character-based tokenization treats each position in a DNA, RNA, or protein sequence as an individual token (e.g., nucleotides A, C, G, T, or amino acids Ala, Cys, Asp, etc.). This results in a straightforward representation that preserves single-base resolution and simplifies certain tasks (e.g., per-nucleotide analyses) where interpretability is key [37]. However, models must then work harder to capture higher-order patterns from single tokens, and the lack of compression in character-tokenized sequences increases computational cost and reduces the potential context window (which may be a notable limitation in some models) [38]. Examples are Hyenadna and ESM 2 (Table 1).

Table 1 Comparison of tokenization methods for biological sequences. Character-based, k-mer, and BPE tokenization represent different strategies for capturing sequence information at varying levels of granularity

Tokenization method	Tokenization rule	Example input (ACGTGCAA)	Tokenized output
Character-based	Each character is a token	ACGTGCAA	A, C, G, T, G, C, A, A
k-mer (k = 3)	Sliding window of length 3	ACGTGCAA	ACG, CGT, GTG, TGC, GCA, CAA
Byte Pair Encoding (BPE)	Iteratively merge frequent character pairs	ACGTGCAA	AC, GTG, CAA

Alternatively, in the biological sequence context, subsequence chunks called *k-mers* can be considered as words belonging to a biological sentence. It is then intuitive to tokenize at this level - for example, given k = 3, the sequence 'ACGTG' can be tokenized as <ACG><CGT><GTG>. Varying the k-value as well as the degree of overlap then produces different tokenization schemes - tokenizing the previous example utilizing a non-overlapping scheme along with k = 2 produces the tokens <AC><GT><G>. This higher-level tokenization can result in significantly higher compression than character-based representations (this is particularly the case with non-overlapping schemes) [38] and encodes local context into the tokenization itself, which may assist a model in capturing short motifs or sequence patterns [26]. However, overlapping k-mers can result in poor generalization due to information leakage, while non-overlapping k-mers disrupt sequence continuity [39]. Furthermore, flexibility means that determining the optimal parameters for this tokenization scheme must often be performed experimentally (e.g., DNABERT1, Nucleotide Transformer) [26].

Experiments have shown varying levels of success in utilizing k-mer tokenization, and several groups have demonstrated improved performance resulting from a switch to byte-pair encoding (BPE) [36, 39]. BPE [17] is a popular subword tokenization scheme in natural language processing that has since been adapted for GLMs, and functions by iteratively merging frequently occurring characters within a sequence until a target vocabulary size is reached. This results in a variable-length tokenization that does not solely rely on traditional biological structures, offering a balance between granularity (small tokens) and vocabulary size (larger tokens) [40]. Furthermore, utilizing BPE can reduce redundant computations by eliminating frequently seen k-mers, and may result in similar compression to k-mer tokenization, leveraging that same benefit to enable larger context lengths in transformer models [39]. However, it is a concern that leveraging this technique often may introduce artificial "words" into the corpus that do not correspond to meaningful biological units, which can reduce model interpretability [23, 38].

2.2 Model Architectures

The advent of the attention-based transformer model in 2017 by Google researchers [41] revolutionized language processing, shifting the field away from the recurrent neural networks (RNNs) [42] and long short-term memory (LSTM) [43] based architectures that had dominated language modeling in the past. Unlike RNNs, which process tokens sequentially, or convolutional neural networks, which use sliding windows over tokens, transformers rely on stacked self-attention mechanisms to process an entire sequence in parallel, enabling models to be trained on much larger datasets. Genomic sequence processing has progressed similarly in keeping with these developments, and a number of transformer-based foundation models have been developed in recent years [44], leveraging the power of attention to process long-range genomic data with greater precision.

The most standard transformer architecture in language processing is known as the encoder-decoder, which utilizes a series of stacked transformer layers that input sequence tokens (and positional embeddings) and iteratively transform and refine them into higher-level abstract embedding representations—this is known as the encoder. Then, a decoder mirrors that process to generate an output sequence based on the encoded embeddings. However, most frequently in GLMs, only the encoder block is utilized as the decoder process is unnecessary, and instead oftentimes a more basic output head converts the hidden embedding representations into a desired output format, such as projecting them into a vector the size of the vocabulary for next-word prediction, or into a classification head to predict class probabilities.

Key to the operation of the transformer layer is the attention function. As the transformer layers receive data, it is first passed through an embedding layer to transform the token sequence into a fixed-size vector representation (along with a positional embedding, which is necessary as transformers do not inherently capture positional information from a sequence). The transformer then utilizes the self-attention mechanism to allow each token in a sequence to directly attend to every other token, capturing relationships over arbitrary distances, enabling the embedding vectors to influence each other to capture contextual information. This is calculated as follows:

$$Attention(Q, K, V) = softmax(\frac{QK^T}{\sqrt{d_k}})V$$

Input sequences are transformed into queries (Q), keys (K), and values (V) d_k is simply the dimensionality of the key vectors. The scaled dot product between the query and the key then contains the information as to which tokens in a sequence attend to each other. Most commonly this process is done in parallel, in a process dubbed multi-headed attention, leveraging multiple attention heads each focusing on different subspaces of the token embeddings. Skip (residual) connections are then added to the output of the multi-layer attention function to help gradients flow more effectively, and to mitigate vanishing or exploding gradient issues. Layer normalization is also performed at this step to stabilize training by normalizing intermediate activations. Next, each token embedding is passed through feed-forward fully connected layers applied independently to every position, which adds non-linearity and helps the model learn complex transformations. By stacking these several transformer layers in series, the embedded output becomes more and more refined, improving the contextual information encoded in the vector positions and more accurately identifying underlying biological signals for each token. These contextual embeddings are far more powerful than the much more basic input vector representation, and can be leveraged for various downstream tasks including variant effect prediction, regulatory motif detection, and protein secondary structure prediction.

However, the nature of the transformer requires a large amount of VRAM, as the self-attention mechanism requirements scale quadratically with sequence length. As a result, processing long sequences becomes untenable, a notable downside in genomic language modeling, where some very long-range dependencies can fall outside the context window of transformer models. To combat this, some models have leveraged implicit convolutions and state-space modeling to enable long-range sequence modeling.

At a high level, a majority of foundation GLMs are built using this sort of architecture, with small adjustments to the core transformer architecture (e.g., tweaking self-attention, positional encodings, and fine-tuning strategies). For example, DNABERT2 [39] is built on the Bidirectional Encoder Representations from Transformers (BERT) architecture [45], an encoder-only transformer framework that leverages information from both directions in a sequence. This is enhanced by the addition of Leveraging Attention with Linear Biases (ALiBi) [46], flash attention [47], and low-rank adaptation (LoRA) [48]—these allow it respectively to handle arbitrarily long inference and fine-tuning sequences, to more efficiently calculate attention, and to reduce the computational expense of fine-tuning.

Implicit convolutions, as used by Hyena [49] and applied to genomics through HyenaDNA [24], utilize shallow feed-forward networks to learn parameterizations of convolutional kernels. Standard convolutional neural networks (CNNs) utilize fixed kernels with learnable weights, but despite their high performance in many tasks, these explicitly defined kernels result in a large bottleneck, as the computational and memory costs scale at an O(Lk) (k is the kernel size of the filter), limiting their efficacy in scenarios where larger kernels may be beneficial. By dynamically generating these filters through an MLP, which learns to produce filter parameters through training, Hyena achieves improved performance in identifying long-range dependencies without the need to utilize sizable explicit kernels, achieving subquadratic computational costs with respect to sequence length.

Classical state-space models [50] analyze system dynamics by describing a system through a series of differential equations that define the evolution of state variables over time. These models are widely used in control systems and signal processing, where system states evolve based on governing equations. Leveraging this concept, the State Space model (SSM) [51] considers a pair of linear differential equations that map a given input sequence to some output, replacing transformer attention mechanisms with structured recurrence and implicit long-range dependencies. SSMs enable high-performance sequence modeling by learning the parameters of the state-space differential equations, efficiently capturing both local and global dependencies across long sequences. This is improved by selectively considering input tokens to influence state transition parameters, maintaining computational efficiency while achieving performance comparable to or exceeding that of transformers [52].

2.3 Training Strategies

Nearly all state-of-the-art GLMs are in some form pretrained. By first training a model on a general task, the intermediate contextual embeddings (those produced by the encoder) become highly attuned toward both short- and long-range biological patterns. These learned features, which contain a wealth of biological information, can then be leveraged for diverse downstream tasks by swapping the language modeling head for a standard classification head and training only a subset of the learned transformer layers. This produces a final fine-tuned model without the need to fully retrain the base model, reducing the computational complexity significantly.

There are two predominant methods of pretraining: **Masked Language Modeling** (MLM), and **Next Token Prediction** (NTP) (see Fig. 2). Both are self-supervised

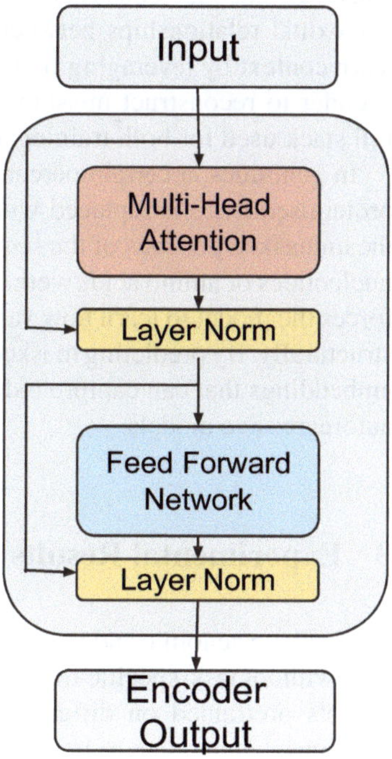

Fig. 1 Schematic of a standard Transformer encoder (Vanilla) used in language modeling

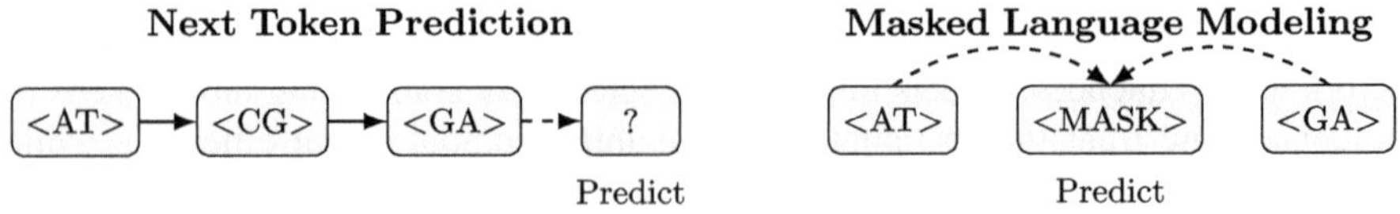

Fig. 2 Comparison between NTP and MLM

learning techniques (meaning they require no labels, and are highly scalable) that rely on using the language model to predict an unknown token; however, there are key differences.

In **autoregressive** models such as GPT [53] that are designed to predict tokens sequentially, using past tokens to infer future ones, NTP is the primary tool for training. This is a fundamental method for training generative LLMs, although it is relatively less common for GLMs where producing high-quality sequence-level embeddings is often the primary goal. However, NTP aligns well with the inherent biology of DNA, which by nature is transcribed sequentially, and has proven highly effective in training state-of-the-art models such as HyenaDNA [24].

In MLM, a portion of the input tokens is randomly "masked" or hidden, and the model is trained to predict the original tokens in these masked positions. This is primarily used in training bidirectional **autoencoder** models such as BERT[45], where an encoder constructs a global representation of a sequence by encoding contextual relationships between all tokens. Here MLM encourages the model to learn context by leveraging both the left and right neighboring tokens, allowing the decoder to reconstruct missing information from the encoded representation. The full stack used for both training and testing in this setup is illustrated in Fig. 3.

In genomics, a certain percentage (e.g., 10–15%) of tokens in a DNA, RNA, or protein sequence is replaced with a special "[MASK]" token. The model analyzes the unmasked portions of the sequence (upstream and downstream) to predict which nucleotides or amino acids were originally present in the masked locations [45]. This forces the model to learn how various parts of the sequence interact functionally and structurally. By predicting masked segments, the model develops robust, context-rich embeddings that can capture bidirectional dependencies [39] that may be missed in autoregressive models.

3 Experimental Results

We evaluate genomic and protein language models on the Scorpio-Gene-Taxa dataset [61] without task-specific fine-tuning. This direct inference setup assesses how well models pretrained on different biological corpora—protein, bacterial, or human genomes—generalize to both gene-level and taxonomic classification. While some pretraining data may overlap, our focus is on relative transfer performance across models.

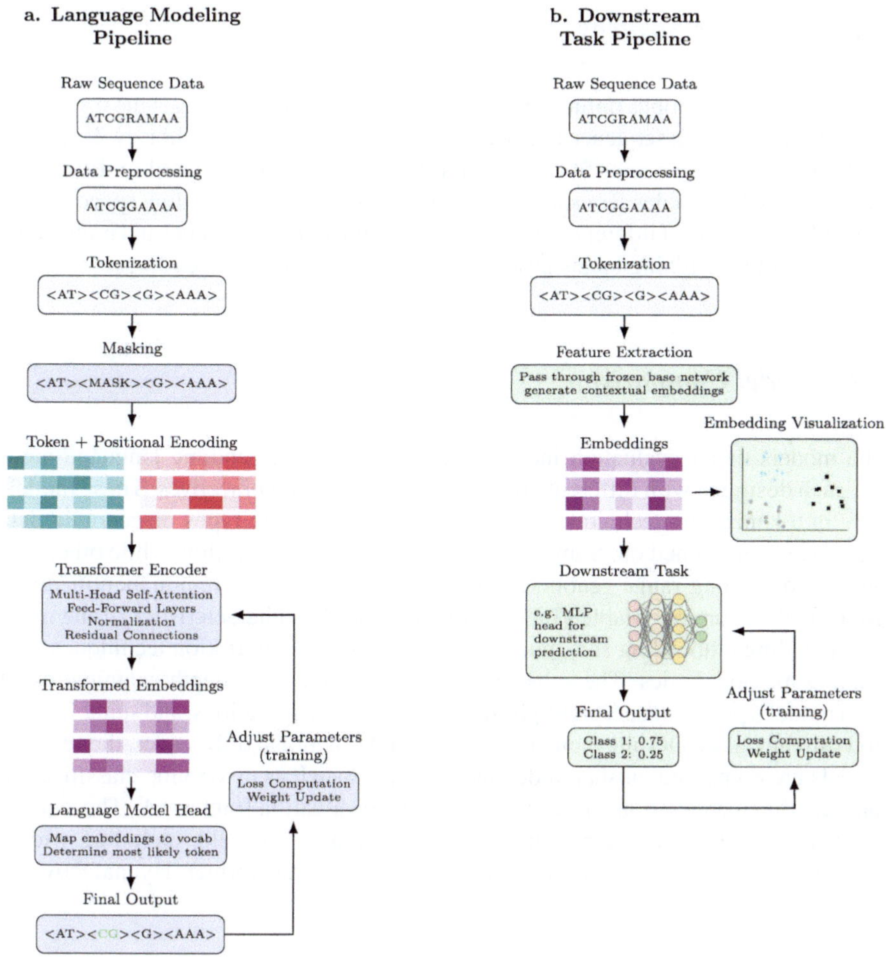

Fig. 3 **Left**: End-to-end masked language modeling (MLM) pipeline using a Transformer architecture. The process includes data preprocessing, tokenization into k-mers (Nucleotide), random token masking, token and positional embedding generation, and passing the embeddings through a Transformer encoder. The model learns to predict the masked tokens and updates its parameters via backpropagation. **Right**: Inference and downstream pipeline. Preprocessed sequences are tokenized and passed through a frozen pretrained Transformer to obtain contextual embeddings. These embeddings are then used in downstream tasks

3.1 Inference Strategy

For inference, we adopt a unified pipeline across all models. We extract the final-layer embeddings for each sequence and use FAISS—a similarity search library developed by Facebook AI Research [54]—for nearest-neighbor retrieval based on Euclidean distance. FAISS enables fast and scalable search over large embedding spaces, making it ideal for best-hit inference. This setup supports both single-label and multi-label predictions by assigning labels based on the closest training embeddings.

3.2 Model Comparison

The models under evaluation include a diverse range of genomic language models, each designed with different architectures and training objectives. These models vary in tokenization techniques, sequence lengths, and the types of genomic data they have been trained on. Some focus on taxonomic classification, while others are optimized for long-range genomic dependencies or regulatory region identification. We provide a summary table (Table 2) that outlines key characteristics of the models, including embedding size, number of parameters, tokenization techniques, and training methodologies. The subsequent sections offer an in-depth examination of each model's performance, highlighting their strengths, limitations, and suitability for downstream tasks. Despite their architectural and functional differences, the selected models are compared on shared downstream tasks such as taxonomic classification and functional annotation, allowing for a unified evaluation framework. These models were chosen to represent a broad cross-section of the field, including widely adopted baselines like DNABERT2 and Nucleotide Transformer. HyenaDNA and

Table 2 Summary of model characteristics, including embedding size, number of parameters, tokenization methods, and training objectives

Model name	Embedding size	Parameters	Tokenization	Training type
DNABERT2	768	117M	Byte-pair encoding	MLM
Nucleotide Transformer	2560	2.5B	6-mer	MLM
HyenaDNA	256	14.2M	Nucleotide	NTP
gLM2	1280	650M	Amino Acid/Nucleotide	NTP
CADUCEUS	256	7.7M	Nucleotide	MLM
BERTAX	250	9.3M	3-mer	MLM
MetaBERTa	1024	35.8M	6-mer	MLM
ESM 2	640	35/150M	Amino Acid	MLM

gLM2 are recent state-of-the-art innovations. Some models such as CADUCEUS and MetaBERTa have emerging specialized architectures to ensure relevance and diversity.

3.2.1 CADUCEUS

CADUCEUS [25] is a bidirectional genomic model built on the Mamba [52] architecture, a recent state-space model (SSM) framework designed to efficiently handle long sequences. By incorporating reverse complement equivariance, CADUCEUS ensures that DNA sequences and their reverse complements are treated identically, improving modeling accuracy. It is trained using masked language modeling (MLM) on the human reference genome (hg38) and supports long-range input sequences up to 131,072 base pairs. CADUCEUS outperforms larger transformer-based models in long-range genomic modeling while maintaining a smaller parameter footprint. Its state-space foundation makes it both scalable and interpretable—well-suited for tasks such as variant effect prediction and regulatory region identification, implementation details can be found in the CADUCEUS documentation on Hugging Face [55].

3.2.2 HyenaDNA

HyenaDNA [24] is a convolution-based genomic model that replaces transformer self-attention with implicit convolutions, allowing it to process long DNA sequences efficiently. Unlike traditional attention-based models, HyenaDNA scales sub-quadratically, enabling faster training while maintaining high accuracy. It is trained using next-token prediction (NTP) on the human reference genome (hg38) at single-nucleotide resolution, with a maximum context length of 1 million tokens. This extended sequence processing ability makes it ideal for tasks such as regulatory element prediction, enhancer identification, and species classification. By capturing long-range dependencies efficiently, HyenaDNA achieves state-of-the-art results while being computationally efficient. Additional details can be found in the HyenaDNA documentation available on Hugging Face [56].

3.2.3 BERTax

BERTax [57] is a masked language model (MLM) designed specifically for taxonomic classification of DNA sequences. It employs 3-mer tokenization, which enables it to capture short-range sequence dependencies more effectively than traditional nucleotide-level tokenization. The model is trained on a diverse dataset containing DNA fragments from Archaea, Bacteria, and Eukarya (eukaryotes) domains, and viruses, ensuring it learns taxonomically relevant features across different domains of life. The dataset was curated at the order taxonomic level, allowing BERTax to generalize well across various taxonomic ranks. Additionally, the model was fine-tuned on taxonomic classification tasks, evaluating its performance on both similar and non-similar test sets. Its ability to classify sequences without closely related references makes it particularly useful for species identification, microbiome analysis, and phylogenetic studies. Further technical information can be found in the BERTax documentation on GitHub [58].

3.2.4 MetaBERTa

MetaBERTa [59, 60] is a masked language model (MLM) trained with 6-mer tokenization, allowing it to capture genomic sequence patterns more effectively than character-level models. Unlike other genomic language models, MetaBERTa is specifically trained on the Scorpio-Gene-Taxa dataset [61], which includes bacterial and archaeal genes. This dataset is divided into functional and taxonomic categories, enabling the model to learn representations that capture both gene function and evolutionary relationships. This dual-focus training makes MetaBERTa particularly effective for functional annotation, microbial taxonomic classification, and gene clustering tasks. Additional details can be found in the MetaBERTa documentation on Hugging Face [61].

3.2.5 gLM2

gLM2 [62] is a mixed-modality genomic language model designed to capture functional and evolutionary patterns in genomic sequences. Unlike conventional genomic models that focus solely on either protein or nucleotide sequences, gLM2 integrates both representations by training on the Open MetaGenomic (OMG) dataset, a large-scale corpus comprising 3.1 trillion base pairs and 3.3 billion protein-coding sequences. The dataset is sourced from JGI's IMG and EMBL's MGnify repositories, ensuring extensive genomic diversity. gLM2 leverages a context-aware learning approach, incorporating both coding (CDS) and intergenic (IGS) sequences to model functional relationships and genomic regulatory syntax. This dual-modality training enables the model to perform well on both protein function prediction and microbial taxonomic classification tasks. Additionally, gLM2's ability to process multi-protein contexts allows it to learn co-evolutionary signals, enhancing its performance in tasks such as protein-protein interaction prediction. By employing a semantic deduplication strategy, gLM2 effectively reduces dataset redundancy, improving generalization and representation learning. As a result, gLM2 demonstrates superior performance on diverse biological benchmarks, making it a powerful tool for functional annotation, genomic sequence classification, and structural biology applications. Additional details can be found in the glm2 documentation on Hugging Face [63].

3.2.6 DNABERT2

DNABERT-2 [39] is an efficient genomic language model designed for multi-species genome analysis. Unlike its predecessor, DNABERT, which relied on k-mer tokenization, DNABERT-2 employs Byte Pair Encoding (BPE) to enhance computational efficiency and reduce tokenization redundancy. This approach allows DNABERT-2 to model genomic sequences more effectively while significantly lowering GPU and memory requirements. Trained on a diverse dataset spanning 135 species, DNABERT-2 captures evolutionary and functional patterns in DNA sequences. DNABERT-2 incorporates architectural improvements such as Attention with Linear Biases (ALiBi) and Flash Attention, enabling it to process long genomic sequences without strict length limitations. Additionally, DNABERT-2 introduces a comprehensive benchmark, Genome Understanding Evaluation (GUE), which standardizes multi-species genome classification tasks across 36 datasets. Through extensive evaluation, DNABERT-2 demonstrates superior performance over DNABERT and competitive results against larger models, all while being 21× smaller and using 92× less GPU time. These advancements establish DNABERT-2 as a highly efficient and scalable model for genomic sequence analysis. Additional details can be found in the DNABERT-2 documentation on Hugging Face [64].

3.2.7 Nucleotide Transformer

Nucleotide Transformer [65] is a large-scale genomic language model designed to capture evolutionary and functional relationships across diverse species. Unlike traditional DNA language models that rely on k-mer tokenization, Nucleotide Transformer employs non-overlapping tokenization to improve computational efficiency and reduce sequence redundancy. Pre-trained on genomic data from 850 species, the model consists of 2.5 billion parameters, enabling it to generalize across a wide range of genomic tasks. The model's architecture is optimized for long-sequence processing, making it suitable for complex applications such as promoter prediction, variant effect analysis, and transcription factor binding site identification.

By leveraging transformer-based architectures, Nucleotide Transformer achieves state-of-the-art performance across 27 genome analysis tasks. Its scalability and efficiency make it a powerful tool for genomic sequence modeling and functional annotation. Additional details can be found in the Nucleotide Transformer documentation on Hugging Face [66].

3.2.8 ESM-2

Evolutionary Scale Modeling 2 (ESM-2) [67] is a family of protein encoders with various protein sizes: 8M, 35M, 150M, 650M, 3B, and 15B. The larger model sizes yield better performances but require more computational resources unnecessary. The ESM-2 models utilize a standard bidirectional transformer encoder architecture.

The input protein sequence is tokenized in such a way that each amino acid is mapped to a given discrete token representation and then processed by the ESM-2 encoder. The ESM models were trained on millions of protein sequences from the UniRef50 database using the masked language modeling (MLM) training paradigm. This brief description of MLM should probably move to the first model (CADUCEUS) that uses the same process. ESM-2 [67] models can be fine-tuned for various tasks such as classification and even protein folding as seen in ESM-Fold, or simply used as is for the protein embedding representations that the encoder creates.

3.3 Dataset

We utilized the Scorpio-Gene-Taxa dataset [59], a comprehensive collection of 2046 genomes, with each genome uniquely representing a single genus. This dataset was meticulously curated to focus on Bacteria and Archaea, excluding viral and fungal genomes, which often lack sufficient and reliable gene information. All coding sequences (CDS) were derived from the NCBI database, resulting in a dataset comprising 800,318 gene sequences across 497 distinct gene types.

To ensure high annotation accuracy, hypothetical proteins, uncharacterized proteins, and sequences with unreliable or inconsistent gene labels were excluded [59]. Additional filters were applied to retain genes with over 1000 samples and to ensure a minimum representation of 350 sequences per phylum, bolstering the dataset's integrity and relevance. As a result, the gene-level dataset is balanced, with most classes containing between 1,000 and 2,000 sequences, supporting fair evaluation of classification performance.

The dataset was split into distinct subsets, including a training set, a testing set, and specialized holdout sets to assess the model's generalizability. These holdout sets included the Taxa Out Set, which excluded phyla present in the training data but retained overlapping genes, and the Gene Out Set, which excluded genes in the training data while retaining overlapping phyla. Furthermore, sequences with multiple representations of the same gene within a genome were removed to prevent redundancy and ensure consistency. A detailed breakdown of the number of classes at each taxonomic level and the sample count per dataset split is provided in Table 3.

3.4 Results

The evaluation of models on between Scorpio and Gene-Taxa dataset revealed notable differences in classification performance across gene and taxonomic levels. The results for the sequence-level test, where the test set contains sequences from the same genus and gene as the training set but consists of different sequences, are shown in Table 4. The ESM2-150M model achieved the highest accuracy across all taxonomic levels, particularly excelling at the gene level with an impressive

Table 3 Number of gene-level and taxonomic-level classes, along with total sample counts for each dataset split. Each row reports the number of unique classes at a specific level

Level	Train set (n = 547,523)	Test set (n = 140,524)	Taxa-out set (n = 11,800)	Gene-out set (n = 43,417)
Gene	437	437	436	60
Kingdom	2	2	1	2
Phylum	27	27	18	27
Class	83	83	25	83
Order	204	204	27	204
Family	505	505	33	505
Genus	1929	1929	39	1928

Table 4 *Evaluation of sequence-level generalization to classification tasks*: identification of additional taxonomy and genes known from training data (test set)

	Accuracy (%)					F1-macro (%)				
	Phylum	Class	Order	Family	Gene	Phylum	Class	Order	Family	Gene
BERTax	77.4	63.0	44.7	34.3	11.6	60.6	**50.7**	**41.6**	**33.0**	11.0
MetaBERTa(Bigbird)	71.2	58.4	42.6	32.2	28.2	48.4	37.3	31.8	25.6	27.4
DNABERT2	70.2	57.1	38.7	27.2	35.0	42.5	31.2	25.8	19.7	34.5
Nucleotide Transformer	78.4	66.8	49.7	36.8	47.8	56.0	*44.6*	*37.1*	*28.9*	47.3
HyenaDNA	53.4	42.3	26.4	16.9	34.3	20.9	16.3	12.9	9.0	33.3
CADUCEUS	46.7	28.4	14.1	7.8	8.0	16.1	9.4	6.9	4.5	7.9
gLM2	61.0	44.8	27.0	16.8	32.6	28.0	17.0	13.4	9.1	31.7
ESM2-35M	*86.8*	*78.4*	*61.1*	*42.5*	*99.7*	*59.6*	36.2	28.1	17.1	*99.7*
ESM2-150M	**88.7**	**80.8**	**63.7**	**44.9**	**99.8**	**63.3**	38.5	30.2	18.3	**99.8**

99.8% accuracy. In terms of F1-macro scores, ESM2 consistently outperformed other models, achieving 99.8% at the gene level while maintaining strong performance across higher taxonomic levels.

BERTax also demonstrated competitive results, particularly at the class, order, and family levels, where it outperformed other models. These observations suggest that BERTax is well-suited for taxonomic classification at finer taxonomic levels. The Nucleotide Transformer model also showed promising results for taxonomic classification, achieving an F1-score of 28.9% at the order level.

While these results highlight the superior performance of ESM2, it is important to consider potential explanations. Both ESM2 models are protein language models, which may give them an advantage as they primarily process amino acid sequences rather than nucleotide sequences. Codon usage varies among species, even when encoding the same amino acids, which might explain some performance differences between genomic language models and protein language models. The latter may encounter less diverse sequences, making them more robust in some cases.

Another potential reason for these performance differences could be the impact of Open Reading Frames (ORFs). As shown in Table 4, MetaBERTa, which is trained on coding sequences (CDS), performs significantly better than other models in gene classification. This observation suggests that genomic language models might struggle when dealing with randomly selected sequences, which could be one of the reasons for their lower performance in gene and functional classification tasks. Additionally, it is important to mention that HyenaDNA and CADUCEUS were not originally designed for this specific genomic task, which may have contributed to their lower performance. These models were trained on human datasets, while between Scorpio and Gene dataset includes a higher proportion of bacterial and archaeal sequences. Another possible reason is that these models are optimized for very long-context dependencies, which may not be ideal for short genomic sequences.

The benchmarking results for taxonomy and gene label generalization highlight significant performance differences among the evaluated models. As shown in Table 5, BERTax achieved the highest accuracy in taxonomic classification, with MetaBERTa being the second-best model, trailing by just over 10%. However, in gene classification, BERTax performed poorly, similar to the results observed in the test set.

In contrast, ESM2 demonstrated exceptional performance in gene label classification, achieving an impressive 98.9% accuracy and a 98.8% F1-macro score, significantly outperforming all other models. The superior performance of the ESM2 use "models" since both ESM2-35M and -150M perform essentially identical. compared to other genomic language models could be attributed to similar factors observed in the test set results, where protein language models may encounter less sequence diversity and variation compared to genomic sequences. As shown in Table 5, in comparison between genomic and protein language models, taxonomic signals may

Table 5 *Evaluation of generalization to gene and taxa level classification tasks*: taxonomy generalization (genes-out set) and gene label generalization (taxa-out set)

	Taxonomy generalization								Gene label generalization	
	Accuracy (%)				F1-macro (%)				Accuracy (%)	F1-macro (%)
Model	phylum	class	order	family	phylum	class	order	family	gene	gene
BERTax	**76.2**	**60.2**	**40.4**	**30.3**	**56.5**	**46.2**	**37.5**	**29.7**	4.1	3.6
MetaBERTa(Bigbird)	*64.0*	*47.1*	*29.0*	*20.4*	*36.7*	*27.6*	*22.4*	*17.8*	7.4	7.1
DNABERT2	22.1	8.9	2.5	0.8	2.1	0.9	0.4	0.1	0.2	0.2
Nucleotide Transformer	22.2	9.2	2.5	0.7	2.2	0.9	0.4	0.1	0.1	0.2
HyenaDNA	45.4	25.3	10.2	4.9	12.0	6.4	4.1	2.7	8.9	8.1
CADUCEUS	45.4	25.5	11.0	5.5	12.3	6.6	4.5	3.0	3.8	3.9
gLM2	53.3	32.4	15.2	8.2	19.9	11.5	8.0	5.5	15.7	14.9
ESM2-35M	47.2	26.4	9.2	4.0	13.7	6.7	3.7	2.0	*98.8*	*98.8*
ESM2-150M	53.9	32.5	12.4	5.8	19.3	10.6	6.1	3.6	**98.9**	**98.8**

be poorly distinguished from functional signals, highlighting the challenges of protein language models in taxonomic classification.

On the other hand, DNABERT2 and Nucleotide Transformer exhibited poor performance across all taxonomy levels. HyenaDNA, CADUCEUS, and gLM2 also performed poorly, although their results differed from those observed in the test set. Interestingly, these models performed better than the Nucleotide Transformer when dealing with embeddings that are significantly different from the training set. Among the results for taxonomy and gene label generalization, gLM2's performance stands out compared to other genomic models in gene classification. This could be due to gLM2 having been trained on both nucleotide and amino acid sequences, making it more effective at recognizing gene-related information than other models. This suggests that gLM2 may be a promising approach for functional and gene classification.

While BERTax shows promising performance, it benefits from having more information during training compared to other genomic language models. Based on their study [57], the authors also introduced label information into the embeddings, which improved BERTax's performance. However, in this work, we use BERTax embeddings instead of classification. This approach can be transferred to other tasks, but it is possible that other models incorporating label classification might achieve even better classification results.

Figure 4 presents a t-SNE visualization comparing the distribution of the 10 most common genes in the training set of the dash between Scorpio and Gene dataset. The clustering results align well with the findings in Table 4. Among the models, Nucleotide Transformer demonstrates the highest ability to identify and cluster the most common genes, followed by HyenaDNA, BERTax, and gLM2.

A notable observation is that some genes, such as *tuf*, appear more distinct from others across different models. *tuf* is a ubiquitous and highly evolutionarily conserved gene, forming part of the core genome. It is also more discriminative than the 16S rRNA gene for identifying bacterial strains from genera such as *Enterococcus*, *Staphylococcus*, and *Streptococcus* [68]. As a result, most models can effectively recognize *tuf* as a highly conserved gene. Similarly, the *groL* gene, which encodes the chaperonin GroEL, is one of the most highly conserved genes across diverse organisms [69]. Another gene, *ftsH*, also exhibits a distinct clustering pattern across multiple models.

An interesting additional observation is the clustering similarity between *fusA* and *ftsH* in certain models such as CADUCEUS and HyenaDNA. These models, which are designed to capture long-context features, reveal a significant overlap between these genes. When examined using STRING-db, we observed that *fusA* and *ftsH* have a co-occurrence score of 0.314 across genomes, indicating a biologically meaningful relationship. This suggests that the embeddings generated by these models not only capture sequence-level features but also encode genomic co-occurrence patterns, which are essential for understanding gene expression and operon structures.

Fig. 4 t-SNE visualization of the 10 most common genes in the Scorpio-Gene-Taxa dataset [61] after using various language models to embed the genes

4 Integration of Multi-omics Data with GLMs

Over the past five decades, significant advancements in multimodal, high-throughput biological assay technologies have revolutionized biology, shifting it from a data-limited, qualitative discipline to a highly data-rich and quantitative field. While early breakthroughs focused on DNA sequencing for partial or whole genomes, modern omics technologies have expanded, enabling large-scale biological sample analysis with diverse methodologies.

Most of these multi-omics innovations have been directed toward eukaryotic systems, including humans, model organisms (such as yeast, *C. elegans*, zebrafish, rats, and mice), and plants. The complexity of eukaryotic cellular and genomic structures—featuring multiple chromosomes, exon-intron arrangements, and intricate regulatory networks—has led to specialized research fields aimed at understanding how molecular components influence phenotypic traits. This broad concept of phenotype is particularly relevant in human health, encompassing both non-communicable diseases (e.g., cancer, diabetes, neurological disorders) and infectious diseases (bacterial, viral, fungal, or protist-related) [70]. Additionally, the challenge of linking genomes to phenotypes extends to metagenomic studies in agricultural and environmental contexts.

In general, omics technologies are directed toward a better mechanistic understanding of the central dogma of biology, the progression from encoding DNA to mRNAs to proteins then the resultant phenotype. However, this path is highly nuanced with multiple regulatory mechanisms and cellular interactions. For example, genomics has evolved beyond just sequencing genomes toward a deeper understanding of regulatory mechanisms such as chromosome accessibility using ATAC sequencing and epigenetic modifications. Transcriptomics, where bulk RNA-seq is still a key technology, now adds single cell RNA (scRNA-seq) and spatial genomics. Importantly, new studies merging genomics with global transcriptomics and proteomics are leading to the identification and functional interpretation of so-called encoded "dark-matter"—those genomic regions that encode novel regulatory microRNAs (miRNAs), long non-coding RNAs (lncRNAs), and small proteins [71].

High throughout global proteomics is now capable of detecting the abundances of thousands of proteins at the cellular or tissue level using either untargeted mass spectrometry (MS) or multi-plex affinity assays built from protein-specific aptamers (Somalogic™) or antibodies (O-link™) [72, 73]. Secondary or post-translational modifications (PTMs) of proteins such as phosphorylation, glycosylation, and protein degradation can also be determined via targeted proteomics assays [74]. A key computational challenge is that while these proteomics platforms have some overlap in the kinds of proteins detected, there are many proteins which are specific to the platform. Therefore, the datasets need to be treated separately or normalized.

Large-scale proteomics and genetics analysis projects are now underway such as the recent characterization of plasma proteomics in thousands of human samples by the UK Biobank Pharma Proteomics Project [72]. Next-generation proteomics, involving direct amino acid sequencing of proteins, has the potential to characterize thousands of novel proteins as well as isoformic variants of known proteins, are still several years away from widespread deployment.

Metabolomics is the detection of endogenous metabolites. Metabolites are the "currency" of communication between the microbiome and its host. In the human gut, metabolites play a crucial role in nutrition extraction, energy transfer, and maintenance of gut immunity homeostasis [75]. The identification of metabolites relies upon matching MS or Nuclear Magnetic Resonance (NMR) spectral profiles to a reference database of known metabolite profiles. The Human Metabolome Database [76] is a repository of both metabolites produced by human enzymes and endogenous microbial species. However, the spectrum of detected metabolites is dependent upon the quality of the reference database which can vary between labs, further complicating meta-analyses of metabolomics datasets.

4.1 Microbiome Multi-omics

In contrast to eukaryotes (i.e., humans, mice, yeast), a much smaller suite of omics technologies is applicable to prokaryotic or bacterial communities. Partially, this is due to the simpler genomic complexity of prokaryotic cells. Besides metagenomics, as discussed previously in this chapter, metabolomics is the most widely used omics technology in microbiome studies, particularly those involving the human stool or gut samples.

Metatranscriptomics is still an evolving technology with multiple challenges such as the isolation and purification of microbial RNAs and the necessary reduction of highly abundant ribosomal RNAs (rRNAs) which can mask the signal from protein-coding mRNAs [77]. Single cell genomics for bacterial cells is being developed; however, challenges here include low biomass resulting in low detection of mRNA and the technical hurdles to isolate single cells [77].

4.2 Current Methods for Multi-omics Analysis

While each of the omics technologies described above has well-established analysis workflows, integrating their outputs into a unified multi-omics framework remains an open challenge. Tools such as OmicsAnalyst [78] and DIABLO [79] provide statistical approaches like PCA, PLS-DA, and clustering to explore cross-omics patterns. OmicsAnalyst enables interactive, web-based exploration, while DIABLO uses supervised multi-block PLS modeling to identify correlated signatures across data types. However, these methods often result in isolated feature lists or correlation networks, with limited synthesis of underlying biological modules. More recent frameworks, such as MintTea [80], use sparse generalized canonical correlation analysis (sgCCA) to uncover disease-associated modules across metagenomic and metabolomic data. Though powerful, such approaches still rely on predefined statistical assumptions, motivating the need for more flexible, data-driven representation learning strategies.

New approaches are necessary to identify higher-level feature structures from multi-omics data. This is where innovative deep learning strategies, particularly LLMs, could be highly impactful by learning shared representations across diverse biological modalities. For example, Tataru et al. applied Latent Dirichlet Allocation (LDA)—a probabilistic topic model from NLP—to microbiome multi-omics data from stool samples of 81 children with and without autism, uncovering coherent cross-omics topics related to metabolism and transcriptional regulation [81]. Zhao et al. introduced CLCLSA, a contrastive-learning self-attention autoencoder trained on mRNA, miRNA, and DNA methylation data, achieving robust classification even with missing modalities through adaptive weighting and mutual information maximization [82]. In the protein domain, Shu et al. aligned LLMs and geometric deep models (GDMs) like ScanNet and GearNet, using contrastive learning to unify sequence and structure representations in a shared space [83]. Their study demonstrated that better alignment—achieved via larger models, projection head tuning, and domain-specific fine-tuning—improves downstream performance and reduces hallucinations. These findings support a broader trend: multimodal representation learning is key for integrating genomics, transcriptomics, proteomics, and structural data. Establishing a unified latent space for biological entities—whether DNA sequences, protein structures, or mass spectrometry features—opens the door to generalist models capable of scalable, interpretable, and robust biological inference [84].

Despite the promise of deep learning and language model-based frameworks, several fundamental challenges limit the development and deployment of unified multi-omics models. First, multi-omics datasets often suffer from the "large p, small n" problem—thousands of molecular features measured across a relatively small number of samples—leading to high-dimensional, low-sample-size regimes that hinder generalization and model stability [85]. Second, there is a scarcity of large-scale, well-annotated datasets that contain matched samples across multiple omics platforms, which are essential for training, validating, and benchmarking multimodal representation learning frameworks. A notable exception is the Inflammatory Bowel Disease (IBD) dataset from the Integrative Human Microbiome Project (HMP2/iHMP) [86], which includes longitudinal multi-omics profiles from 132 individuals—collected over a year—diagnosed with Crohn's disease, ulcerative colitis, or non-IBD. The dataset includes metagenomics, metatranscriptomics, viromes, proteomics, and metabolomics from stool samples, along with human RNA-seq data from tissue biopsies. This rich, multi-layered data enabled non-deep learning statistical models to identify key host-microbiome associations linked to disease state. However, due to missing data across some modalities and time points, significant sparsity remains, complicating the creation of balanced and complete training, validation, and test splits for LLM-based models. Overcoming these limitations will require both method development—e.g., for contrastive and self-supervised training under partial observations—and community-wide efforts to collect more standardized, multimodal biological datasets.

4.3 Host-Microbe Interactions: Opportunities for GLMs and DL

LLMs and other AI-driven methods can help in combating infectious diseases and antimicrobial resistance (AMR) as discussed earlier and elsewhere [59, 87]. Metagenomic GLMs are mainly focused on the detection of known genetic determinants of AMR such as the presence of specific genes and DNA variants. However, microbial drug-resistance is also facilitated by non-genetic factors such as the formation of biofilms [88], environmentally driven persistence [89], and evasion of the immune system via metabolic dormancy [90]. Therefore, deeper analyses of transcriptomic and metabolic data along with phenotypic data are necessary for the development of better AMR detection models.

Understanding the roles of host-microbe interactions in both communicable and non-communicable diseases has strong justifications for disease understanding and developing new therapeutic modalities. Many pathogenic bacteria and viruses modulate host pathways. *Mycobacterium tuberculosis*, the causative agent of TB, evades destruction by the immune system by entering a dormancy phase in host respiratory immunity cell-types called macrophages [91]. Dysbiosis of the gut microbiome is associated with changes in host gene expression that can lead to dysregulation of host metabolic and immune pathways [92]. Understanding these interactions is critical for the development of new therapeutics, vaccines, and diagnostic tools.

From a multi-omics perspective, the initiation of host-microbiome crosstalk occurs when environmental factors, such as dietary intake, alter the activity and composition of the microbiome (metagenomic and metatranscriptomics), leading to the production of certain metabolites (metabolomics) which have evolved tight affinities for specific human receptors (host transcriptomics, metabolomics, and proteomics). Activation or inhibition of these host receptors by microbial metabolites can modulate tissue-specific pathways in ways which potentially manifested as either pro-health or disease phenotypes (Fig. 5).

Increasingly, clinical studies involved with exploratory medicine or drug development employ some sort of multi-omics profiling. As shown in Fig. 5, to best understand the microbiome-host response axis, the optimal strategy would be to mea-

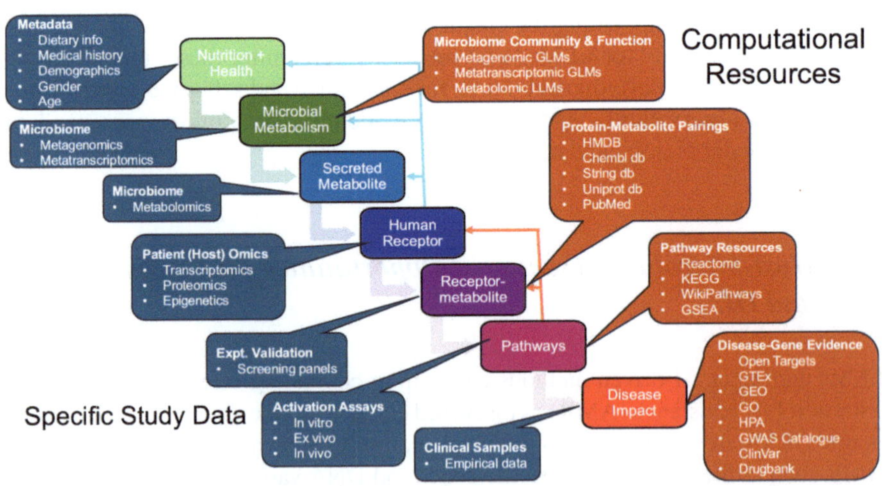

Fig. 5 Overview of proposed workflow for understanding host-microbe interactions. computational analysis. The central box cascade represents high-level biological processes involved with host-microbiome crosstalk. Left-side callouts represent potential multi-omics, clinical metadata, and de novo assay data. Right-side callouts are useful computational resources. For expansions of the abbreviations see Table 6

Table 6 Public databases relevant for the integrative analyses of host-microbe interactions, which are also shown in Fig. 1

Resource	Full name	Purpose	URL
Chembl	ChEMBL	Bioactive molecules	https://www.ebi.ac.uk/chembl/
ClinVar	ClinVar	Genomic variation and human health	https://www.ncbi.nlm.nih.gov/clinvar/
DrugBank	Drug bank	Drug-target associations	https://go.drugbank.com/
GEO	Gene expression omnibus	Gene expression	https://www.ncbi.nlm.nih.gov/gds
GO	Gene ontology resource	Disease-associated gene sets	https://geneontology.org/
GSEA	Gene set enrichment analysis	Pathway enrichment	https://www.gseamsigdb.org/gsea/index.jsp
GTEx	Genotype-tissue expression	Tissue and cell-specific gene expression	https://gtexportal.org/home/
GWAS Catalogue	Human Genome-Wide Association Studies	Human genetic-phenotypic associations	https://www.ebi.ac.uk/gwas/home
HMDB	The human metabolome database	Protein-metabolite interactions	https://hmdb.ca/
HPA	The human protein atlas	Tissue-specific protein profiles	https://www.proteinatlas.org/
KEGG	Kyoto encyclopedia of genes and genomes	Pathway mapping	https://www.genome.jp/kegg/
Open targets	Open targets	Multi-omics for drug discovery	https://platform.opentargets.org/
PubMed	PubMed—national library of medicine	Literature	https://pubmed.ncbi.nlm.nih.gov/
Reactome	Reactome	Pathway mapping	https://reactome.org/
STRING	STRING: functional protein association networks	Protein-metabolite interactions	https://string-db.org/
UniProt	UniProt: The universal protein knowledgebase	Protein functional annotations	https://www.uniprot.org/
WikiPathways	WikiPathways	Pathway mapping	https://www.wikipathways.org/

sure both the microbiome (i.e., stool samples) and human (blood or tissue samples) components using several multi-omics technologies. The panoply of multi-omics methodologies and possible computational approaches can be readily aligned with key inflection points along the microbiome-host axis of interactions. However, limits on the type and quantity of multi-omics analyses are often imposed by logistical and financial constraints.

Computational resources are essential for processing the multi-omics data and should add value in terms of improved mechanistic understanding of the disease or treatment, as well as identifying potential inflection points for therapeutic intervention, such as the agonism or antagonism of specific metabolite-protein interactions. At present, microbiome-centric GLMs are mostly focused on predictions of microbial taxonomy, genes, and proteins [93]. Human genome foundational models are also extendable to RNA and proteins [29].

With information on differentially abundant metabolites (DAMs) from stool samples as well as differentially expressed genes (DEGs) from host blood or tissues, public databases such as the Human Metabolome Database (HMDB), UniProt, and Chembl can be used to find potential pairings between metabolites and human proteins. Nuzzo et al. used this approach to identify host protein target-metabolite interactions using human transcriptomics and metabolomics from the HMP2 IBD dataset [94]. Two-dimensional character representations of DAM's chemical structures, called SMILES strings, were used to search for metabolite-like molecules in CHEMBL with associated protein targets. SMILES strings embeddings for downstream tasks can also be computed by LLMs such as GPT and LLaMA, which opens further opportunities for integrative analyses [95]. Given that metabolites can be useful chemical scaffolds for drug design, such analyses can accelerate drug discovery and development [96].

As shown in Fig. 5 and Table 6, there are several large, well-structured data repositories, which are potentially useful for placing gene-metabolite pairings in the context of particular disease pathologies using pathway reconstruction (REACTOME, KEGG, WikiPathways, and GSEA), gene/protein expression patterns (GEO, GTEx, and HPA), and human genetics (GWAS Catalogue and ClinVar). Initial foundational models have been constructed using some of these resources such as ProteinBert (GO or gene ontology) [97] and ClinVar-BERT (ClinVar or clinical genetic variants) [98].

In summary, there are many exciting opportunities for the application of deep learning methodologies, especially LLMs and GLMs, in the analysis of multi-omics data.

5 Ethical and Legal Implications of Genome Language Models

The rapid advancement of genomic sequencing technologies has opened up unprecedented opportunities for understanding human biology and advancing personalized medicine. As we discuss in this paper, the plethora of available sequencing data has enabled the development of GLMs. However, even as the widespread availability of genome sequencing technology has raised a host of ethical, legal, and social implications (ELSI), there are specific challenges associated with GLMs. At the heart of these concerns lies the sensitive and personal nature of genomic information. Unlike other forms of data, our genome is a unique and immutable part of our identity, with the potential to reveal intimate details about our health, ancestry, and even family relationships. The ability to access and analyze genomic data at scale has raised critical questions about privacy, consent, and the potential for misuse or discrimination [99–101].

Legal structures have been put in place to prevent discrimination on the basis of genetic information. For example, in Canada, the Genetic Non-Discrimination Act prohibits requiring genetic tests or the disclosure of genetic test results in specific contexts, such as entering into a contract or agreement [102]. In the United States, the Genetic Information Nondiscrimination Act (GINA) prohibits discrimination based on genetic information in health insurance and employment [103–106]. GINA also restricts employers and health insurers from requesting or requiring genetic information, with some exceptions. Additionally, the Health Insurance Portability and Accountability Act (HIPAA) in the US provides privacy protections for personal health information, including genetic data, by limiting the disclosure and use of such information by certain entities, such as healthcare providers and health plans, without an individual's consent [107–109]. While these legal frameworks offer some protections, they may not fully address all the ethical and privacy concerns raised by the increasing availability and use of genomic data, particularly in the context of GLMs.

The rise of consumer genomics companies exposes governance weaknesses, particularly in the U.S., where data protection depends on private companies and informed consent agreements [110]. The 23andMe case highlights these risks, as financial struggles raise concerns about whether privacy agreements will safeguard user data in cases of insolvency or acquisition [111]. This issue stems from weaknesses in information governance, not hacking, leaving users vulnerable to policy changes. While Europe's GDPR provides stronger data protection, it may not cover consumers engaging with foreign providers that bypass local regulations [112, 113]. Similarly, genetic diagnostics remain a regulatory challenge in countries like China [114].

Similarly, the Ubiome scandal revealed the dangers of unchecked ambition in a rapidly developing field [115, 116]. Ubiome was a startup company founded with the promise of sequencing the human microbiome and using that information to diagnose health conditions. Faced with intense pressure to demonstrate growth, the founders allegedly implemented fraudulent billing practices, falsified "doctor's notes," and misled both investors and insurers. This case highlighted the problematic incentives that can arise when venture capital is prioritized over scientific integrity, forcing rapid business development and corner-cutting, with potentially devastating consequences for patients and the integrity of genomic research.

As GLMs become increasingly sophisticated and influential in shaping our understanding of human genetics, it is necessary to accompany GLM research with the development of robust governance frameworks to address emerging ELSI challenges, as further discussed in the following section. The research community should engage with stakeholders and communities to maximize the transformative potential of these technologies while safeguarding the rights and interests of individuals and society as a whole. As further described in this section, such an effort includes grappling with complex issues such as data privacy, intellectual property, consent, and community engagement, while also ensuring that GLMs realize their potential by being used in ways that promote the public good.

To address the ethical and legal implications of GLMs, stakeholders must understand and confront the reality that GLMs implicate two emerging technical challenges: genomics and artificial intelligence. As a result, no single established ethical or regulatory framework can handle all of the myriad issues of GLMs. Instead, we frame our discussion of ethical and legal challenges in genomic language models (GLMs) by drawing on established frameworks from multiple disciplines: the Responsible AI principles (e.g., OECD, EU High-Level Expert Group), bioethics (e.g., autonomy, beneficence, nonmaleficence, and justice), and the FAIR (Fair, Accurate, Interpretable, Reproducible) principles for machine learning and data stewardship [117–119]. We also highlight intellectual property (IP) as an emerging and contested area of concern for GLMs. While traditional ethical frameworks often focus on data privacy and consent, the use of proprietary genomic datasets, unclear licensing of model outputs, and questions around ownership of AI-generated biological insights all raise unresolved legal and ethical questions [120]. Rather than adhering strictly to one framework, we selected six focus areas—privacy, bias and fairness, transparency, consent, governance, and intellectual property (IP)—that represent points of overlap and priority across these ethical approaches and policy issues. This alignment is reflected in Table 7, which maps these categories to mitigation strategies supported by relevant scholarship.

5.1 *Data Privacy, Security, and Re-identification Risks*

Concerns around data privacy and re-identification are foundational across major ethical frameworks. In bioethics, they relate directly to the principle of autonomy and the right to control one's personal information [118], while Responsible AI guidelines stress the need for robust data governance, accountability, and protection against misuse [119]. Genomic data used for pretraining GLMs presents significant privacy risks, even when anonymized, due to its inherently identifiable nature. Pretraining on vast genomic datasets may unintentionally embed sensitive information within model parameters, similar to PII leakage in LLMs[121–123]. While de-identification techniques are commonly applied to genome databases, they may be insufficient for GLMs, as model training can retain patterns that enable re-identification[109, 124, 125]. The risk is heightened in hybrid models that combine genomic and textual data, increasing the likelihood of re-identification attacks and privacy breaches.

Table 7 Summary of ethical and legal challenges in genomic language models

Issue	Concerns	Mitigation strategies
Privacy and security	Genomic data re-identification risks, potential misuse in insurance/employment	Differential privacy, federated learning, encryption, regulatory compliance
Bias and fairness	Overrepresentation of specific populations causing biased predictions	Dataset diversity, fairness-aware training, bias auditing
Model transparency	Black-box behavior and hallucinations can cause erroneous predictions	Explainability techniques, model transparency, interpretability tools
Intellectual property	Unclear ownership of genomic data and model-generated outputs	Ethical licensing, controlled open-source policies
Governance	Lack of global regulatory clarity, limited stakeholder participation in policymaking	Inclusive policy development, stakeholder engagement, international coordination
Consent	Lack of transparency and insufficient community/patient engagement	Mandated community engagement, clear communication protocols

Fine-tuning GLMs on personal genomic data raises ethical and legal concerns due to the immutable and lifelong implications of genetic information. While laws mitigate formal discrimination, informal biases and privacy breaches can still affect individuals and their relatives, as seen in the forensic use of genetic databases, which may lead to false identification and racial disparities [106, 126, 127].

Technical solutions such as machine "unlearning" methods, differential privacy techniques, and synthetic genomic data generation are being explored to address privacy risks [128–136]. Federated learning, which keeps data localized while only sharing model updates, offers another privacy-preserving approach [137–139]. However, no single technique can fully eliminate risks, necessitating a multi-layered approach that integrates privacy-enhancing methods with ethical guidelines, regulatory frameworks, and governance structures to ensure responsible data stewardship [140, 141].

5.2 Preventing Demographic Bias and Ensuring Fairness

Issues of demographic bias and representational fairness are important components of both Responsible AI and bioethical frameworks. They reflect the principles of justice and nonmaleficence in bioethics [118], and are emphasized in OECD and EU AI guidelines, which advocate for fairness-aware design, inclusive datasets, and mitigation of systemic bias in algorithmic outcomes [119]. GLMs, like other LLMs, inherit biases from their training data, which can result in demographic disparities [142–146]. Bias arises when training data over represents populations of higher economic status or European ancestry, limiting the model's applicability to diverse groups [147]. This can lead to inaccurate predictions for under represented populations, exacerbating health disparities and leading to inequities in disease risk assessment and treatment recommendations [148–151]. Attempts to remove demographic features from models do not always resolve bias; in some cases, such as colorectal cancer prediction, excluding race reduced model accuracy, highlighting the complexity of bias mitigation [152–155].

Ensuring fairness in GLMs requires building diverse training datasets and expanding genomic research to include historically under represented populations [156]. However, increasing dataset diversity alone is insufficient; models must also incorporate bias detection and mitigation strategies, such as adversarial debiasing and fairness constraints, to ensure equitable outputs [157–159]. Regular audits, standardized evaluation metrics, and transparent reporting are essential to identifying and addressing disparities [160–164].

Addressing bias in GLMs requires interdisciplinary collaboration between geneticists, machine learning researchers, bioethicists, and social scientists. Engaging underrepresented communities is crucial to understanding their concerns and ensuring equitable benefits from genomic language models.

5.3 Model Transparency and Responsible Use

GLMs offer powerful predictive capabilities but face significant challenges regarding interpretability, potential misuse, and downstream harms. Like other deep learning models, GLMs function as "black boxes," making it difficult to understand how they generate outputs, raising concerns about their reliability and scientific validity [165, 166]. Misinterpretation of GLM outputs can lead to harmful applications, such as clinicians overestimating predictive power in medical decisions or insurance companies using them to discriminate based on genetic risk [167–169].

GLMs, like LLMs, are prone to hallucination, which can sometimes generate novel biological insights but also introduce inaccuracies that undermine genomic analysis and clinical applications [170–172]. Research into mitigating hallucinations in LLMs should be extended to GLMs [157, 173–175].

Efforts to enhance interpretability in GLMs include attention mechanisms, probing classifiers, and input perturbation analysis [149, 167, 169, 176]. While these methods can highlight important sequence motifs or features, they often suffer from limited resolution and inconsistent attributions, which may hinder their usefulness for biological interpretation.

To address these limitations, model-agnostic attribution techniques such as SHAP and Integrated Gradients have been increasingly applied to genomic tasks [177, 178]. For instance, SHapley Additive exPlanations (SHAP) DeepExplainer has been used to identify nucleotide-level contributions to Clustered Regularly Interspaced Short Palindromic Repeats (CRISPR) off-target effects [179], and Integrated Gradients has helped uncover biomarker relevance in cancer gene expression models [180]. These approaches offer more robust and interpretable explanations, making them valuable tools for building trust, supporting experimental validation, and enabling responsible use of GLMs in biological and clinical settings. The need for model transparency and interpretability is an aspect of the Responsible AI, FAIR, and bioethics frameworks. It aligns with the FAIR goal of interpretable and reproducible models [117], and with the bioethical principle of autonomy, which requires that affected individuals and stakeholders can understand how decisions about them are made [118, 119].

5.4 Intellectual Property and Commercialization Issues

GLMs present complex intellectual property (IP) and commercialization challenges. While IP protection incentivizes innovation, excessive restrictions may hinder scientific progress. Many GLMs are open-source, raising questions about data ownership and benefit sharing, necessitating clear data-sharing agreements [181].

Patentability remains uncertain; software and algorithms face strict limitations, and genetic sequences are largely unpatentable in the U.S. [182–185]. Copyright protections for GLMs are also unclear—while model code may be protected, model weights are unlikely to qualify, and IP laws differ across jurisdictions [120, 186–188]. Additionally, multimodal GLMs incorporating text or images may risk infringing on existing copyrights [189].

Ethical licensing has emerged as a strategy to enforce responsible AI use, including restricting discriminatory applications and requiring benefit sharing [190–193]. However, its enforceability is questionable, particularly for open-source models, highlighting the need for strong governance frameworks to regulate GLM usage responsibly [120, 194].

While IP is not a central theme in traditional bioethics or FAIR principles, it is an emerging area of concern in genomic AI due to questions around data ownership, licensing, and commercialization of model-generated outputs. As GLMs blur the lines between input data and intellectual products, their IP issues are informed by recent scholarship on LLMs, which has highlighted how IP is an important mechanism for regulating the degree of accountability, equitable access, and transparency in research and innovation [190].

5.5 Consent and Community Engagement

Informed consent is a cornerstone of bioethics, rooted in the principle of autonomy, and is especially complex in the context of GLMs that may use de-identified or aggregated genomic data [118]. While Responsible AI frameworks increasingly call for transparent and participatory design, applying AI to genomics requires extending these principles to include community-level consent, particularly for underrepresented or indigenous populations whose data may be indirectly included [118].

Ensuring proper consent and community engagement in GLMs is essential to maintaining trust, inclusivity, and ethical research practices. Participants must be fully informed about how their genomic data will be used, including unforeseen applications, requiring dynamic consent models that allow for updates and withdrawal over time [195–200]. However, implementing such consent frameworks at scale remains a challenge.

Public transparency is crucial, involving clear communication on data usage, model development, and potential applications. Initiatives like the Genomics England public dialogue highlight the benefits of stakeholder engagement but require continuous effort to adapt to evolving GLM advancements [201]. Participatory approaches, such as advisory boards and community workshops, further enhance trust and inclusivity, as demonstrated by the H3Africa Consortium's efforts in Africa [202].

Global collaboration in genomic data sharing must also address disparities in representation. Efforts like GISAID promote data inclusivity but face tensions between open science and restricted data-sharing policies, particularly during crises like the COVID-19 pandemic [203–205]. Striking a balance between equity and open-access research remains an ongoing challenge requiring sustained dialogue and adaptive governance.

5.6 Governance Frameworks and Policy Considerations

Governance frameworks for GLMs must address ethical, legal, and regulatory challenges to ensure responsible deployment. Best practices from LLM governance can guide the development of policies for data access, privacy, and compliance [141, 206–208].

Key strategies include establishing data governance mechanisms with controlled access policies, implementing tiered access models to balance privacy with research needs, and promoting transparency in model development through open science and regular audits. Engaging diverse stakeholders is essential to developing ethical guidelines that reflect multiple perspectives [209–213].

International collaboration is needed to harmonize standards across jurisdictions while ensuring adaptability to evolving technological and social landscapes. Regular updates based on stakeholder feedback will help mitigate risks and foster responsible innovation, ultimately enabling GLMs to advance genomics research while protecting individual rights. To further contextualize these concerns, we summarize the key ethical, legal, and technical risks associated with genomic language models—along with potential mitigation strategies—in Table 7.

Best practices from LLM governance can guide the development of policies for data access, privacy, and compliance [141, 206–208]. While Responsible AI frameworks emphasize inclusive, transparent, and accountable oversight [118], genomic data introduces cross-jurisdictional challenges—such as differing privacy laws and uneven regulatory protections, which require going beyond existing regulatory models.

6 Conclusion and Future Directions

Key Points

- **Genomic Language Models (GLMs)** adapt NLP techniques to analyze DNA, RNA, and protein sequences.
- We compare diverse GLMs—including transformer-based and state-space architectures—on a unified benchmark dataset.
- Evaluation is conducted without task-specific model fine-tuning to assess true generalization.
- Insights are provided on model scalability, tokenization strategies, and gene-taxa level classification.
- The chapter also highlights the promise of multi-omics integration and the importance of ethical and legal considerations in GLM development.

Genomic Language Models (GLMs) have revolutionized the field of computational genomics by using deep learning to extract meaningful patterns from genomic sequences. These models leverage advancements in natural language processing to improve taxonomic classification, gene annotation, and functional prediction. However, their effectiveness varies depending on model architecture and dataset characteristics. Our study highlights the strengths and limitations of various GLMs in taxonomic and gene classification. Through an extensive evaluation on the Scorpio Gene-Taxa dataset, we observed that transformer-based models such as ESM2 excelled in gene classification, while BERTax demonstrated superior performance in taxonomic classification. The comparison between genomic and protein language models revealed that while protein models achieved remarkable accuracy for gene labeling, their success may be due to the fact that protein sequences retain information important to function and contain less mutation-dependent information like taxonomy. Such behavior of GLM "specialization" highlights that pretraining/training can be pivotal for the application of these models, for example, the aforementioned examples of training on proteins help gene labeling, or training on taxonomy optimizes taxonomic classification with BERTax. However, "specialization" can also have disadvantages. Models such as HyenaDNA and CADUCEUS, optimized for long-range genomic contexts and trained on human genomes, exhibited limitations when applied to microbial datasets. These findings emphasize the need for task-specific model selection, considering factors such as sequence diversity, taxonomic complexity, and functional annotation requirements. Another interesting observation is the performance of gLM2, which demonstrated promising results in gene classification, suggesting that models trained on both nucleotide and amino acid sequences could be better suited for gene-related tasks.

Looking ahead, integrating multiple biological modalities represents a vital direction for refining genomic language models. While current GLMs primarily operate on nucleotide or amino acid sequences, they often miss the regulatory, structural, and environmental context underlying biological functions. Incorporating transcriptomic, epigenomic, proteomic, and metabolomic data—alongside sequence data—can provide richer biological representations. Advances in contrastive learning and cross-modal alignment offer promising strategies to fuse such heterogeneous data into unified, biologically grounded embeddings. By embedding diverse omics modalities into shared latent spaces, future models—whether GLMs or broader LLMs adapted for biology—will be better equipped for complex tasks like phenotype prediction, disease association, and causal inference. Multi-modal AI is therefore not just an extension but a necessary evolution for unlocking deeper biological insight from genomic data.

Finally, the increasing use of AI in genomics raises important ethical and legal considerations. The application of GLMs to genomic datasets requires careful attention to privacy, data ownership, and potential biases in model training. Ensuring the ethical use of genomic AI involves implementing robust data governance policies, addressing demographic imbalances in training datasets, and maintaining transparency in model decision-making. Additionally, as GLMs become more integrated into clinical and biomedical research, regulatory frameworks must evolve to safeguard data security and prevent misuse. Ethical AI deployment in genomics will be key to fostering trust, ensuring equitable benefits across populations, and aligning computational advancements with real-world applications.

As the field progresses, future research should focus on improving model generalization, reducing computational costs, and enhancing interpretability. Developing hybrid architectures that balance the efficiency of convolutional networks with the contextual learning power of transformers could improve both performance and scalability. Furthermore, advancing techniques for interpreting GLM predictions will be crucial for making these models more useful for biological and medical applications. Ultimately, the next generation of GLMs has the potential to unlock new frontiers in genomic discovery, driving innovations in precision medicine, microbiome research, and evolutionary biology.

Code Availability

To support reproducibility and further research, we have released all models and their tokenization methods, along with instructions for running the evaluations, at the following GitHub repository: https://github.com/EESI/Genomics-Language-Model-Comparison.

Acknowledgements This work is supported by the National Science Foundation under Grant Numbers #1936791, #1919691, and #2107108. We thank the University Research Computing Facility for their paid services.

References

1. Laure Segurel, Perle Guarino-Vignon, Nina Marchi, Sophie Lafosse, Romain Laurent, Céline Bon, Alexandre Fabre, Tatyana Hegay, and Evelyne Heyer. Why and when was lactase persistence selected for? insights from central asian herders and ancient dna. *PLOS Biology*, 18(6):e3000742, 2020.
2. Rasmus Nielsen. Molecular signatures of natural selection. *Annual Review of Genetics*, 39(1):197–218, 2005.
3. Joseph J. Vitti, Sharon R. Grossman, and Pardis C. Sabeti. Detecting natural selection in genomic data. *Annual Review of Genetics*, 47(1):97–120, 2013.
4. Wenqing Fu and Joshua M. Akey. Selection and adaptation in the human genome. *Annual Review of Genomics and Human Genetics*, 14(1):467–489, 2013.
5. Pardis C Sabeti. Natural selection: Uncovering mechanisms of evolutionary adaptation to infectious disease. *Nature Education*, 1(1):13, 2008.
6. Bacem Saada, Tianchi Zhang, Estevao Siga, Jing Zhang, and Maria Malane Magalhães Muniz. Whole-genome alignment: Methods, challenges, and future directions. *Applied Sciences*, 14(11):4837, 2024.
7. Gill Bejerano, Michael Pheasant, Igor Makunin, Stuart Stephen, W. James Kent, John S. Mattick, and David Haussler. Ultraconserved elements in the human genome. *Science*, 304(5675):1321–1325, 2004.
8. Jingjing Wang, Xianwen Meng, Hongjun Chen, Chunhui Yuan, Xue Li, Yincong Zhou, and Ming Chen. Exploring the mechanisms of genome-wide long-range interactions: interpreting chromosome organization. *Briefings in Functional Genomics*, 15(5):385–395, 2016.
9. Stefan Schoenfelder and Peter Fraser. Long-range enhancer–promoter contacts in gene expression control. *Nature Reviews Genetics*, 20(8):437–455, 2019.
10. J. Hu, B. Li, and D. Kihara. Limitations and potentials of current motif discovery algorithms. *Nucleic Acids Research*, 33(15):4899–4913, 2005.
11. C. Mathe. Current methods of gene prediction, their strengths and weaknesses. *Nucleic Acids Research*, 30(19):4103–4117, 2002.
12. Jiekun Yang and Mazhar Adli. Mapping and making sense of noncoding mutations in the genome. *Cancer Research*, 79(17):4309–4314, 2019.
13. Lucia A. Hindorff, Praveen Sethupathy, Heather A. Junkins, Erin M. Ramos, Jayashri P. Mehta, Francis S. Collins, and Teri A. Manolio. Potential etiologic and functional implications of genome-wide association loci for human diseases and traits. *Proceedings of the National Academy of Sciences*, 106(23):9362–9367, 2009.
14. Robert J.A. Bell, H. Tomas Rube, Ana Xavier-Magalhães, Bruno M. Costa, Andrew Mancini, Jun S. Song, and Joseph F. Costello. Understanding tert promoter mutations: A common path to immortality. *Molecular Cancer Research*, 14(4):315–323, 2016.
15. Lei Li and Cheng Zhao. Large language model applications in nucleic acid research. *Genomics Communications*, 2(1):0–0, 2025.
16. Ashish Vaswani, Noam Shazeer, Niki Parmar, Jakob Uszkoreit, Llion Jones, Aidan N. Gomez, Łukasz Kaiser, and Illia Polosukhin. Attention is all you need. *arXiv preprint* arXiv:1706.03762, 2017.
17. Philip Gage. A new algorithm for data compression. *The C Users Journal archive*, 12:23–38, 1994.
18. Md Jubaer Hossain, Mohammed Imamul Hassan Bhuiyan, and Zaowad Rahabin Abdullah. Cpg island detection using transformer model with conditional random field. In *2022 IEEE Bombay Section Signature Conference (IBSSC)*, page 1–5. IEEE, 2022.
19. Shuyan Cheng, Yishu Wei, Yiliang Zhou, Zihan Xu, Drew N Wright, Jinze Liu, and Yifan Peng. Deciphering genomic codes using advanced natural language processing techniques: a scoping review. *Journal of the American Medical Informatics Association*, 32(4):761–772, 2025.
20. Mhaned Oubounyt, Zakaria Louadi, Hilal Tayara, and Kil To Chong. Deepromoter: Robust promoter predictor using deep learning. *Frontiers in Genetics*, 10, 2019.

21. Yanrong Ji, Zhihan Zhou, Han Liu, and Ramana V Davuluri. Dnabert: pre-trained bidirectional encoder representations from transformers model for dna-language in genome. *Bioinformatics*, 37(15):2112–2120, 2021.
22. Kishore Jaganathan, Sofia Kyriazopoulou Panagiotopoulou, Jeremy F. McRae, Siavash Fazel Darbandi, David Knowles, Yang I. Li, Jack A. Kosmicki, Juan Arbelaez, Wenwu Cui, Grace B. Schwartz, Eric D. Chow, Efstathios Kanterakis, Hong Gao, Amirali Kia, Serafim Batzoglou, Stephan J. Sanders, and Kyle Kai-How Farh. Predicting splicing from primary sequence with deep learning. *Cell*, 176(3):535–548.e24, 2019.
23. Hugo Dalla-Torre, Liam Gonzalez, Javier Mendoza-Revilla, Nicolas Lopez Carranza, Adam Henryk Grzywaczewski, Francesco Oteri, Christian Dallago, Evan Trop, Bernardo P de Almeida, Hassan Sirelkhatim, et al. Nucleotide transformer: building and evaluating robust foundation models for human genomics. *Nature Methods*, pages 1–11, 2024.
24. Eric Nguyen, Michael Poli, Marjan Faizi, Armin Thomas, Michael Wornow, Callum Birch-Sykes, Stefano Massaroli, Aman Patel, Clayton Rabideau, Yoshua Bengio, et al. Hyenadna: Long-range genomic sequence modeling at single nucleotide resolution. *Advances in neural information processing systems*, 36:43177–43201, 2023.
25. Yair Schiff, Chia-Hsiang Kao, Aaron Gokaslan, Tri Dao, Albert Gu, and Volodymyr Kuleshov. Caduceus: Bi-directional equivariant long-range dna sequence modeling. *arXiv preprint* arXiv:2403.03234, 2024.
26. Yanrong Ji, Zhihan Zhou, Han Liu, and Ramana V Davuluri. Dnabert: pre-trained bidirectional encoder representations from transformers model for dna-language in genome. *Bioinformatics*, 37(15):2112–2120, 2021.
27. Eric Nguyen, Michael Poli, Matthew G Durrant, Brian Kang, Dhruva Katrekar, David B Li, Liam J Bartie, Armin W Thomas, Samuel H King, Garyk Brixi, et al. Sequence modeling and design from molecular to genome scale with evo. *Science*, 386(6723):eado9336, 2024.
28. Garyk Brixi, Matthew G Durrant, Jerome Ku, Michael Poli, Greg Brockman, Daniel Chang, Gabriel A Gonzalez, Samuel H King, David B Li, Aditi T Merchant, et al. Genome modeling and design across all domains of life with evo 2. *BioRxiv*, pages 2025–02, 2025.
29. Zhihan Zhou, Robert Riley, Satria Kautsar, Weimin Wu, Rob Egan, Steven Hofmeyr, Shira Goldhaber-Gordon, Mutian Yu, Harrison Ho, Fengchen Liu, et al. Genomeocean: An efficient genome foundation model trained on large-scale metagenomic assemblies. *bioRxiv*, pages 2025–01, 2025.
30. Wei Lan, Haibo Liao, Qingfeng Chen, Lingzhi Zhu, Yi Pan, and Yi-Ping Phoebe Chen. Deepkegg: a multi-omics data integration framework with biological insights for cancer recurrence prediction and biomarker discovery. *Briefings in bioinformatics*, 25(3):bbae185, 2024.
31. Chen Zhao, Anqi Liu, Xiao Zhang, Xuewei Cao, Zhengming Ding, Qiuying Sha, Hui Shen, Hong-Wen Deng, and Weihua Zhou. Clclsa: Cross-omics linked embedding with contrastive learning and self attention for multi-omics integration with incomplete multi-omics data. *bioRxiv*, 2023.
32. Yunha Hwang, Andre L Cornman, Elizabeth H Kellogg, Sergey Ovchinnikov, and Peter R Girguis. Genomic language model predicts protein co-regulation and function. *Nature communications*, 15(1):2880, 2024.
33. Yaniv Erlich and Arvind Narayanan. Genome privacy: challenges, technical approaches to mitigate risk, and ethical considerations in the united states. *The Annals of the New York Academy of Sciences*, 1387(1):73–83, 2017.
34. Amanda K Sarata and Jody Feder. *The Genetic Information Nondiscrimination Act of 2008 (GINA)*. Congressional Research Service Washington, 2015.
35. Sandra Soo-Jin Lee. The ethics of consent in a shifting genomic ecosystem. *Annual review of biomedical data science*, 4(1):145–164, 2021.
36. Edo Dotan, Gal Jaschek, Tal Pupko, and Yonatan Belinkov. Effect of tokenization on transformers for biological sequences. *Bioinformatics*, 40(4):btae196, 2024.
37. Eric Nguyen, Michael Poli, Marjan Faizi, Armin Thomas, Michael Wornow, Callum Birch-Sykes, Stefano Massaroli, Aman Patel, Clayton Rabideau, Yoshua Bengio, et al. Hyenadna: Long-range genomic sequence modeling at single nucleotide resolution. *Advances in neural information processing systems*, 36, 2024.

38. LeAnn M Lindsey, Nicole L Pershing, Anisa Habib, W Zac Stephens, Anne J Blaschke, and Hari Sundar. A comparison of tokenization impact in attention based and state space genomic language models. *bioRxiv*, pages 2024–09, 2024.
39. Zhihan Zhou, Yanrong Ji, Weijian Li, Pratik Dutta, Ramana Davuluri, and Han Liu. Dnabert-2: Efficient foundation model and benchmark for multi-species genome. *arXiv preprint* arXiv:2306.15006, 2023.
40. Rico Sennrich. Neural machine translation of rare words with subword units. *arXiv preprint* arXiv:1508.07909, 2015.
41. Noam Shazeer Ashish Vaswani, Jakob Uszkoreit Niki Parmar, Llion Jones, and Lukasz Kaiserand Illia Polosukhin Aidan N. Gomez. Attention is all you need. *aRxiv*, 2017.
42. John J Hopfield. Neural networks and physical systems with emergent collective computational abilities. *Proceedings of the national academy of sciences*, 79(8):2554–2558, 1982.
43. Sepp Hochreiter and Jürgen Schmidhuber. Long short-term memory. *Neural computation*, 9(8):1735–1780, 1997.
44. Micaela E Consens, Cameron Dufault, Michael Wainberg, Duncan Forster, Mehran Karimzadeh, Hani Goodarzi, Fabian J Theis, Alan Moses, and Bo Wang. To transformers and beyond: large language models for the genome. *arXiv preprint* arXiv:2311.07621, 2023.
45. Jacob Devlin, Ming-Wei Chang, Kenton Lee, and Kristina Toutanova. Bert: Pre-training of deep bidirectional transformers for language understanding. In *Proceedings of the 2019 conference of the North American chapter of the association for computational linguistics: human language technologies, volume 1 (long and short papers)*, pages 4171–4186, 2019.
46. Ofir Press, Noah A Smith, and Mike Lewis. Train short, test long: Attention with linear biases enables input length extrapolation. *arXiv preprint* arXiv:2108.12409, 2021.
47. Tri Dao, Dan Fu, Stefano Ermon, Atri Rudra, and Christopher Ré. Flashattention: Fast and memory-efficient exact attention with io-awareness. *Advances in neural information processing systems*, 35:16344–16359, 2022.
48. Edward J Hu, Yelong Shen, Phillip Wallis, Zeyuan Allen-Zhu, Yuanzhi Li, Shean Wang, Lu Wang, Weizhu Chen, et al. Lora: Low-rank adaptation of large language models. *ICLR*, 1(2):3, 2022.
49. Michael Poli, Stefano Massaroli, Eric Nguyen, Daniel Y Fu, Tri Dao, Stephen Baccus, Yoshua Bengio, Stefano Ermon, and Christopher Ré. Hyena hierarchy: Towards larger convolutional language models. In *International Conference on Machine Learning*, pages 28043–28078. PMLR, 2023.
50. Rudolph Emil Kalman. A new approach to linear filtering and prediction problems. 1960.
51. Albert Gu, Karan Goel, and Christopher Ré. Efficiently modeling long sequences with structured state spaces. *arXiv preprint* arXiv:2111.00396, 2021.
52. Albert Gu and Tri Dao. Mamba: Linear-time sequence modeling with selective state spaces. *arXiv preprint* arXiv:2312.00752, 2023.
53. Alec Radford, Karthik Narasimhan, Tim Salimans, Ilya Sutskever, et al. Improving language understanding by generative pre-training. 2018.
54. Matthijs Douze, Alexandr Guzhva, Chengqi Deng, Jeff Johnson, Gergely Szilvasy, Pierre-Emmanuel Mazaré, Maria Lomeli, Lucas Hosseini, and Hervé Jégou. The faiss library. *arXiv preprint* arXiv:2401.08281, 2024.
55. Kuleshov Group. Caduceus models on hugging face. https://huggingface.co/collections/kuleshov-group/caduceus-65dcb89b4f54e416ef61c350, 2024.
56. LongSafari Group. Hyenadna models on hugging face. https://huggingface.co/collections/LongSafari/hyenadna-models-654d0cbbe113b04ba5a0f638, 2024.
57. Florian Mock, Fleming Kretschmer, Anton Kriese, Sebastian Böcker, and Manja Marz. Bertax: taxonomic classification of dna sequences with deep neural networks. *BioRxiv*, pages 2021–07, 2021.
58. Florian Mock, Fleming Kretschmer, Anton Kriese, Sebastian Böcker, and Manja Marz. Bertax: Taxonomic classification of dna sequences. https://github.com/rnajena/bertax, 2020.
59. Mohammadsaleh Refahi, Bahrad A Sokhansanj, Joshua C Mell, James R Brown, Hyunwoo Yoo, Gavin Hearne, and Gail L Rosen. Enhancing nucleotide sequence representations in genomic analysis with contrastive optimization. *Communications biology*, 8(1):517, 2025.

60. MS Refahi, BA Sokhansanj, and GL Rosen. Leveraging large language models for metagenomic analysis. In *2023 IEEE signal processing in medicine and biology symposium (SPMB)*, pages 1–6. IEEE, 2023.
61. Mohammadsaleh Refahi. Metaberta-bigbird-gene. https://huggingface.co/MsAlEhR/MetaBERTa-bigbird-gene, 2024.
62. Andre Cornman, Jacob West-Roberts, Antonio Pedro Camargo, Simon Roux, Martin Beracochea, Milot Mirdita, Sergey Ovchinnikov, and Yunha Hwang. The omg dataset: An open metagenomic corpus for mixed-modality genomic language modeling. In *Proceedings of the International Conference on Learning Representations (ICLR)*. ICLR, 2025.
63. tattabio. glm2_650m. https://huggingface.co/tattabio/gLM2_650M, 2025.
64. Zhihan Zhou. Dnabert-2-117m. https://huggingface.co/zhihan1996/DNABERT-2-117M, 2023.
65. Hugo Dalla-Torre, Liam Gonzalez, Javier Mendoza-Revilla, Nicolas Lopez Carranza, Adam Henryk Grzywaczewski, Francesco Oteri, Christian Dallago, Evan Trop, Bernardo P. de Almeida, Hassan Sirelkhatim, Guillaume Richard, Marcin Skwark, Karim Beguir, Marie Lopez, and Thomas Pierrot. Nucleotide transformer: Building and evaluating robust foundation models for human genomics. *Nature Methods*, 2024.
66. InstaDeepAI. Nucleotide transformer collection. https://huggingface.co/collections/InstaDeepAI/nucleotide-transformer-65099cdde13ff96230f2e592, 2024.
67. Zeming Lin, Halil Akin, Roshan Rao, Brian Hie, Zhongkai Zhu, Wenting Lu, Nikita Smetanin, Robert Verkuil, Ori Kabeli, Yaniv Shmueli, et al. Evolutionary-scale prediction of atomic-level protein structure with a language model. *Science*, 379(6637):1123–1130, 2023.
68. Xuerui Li, Juan Xing, Baoyu Li, Pu Wang, and Jixing Liu. Use of *tuf* as a target for sequence-based identification of gram-positive cocci of the genus enterococcus, streptococcus, coagulase-negative staphylococcus, and lactococcus. *Annals of Clinical Microbiology and Antimicrobials*, 11:1–6, 2012.
69. Gil Segal and Eliora Z Ron. Regulation and organization of the groe and dnak operons in eubacteria. *FEMS microbiology letters*, 138(1):1–10, 1996.
70. Zahra Safaei and Gary L Thompson. Histone deacetylase 4 and 5 translocation elicited by microsecond pulsed electric field exposure is mediated by kinase activity. *Frontiers in Bioengineering and Biotechnology*, 10:1047851, 2022.
71. Huifen Cao and Philipp Kapranov. Methods to analyze the non-coding rna interactome–recent advances and challenges. *Frontiers in Genetics*, 13:857759, 2022.
72. Grimur Hjorleifsson Eldjarn, Egil Ferkingstad, Sigrun H Lund, Hannes Helgason, Olafur Th Magnusson, Kristbjorg Gunnarsdottir, Thorunn A Olafsdottir, Bjarni V Halldorsson, Pall I Olason, Florian Zink, et al. Large-scale plasma proteomics comparisons through genetics and disease associations. *Nature*, 622(7982):348–358, 2023.
73. Miao Cui, Chao Cheng, and Lanjing Zhang. High-throughput proteomics: a methodological mini-review. *Laboratory investigation*, 102(11):1170–1181, 2022.
74. Lijun Dou, Fenglong Yang, Lei Xu, and Quan Zou. A comprehensive review of the imbalance classification of protein post-translational modifications. *Briefings in Bioinformatics*, 22(5):bbab089, 2021.
75. J.-Y. Lee, D. J. Bays, H. P. Savage, and A. J. Bäumler. The human gut microbiome in health and disease: time for a new chapter? *Infection and Immunity*, 92:e0030224, 2024.
76. David S Wishart, AnChi Guo, Eponine Oler, Fei Wang, Afia Anjum, Harrison Peters, Raynard Dizon, Zinat Sayeeda, Siyang Tian, Brian L Lee, et al. Hmdb 5.0: the human metabolome database for 2022. *Nucleic acids research*, 50(D1):D622–D631, 2022.
77. T. Ojala, A.-E. Häkkinen, E. Kankuri, and M. Kankainen. Current concepts, advances, and challenges in deciphering the human microbiota with metatranscriptomics. *Trends in Genetics*, 39:686–702, 2023.
78. Guangyan Zhou, Jessica Ewald, and Jianguo Xia. Omicsanalyst: a comprehensive web-based platform for visual analytics of multi-omics data. *Nucleic Acids Research*, 49(W1):W476–W482, 2021.

79. Amrit Singh, Casey P Shannon, Benoît Gautier, Florian Rohart, Michaël Vacher, Scott J Tebbutt, and Kim-Anh Lê Cao. Diablo: an integrative approach for identifying key molecular drivers from multi-omics assays. *Bioinformatics*, 35(17):3055–3062, 2019.
80. Efrat Muller, Itamar Shiryan, and Elhanan Borenstein. Multi-omic integration of microbiome data for identifying disease-associated modules. *Nature Communications*, 15(1):2621, 2024.
81. Christine Tataru, Marie Peras, Erica Rutherford, Kaiti Dunlap, Xiaochen Yin, Brianna S Chrisman, Todd Z DeSantis, Dennis P Wall, Shoko Iwai, and Maude M David. Topic modeling for multi-omic integration in the human gut microbiome and implications for autism. *Scientific Reports*, 13(1):11353, 2023.
82. Chen Zhao, Anqi Liu, Xiao Zhang, Xuewei Cao, Zhengming Ding, Qiuying Sha, Hui Shen, Hong-Wen Deng, and Weihua Zhou. Clclsa: Cross-omics linked embedding with contrastive learning and self attention for integration with incomplete multi-omics data. *Computers in biology and medicine*, 170:108058, 2024.
83. Dong Shu, Bingbing Duan, Kai Guo, Kaixiong Zhou, Jiliang Tang, and Mengnan Du. Aligning large language models and geometric deep models for protein representation. *Patterns*, 6(5), 2025.
84. Gonzalo Benegas, Chengzhong Ye, Carlos Albors, Jianan Canal Li, and Yun S Song. Genomic language models: opportunities and challenges. *Trends in Genetics*, 2025.
85. Dylan Feldner-Busztin, Panos Firbas Nisantzis, Shelley Jane Edmunds, Gergely Boza, Fernando Racimo, Shyam Gopalakrishnan, Morten Tønsberg Limborg, Leo Lahti, and Gonzalo G de Polavieja. Dealing with dimensionality: the application of machine learning to multi-omics data. *Bioinformatics*, 39(2):btad021, 2023.
86. Jason Lloyd-Price, Cesar Arze, Ashwin N Ananthakrishnan, Melanie Schirmer, Julian Avila-Pacheco, Tiffany W Poon, Elizabeth Andrews, Nadim J Ajami, Kevin S Bonham, Colin J Brislawn, et al. Multi-omics of the gut microbial ecosystem in inflammatory bowel diseases. *Nature*, 569(7758):655–662, 2019.
87. Angela Cesaro, Samuel C Hoffman, Payel Das, and Cesar de la Fuente-Nunez. Challenges and applications of artificial intelligence in infectious diseases and antimicrobial resistance. *npj Antimicrobials and Resistance*, 3(1):2, 2025.
88. Javairia Khan, Sumbal Mudassar Tarar, Iram Gul, Uzam Nawaz, and Muhammad Arshad. Challenges of antibiotic resistance biofilms and potential combating strategies: a review. *3 Biotech*, 11:1–15, 2021.
89. Tahmina Hossain, Abhyudai Singh, and Nicholas C Butzin. Escherichia coli cells are primed for survival before lethal antibiotic stress. *Microbiology Spectrum*, 11(5):e01219–23, 2023.
90. Ali Akbar Velayati, Satoshi Mitaria, Parissa Farnia, Poopak Farnia, Jaladein Ghanavi, Mohsan Maleknia, Yoshiro Murase, and Sven Hoffner. Exploring the genetic landscape of mycobacterium tuberculosis: Unlocking the differences in between latent and active tuberculosis. *The International Journal of Mycobacteriology*, 13(4):410–419, 2024.
91. Zhang Wang, Seda Arat, Michal Magid-Slav, and James R Brown. Meta-analysis of human gene expression in response to mycobacterium tuberculosis infection reveals potential therapeutic targets. *BMC systems biology*, 12:1–18, 2018.
92. Christoph A Thaiss, Niv Zmora, Maayan Levy, and Eran Elinav. The microbiome and innate immunity. *Nature*, 535(7610):65–74, 2016.
93. Binghao Yan, Yunbi Nam, Lingyao Li, Rebecca A Deek, Hongzhe Li, and Siyuan Ma. Recent advances in deep learning and language models for studying the microbiome. *Frontiers in genetics*, 15:1494474, 2025.
94. Andrea Nuzzo, Somdutta Saha, Ellen Berg, Channa Jayawickreme, Joel Tocker, and James R Brown. Expanding the drug discovery space with predicted metabolite–target interactions. *Communications biology*, 4(1):288, 2021.
95. Shaghayegh Sadeghi, Alan Bui, Ali Forooghi, Jianguo Lu, and Alioune Ngom. Can large language models understand molecules? *BMC bioinformatics*, 25(1):225, 2024.
96. Somdutta Saha, Deepak K Rajpal, and James R Brown. Human microbial metabolites as a source of new drugs. *Drug discovery today*, 21(4):692–698, 2016.

97. Nadav Brandes, Dan Ofer, Yam Peleg, Nadav Rappoport, and Michal Linial. Proteinbert: a universal deep-learning model of protein sequence and function. *Bioinformatics*, 38(8):2102–2110, 2022.
98. Weijiang Li, Xiaomin Li, Ethan Lavallee, Alice Saparov, Marinka Zitnik, and Christopher Cassa. From text to translation: Using language models to prioritize variants for clinical review. *medRxiv*, pages 2024–12, 2024.
99. Jane Kaye, Eric M. Meslin, Bartha M. Knoppers, Eric T. Juengst, Mylène Deschênes, Anne Cambon-Thomsen, Donald Chalmers, Jantina De Vries, Kelly Edwards, Nils Hoppe, Alastair Kent, Clement Adebamowo, Patricia Marshall, and Kazuto Kato. ELSI 2.0 for Genomics and Society. *Science*, 336(6082):673–674, 2012.
100. Bartha Maria Knoppers, Ma'n H. Zawati, and Emily S. Kirby. Sampling Populations of Humans Across the World: ELSI Issues. *Annual Review of Genomics and Human Genetics*, 13(Volume 13, 2012):395–413, 2012.
101. Go Yoshizawa, Calvin Wai-Loon Ho, Wei Zhu, Chingli Hu, Yoni Syukriani, Ilhak Lee, Hannah Kim, Daniel Fu Chang Tsai, Jusaku Minari, and Kazuto Kato. ELSI practices in genomic research in East Asia: Implications for research collaboration and public participation. *Genome Medicine*, 6(5):39, 2014.
102. Victoria Di Felice. Canada's Genetic Non-Discrimination Act: False Security for Customers of Direct-to-Consumer Genetic Testing Services. *McGill Journal of Law and Health*, 16(1):1–36, 2024.
103. Yann Joly and Gratien Dalpe. Genetic discrimination still casts a large shadow in 2022. *European Journal of Human Genetics*, 30(12):1320–1322, 2022.
104. Andrea Lenartz, Aaron M. Scherer, Wendy R. Uhlmann, Sonia M. Suter, Colleen Anderson Hartley, and Anya E. R. Prince. The persistent lack of knowledge and misunderstanding of the Genetic Information Nondiscrimination Act (GINA) more than a decade after passage. *Genetics in Medicine*, 23(12):2324–2334, 2021.
105. Jessica L. Roberts. Preempting Discrimination: Lessons from the Genetic Information Nondiscrimination Act. *Vanderbilt Law Review*, 63(2):439–490, 2010.
106. Bahrad A. Sokhansanj. Beyond Protecting Genetic Privacy: Understanding Genetic Discrimination Through Its Disparate Impact on Racial Minorities. *Columbia Journal of Race and Law*, 2(2):279–309, 2012.
107. I. Glenn Cohen and Michelle M. Mello. HIPAA and Protecting Health Information in the 21st Century. *JAMA*, 320(3):231–232, 2018.
108. Jitendra Jonnagaddala and Zoie Shui-Yee Wong. Privacy preserving strategies for electronic health records in the era of large language models. *npj Digital Medicine*, 8(1):1–3, 2025.
109. Jennifer K Wagner, Juhi K Tanniru, Courtney A Chane, and Michelle N Meyer. Exploring access to genomic risk information and the contours of the HIPAA public health exception. *Journal of Law and the Biosciences*, 9(2):lsac034, 2022.
110. Katherine Drabiak. Caveat Emptor: How the Intersection of Big Data and Consumer Genomics Exponentially Increases Informational Privacy Risks. *Health Matrix: Journal of Law-Medicine*, 27:143–184, 2017.
111. Diana Kwon. What went wrong at 23andMe? Why the genetic-data giant risks collapse. *Nature*, 638(8049):14–15, 2025.
112. Kristi Harbord. Genetic Data Privacy Solutions in the GDPR Comment. *Texas A&M Law Review*, 7(1):[i]–298, 2019/2020.
113. Miriam C. Buiten. 'Your DNA Is One Click Away': The GDPR and Direct-to-Consumer Genetic Testing. In Klaus Mathis and Avishalom Tor, editors, *Consumer Law and Economics*, pages 205–223, Cham, 2021. Springer International Publishing.
114. Li Du and Meng Wang. Genetic Privacy and Data Protection: A Review of Chinese Direct-to-Consumer Genetic Test Services. *Frontiers in Genetics*, 11, 2020.
115. Northern District of California | uBiome Co-Founders Charged With Federal Securities, Health Care Fraud Conspiracies | United States Department of Justice. https://www.justice.gov/usao-ndca/pr/ubiome-co-founders-charged-federal-securities-health-care-fraud-conspiracies, 2021.

116. Matthew S. Wood, David J. Scheaf, and Sean M. Dwyer. Fake it 'til you make it: Hazards of a cultural norm in entrepreneurship. *Business Horizons*, 65(5):681–696, 2022.
117. EA Huerta, Ben Blaiszik, L Catherine Brinson, Kristofer E Bouchard, Daniel Diaz, Caterina Doglioni, Javier M Duarte, Murali Emani, Ian Foster, Geoffrey Fox, et al. Fair for ai: An interdisciplinary and international community building perspective. *Scientific data*, 10(1):487, 2023.
118. Emily R Winslow and Ira J Kodner. Ethics and genetic testing. In *Seminars in Colon and Rectal Surgery*, volume 15, pages 186–190. Elsevier, 2004.
119. Mark D Cole. AI regulation and governance on a global scale: An overview of international, regional and national instruments. *Journal of AI Law and Regulation*, 1(1):126–142, 2024.
120. Mark A. Lemley and Peter Henderson. The Mirage of Artificial Intelligence Terms of Use Restrictions, 2024.
121. Badhan Chandra Das, M. Hadi Amini, and Yanzhao Wu. Security and Privacy Challenges of Large Language Models: A Survey. *ACM Comput. Surv.*, 57(6):152:1–152:39, 2025.
122. Nils Lukas, Ahmed Salem, Robert Sim, Shruti Tople, Lukas Wutschitz, and Santiago Zanella-Béguelin. Analyzing Leakage of Personally Identifiable Information in Language Models. In *2023 IEEE Symposium on Security and Privacy (SP)*, pages 346–363, 2023.
123. Yifan Yao, Jinhao Duan, Kaidi Xu, Yuanfang Cai, Zhibo Sun, and Yue Zhang. A survey on large language model (LLM) security and privacy: The Good, The Bad, and The Ugly. *High-Confidence Computing*, 4(2):100211, 2024.
124. Luca Bonomi, Yingxiang Huang, and Lucila Ohno-Machado. Privacy Challenges and Research Opportunities for Genomic Data Sharing. *Nature genetics*, 52(7):646–654, 2020.
125. Mara Thomas, Nuria Mackes, Asad Preuss-Dodhy, Thomas Wieland, and Markus Bundschus. Assessing Privacy Vulnerabilities in Genetic Data Sets: Scoping Review. *JMIR Bioinformatics and Biotechnology*, 5(1):e54332, 2024.
126. Greg Miller. Forensics. Familial DNA testing scores a win in serial killer case. *Science (New York, N.Y.)*, 329(5989):262, 2010.
127. Nina F. de Groot, Britta C. van Beers, Lieven Decock, and Gerben Meynen. Accessing medical biobanks to solve crimes: Ethical considerations. *Journal of Medical Ethics*, pages medethics–2020–106133, 2020.
128. Sungmin Cha, Sungjun Cho, Dasol Hwang, and Moontae Lee. Towards Robust and Cost-Efficient Knowledge Unlearning for Large Language Models, 2024.
129. Robert Dilworth. Privacy Preservation through Practical Machine Unlearning, 2025.
130. Debdeep Sanyal and Murari Mandal. ALU: Agentic LLM Unlearning, 2025.
131. Bozhong Tian, Xiaozhuan Liang, Siyuan Cheng, Qingbin Liu, Mengru Wang, Dianbo Sui, Xi Chen, Huajun Chen, and Ningyu Zhang. To Forget or Not? Towards Practical Knowledge Unlearning for Large Language Models, 2024.
132. Rouzbeh Behnia, Mohammadreza Reza Ebrahimi, Jason Pacheco, and Balaji Padmanabhan. EW-Tune: A Framework for Privately Fine-Tuning Large Language Models with Differential Privacy. In *2022 IEEE International Conference on Data Mining Workshops (ICDMW)*, pages 560–566, 2022.
133. Tanmay Singh, Harshvardhan Aditya, Vijay K. Madisetti, and Arshdeep Bahga. Whispered Tuning: Data Privacy Preservation in Fine-Tuning LLMs through Differential Privacy. *Journal of Software Engineering and Applications*, 17(1):1–22, 2024.
134. Taslim Murad, Sarwan Ali, and Murray Patterson. Exploring the Potential of GANs in Biological Sequence Analysis. *Biology*, 12(6):854, 2023.
135. Bristena Oprisanu, Georgi Ganev, and Emiliano De Cristofaro. On Utility and Privacy in Synthetic Genomic Data, 2022.
136. Xinyue Wang, Sitao Min, and Jaideep Vaidya. Descriptor: Synthetic Genomic Dataset With Diverse Ancestry (SynGen6). *IEEE Data Descriptions*, 2:1–7, 2025.
137. Charlie Hou, Akshat Shrivastava, Hongyuan Zhan, Rylan Conway, Trang Le, Adithya Sagar, Giulia Fanti, and Daniel Lazar. PrE-Text: Training Language Models on Private Federated Data in the Age of LLMs, 2024.

138. Feijie Wu, Zitao Li, Yaliang Li, Bolin Ding, and Jing Gao. FedBiOT: LLM Local Fine-tuning in Federated Learning without Full Model. In *Proceedings of the 30th ACM SIGKDD Conference on Knowledge Discovery and Data Mining*, KDD '24, pages 3345–3355, New York, NY, USA, 2024. Association for Computing Machinery.
139. JiaYing Zheng, HaiNan Zhang, LingXiang Wang, WangJie Qiu, HongWei Zheng, and ZhiMing Zheng. Safely Learning with Private Data: A Federated Learning Framework for Large Language Model. In *Proceedings of the 2024 Conference on Empirical Methods in Natural Language Processing*, pages 5293–5306, 2024.
140. Natalia Díaz-Rodríguez, Javier Del Ser, Mark Coeckelbergh, Marcos López de Prado, Enrique Herrera-Viedma, and Francisco Herrera. Connecting the dots in trustworthy Artificial Intelligence: From AI principles, ethics, and key requirements to responsible AI systems and regulation. *Information Fusion*, 99:101896, 2023.
141. Jason Hausenloy, Duncan McClements, and Madhavendra Thakur. Towards Data Governance of Frontier AI Models, 2024.
142. Daphne Lenders and Anne Oloo. 15 Years of Algorithmic Fairness – Scoping Review of Interdisciplinary Developments in the Field, 2024.
143. Noel F. Ayoub, Karthik Balakrishnan, Marc S. Ayoub, Thomas F. Barrett, Abel P. David, and Stacey T. Gray. Inherent Bias in Large Language Models: A Random Sampling Analysis. *Mayo Clinic Proceedings: Digital Health*, 2(2):186–191, 2024.
144. Md Meftahul Ferdaus, Mahdi Abdelguerfi, Elias Ioup, Kendall N. Niles, Ken Pathak, and Steven Sloan. Towards Trustworthy AI: A Review of Ethical and Robust Large Language Models, 2024.
145. Anna Kruspe. Towards detecting unanticipated bias in Large Language Models, 2024.
146. Alberto Acerbi and Joseph M. Stubbersfield. Large language models show human-like content biases in transmission chain experiments. *Proceedings of the National Academy of Sciences*, 120(44):e2313790120, 2023.
147. Daniel E. Spratt, Tiffany Chan, Levi Waldron, Corey Speers, Felix Y. Feng, Olorunseun O. Ogunwobi, and Joseph R. Osborne. Racial/Ethnic Disparities in Genomic Sequencing. *JAMA oncology*, 2(8):1070–1074, 2016.
148. Alicia R. Martin, Christopher R. Gignoux, Raymond K. Walters, Genevieve L. Wojcik, Benjamin M. Neale, Simon Gravel, Mark J. Daly, Carlos D. Bustamante, and Eimear E. Kenny. Human Demographic History Impacts Genetic Risk Prediction across Diverse Populations. *The American Journal of Human Genetics*, 100(4):635–649, 2017.
149. Gregory McInnes, Andrew G. Sharo, Megan L. Koleske, Julia E. H. Brown, Matthew Norstad, Aashish N. Adhikari, Sheng Wang, Steven E. Brenner, Jodi Halpern, Barbara A. Koenig, David C. Magnus, Renata C. Gallagher, Kathleen M. Giacomini, and Russ B. Altman. Opportunities and challenges for the computational interpretation of rare variation in clinically important genes. *The American Journal of Human Genetics*, 108(4):535–548, 2021.
150. Noam Barda, Gal Yona, Guy N Rothblum, Philip Greenland, Morton Leibowitz, Ran Balicer, Eitan Bachmat, and Noa Dagan. Addressing bias in prediction models by improving subpopulation calibration. *Journal of the American Medical Informatics Association*, 28(3):549–558, 2021.
151. Anurag Vaidya, Richard J. Chen, Drew F. K. Williamson, Andrew H. Song, Guillaume Jaume, Yuzhe Yang, Thomas Hartvigsen, Emma C. Dyer, Ming Y. Lu, Jana Lipkova, Muhammad Shaban, Tiffany Y. Chen, and Faisal Mahmood. Demographic bias in misdiagnosis by computational pathology models. *Nature Medicine*, 30(4):1174–1190, 2024.
152. Sara Khor, Eric C. Haupt, Erin E. Hahn, Lindsay Joe L. Lyons, Veena Shankaran, and Aasthaa Bansal. Racial and Ethnic Bias in Risk Prediction Models for Colorectal Cancer Recurrence When Race and Ethnicity Are Omitted as Predictors. *JAMA Network Open*, 6(6):e2318495, 2023.
153. Heidi Ledford. Millions of black people affected by racial bias in health-care algorithms. *Nature*, 574(7780):608–609, 2019.
154. Ziad Obermeyer, Brian Powers, Christine Vogeli, and Sendhil Mullainathan. Dissecting racial bias in an algorithm used to manage the health of populations. *Science*, 366(6464):447–453, 2019.

155. Zahra Safaei and Gary L Thompson. Caspase-dependent cell death and hdac4 translocation following microsecond pulsed electric field (μspef) exposure in mcf-7 breast cancer cells. *Bioelectromagnetics*, 46(4):e70009, 2025.
156. Alexander G. Bick, Ginger A. Metcalf, Kelsey R. Mayo, Lee Lichtenstein, Shimon Rura, Robert J. Carroll, Anjene Musick, Jodell E. Linder, I. King Jordan, Shashwat Deepali Nagar, Shivam Sharma, Robert Meller, Melissa Basford, Eric Boerwinkle, Mine S. Cicek, Kimberly F. Doheny, Evan E. Eichler, Stacey Gabriel, Richard A. Gibbs, David Glazer, Paul A. Harris, Gail P. Jarvik, Anthony Philippakis, Heidi L. Rehm, Dan M. Roden, Stephen N. Thibodeau, Scott Topper, Ashley L. Blegen, Samantha J. Wirkus, Victoria A. Wagner, Jeffrey G. Meyer, Mine S. Cicek, Donna M. Muzny, Eric Venner, Michelle Z. Mawhinney, Sean M. L. Griffith, Elvin Hsu, Hua Ling, Marcia K. Adams, Kimberly Walker, Jianhong Hu, Harsha Doddapaneni, Christie L. Kovar, Mullai Murugan, Shannon Dugan, Ziad Khan, Eric Boerwinkle, Niall J. Lennon, Christina Austin-Tse, Eric Banks, Michael Gatzen, Namrata Gupta, Emma Henricks, Katie Larsson, Sheli McDonough, Steven M. Harrison, Christopher Kachulis, Matthew S. Lebo, Cynthia L. Neben, Marcie Steeves, Alicia Y. Zhou, Joshua D. Smith, Christian D. Frazar, Colleen P. Davis, Karynne E. Patterson, Marsha M. Wheeler, Sean McGee, Christina M. Lockwood, Brian H. Shirts, Colin C. Pritchard, Mitzi L. Murray, Valeria Vasta, Dru Leistritz, Matthew A. Richardson, Jillian G. Buchan, Aparna Radhakrishnan, Niklas Krumm, Brenna W. Ehmen, Sophie Schwartz, M. Morgan T. Aster, Kristian Cibulskis, Andrea Haessly, Rebecca Asch, Aurora Cremer, Kylee Degatano, Akum Shergill, Laura D. Gauthier, Samuel K. Lee, Aaron Hatcher, George B. Grant, Genevieve R. Brandt, Miguel Covarrubias, Eric Banks, Ashley Able, Ashley E. Green, Robert J. Carroll, Jennifer Zhang, Henry R. Condon, Yuanyuan Wang, Moira K. Dillon, C. H. Albach, Wail Baalawi, Seung Hoan Choi, Xin Wang, Elisabeth A. Rosenthal, Andrea H. Ramirez, Sokny Lim, Siddhartha Nambiar, Bradley Ozenberger, Anastasia L. Wise, Chris Lunt, Geoffrey S. Ginsburg, Joshua C. Denny, The All of Us Research Program Genomics Investigators, Manuscript Writing Group, All of Us Research Program Genomics Principal Investigators, Mayo Biobank, Genome Center: Baylor-Hopkins Clinical Genome Center, Color Genome Center: Broad, and Mass General Brigham Laboratory for Molecular Medicine, Genome Center: University of Washington, Data and Research Center, All of Us Research Demonstration Project Teams, and NIH All of Us Research Program Staff. Genomic data in the All of Us Research Program. *Nature*, 627(8003):340–346, 2024.
157. Zichao Lin, Shuyan Guan, Wending Zhang, Huiyan Zhang, Yugang Li, and Huaping Zhang. Towards trustworthy LLMs: A review on debiasing and dehallucinating in large language models. *Artificial Intelligence Review*, 57(9):243, 2024.
158. Jenny Yang, Andrew A. S. Soltan, David W. Eyre, and David A. Clifton. Algorithmic fairness and bias mitigation for clinical machine learning with deep reinforcement learning. *Nature Machine Intelligence*, 5(8):884–894, 2023.
159. Honggang Zhao and Wenlu Wang. Adversarial Removal of Population Bias in Genomics Phenotype Prediction. In *2022 IEEE International Conference on Data Mining Workshops (ICDMW)*, pages 334–340, 2022.
160. Fatma-Elzahraa Eid, Haitham A. Elmarakeby, Yujia Alina Chan, Nadine Fornelos, Mahmoud ElHefnawi, Eliezer M. Van Allen, Lenwood S. Heath, and Kasper Lage. Systematic auditing is essential to debiasing machine learning in biology. *Communications Biology*, 4(1):1–9, 2021.
161. Jakob Mökander, Jonas Schuett, Hannah Rose Kirk, and Luciano Floridi. Auditing large language models: A three-layered approach. *AI and Ethics*, 4(4):1085–1115, 2024.
162. Charvi Rastogi, Marco Tulio Ribeiro, Nicholas King, Harsha Nori, and Saleema Amershi. Supporting Human-AI Collaboration in Auditing LLMs with LLMs. In *Proceedings of the 2023 AAAI/ACM Conference on AI, Ethics, and Society*, AIES '23, pages 913–926, New York, NY, USA, 2023. Association for Computing Machinery.
163. Jack Gallifant, Majid Afshar, Saleem Ameen, Yindalon Aphinyanaphongs, Shan Chen, Giovanni Cacciamani, Dina Demner-Fushman, Dmitriy Dligach, Roxana Daneshjou, Chrystinne Fernandes, Lasse Hyldig Hansen, Adam Landman, Lisa Lehmann, Liam G. McCoy, Timothy

Miller, Amy Moreno, Nikolaj Munch, David Restrepo, Guergana Savova, Renato Umeton, Judy Wawira Gichoya, Gary S. Collins, Karel G. M. Moons, Leo A. Celi, and Danielle S. Bitterman. The TRIPOD-LLM reporting guideline for studies using large language models. *Nature Medicine*, 31(1):60–69, 2025.

164. Lena Maier-Hein, Annika Reinke, Michal Kozubek, Anne L. Martel, Tal Arbel, Matthias Eisenmann, Allan Hanbury, Pierre Jannin, Henning Müller, Sinan Onogur, Julio Saez-Rodriguez, Bram van Ginneken, Annette Kopp-Schneider, and Bennett A. Landman. BIAS: Transparent reporting of biomedical image analysis challenges. *Medical Image Analysis*, 66:101796, 2020.
165. Pantelis Linardatos, Vasilis Papastefanopoulos, and Sotiris B Kotsiantis. Explainable AI: A Review of Machine Learning Interpretability Methods. *Entropy*, 23, 2021.
166. W James Murdoch, Chandan Singh, Karl Kumbier, Reza Abbasi-Asl, and Bin Yu. Definitions, methods, and applications in interpretable machine learning. *Proceedings of the National Academy of Sciences of the United States of America*, 116(44):22071–22080, 2019.
167. K. D. Ahlquist, Lauren A. Sugden, and Sohini Ramachandran. Enabling interpretable machine learning for biological data with reliability scores. *PLOS Computational Biology*, 19(5):e1011175, 2023.
168. David S Watson. Interpretable machine learning for genomics. *Human genetics*, pages 1–15, 2021.
169. Amanda Momenzadeh, Ali Shamsa, and Jesse G Meyer. Bias or biology? Importance of model interpretation in machine learning studies from electronic health records. *JAMIA Open*, 5(3):ooac063, 2022.
170. Ziwei Xu, Sanjay Jain, and Mohan Kankanhalli. Hallucination is Inevitable: An Innate Limitation of Large Language Models, 2024.
171. Abbi Abdel-Rehim, Hector Zenil, Oghenejokpeme Orhobor, Marie Fisher, Ross J. Collins, Elizabeth Bourne, Gareth W. Fearnley, Emma Tate, Holly X. Smith, Larisa N. Soldatova, and Ross D. King. Scientific Hypothesis Generation by a Large Language Model: Laboratory Validation in Breast Cancer Treatment, 2024.
172. Shuzhou Yuan and Michael Färber. Hallucinations Can Improve Large Language Models in Drug Discovery, 2025.
173. Minda Hu, Bowei He, Yufei Wang, Liangyou Li, Chen Ma, and Irwin King. Mitigating Large Language Model Hallucination with Faithful Finetuning, 2024.
174. Subhabrata Mukherjee, Paul Gamble, Markel Sanz Ausin, Neel Kant, Kriti Aggarwal, Neha Manjunath, Debajyoti Datta, Zhengliang Liu, Jiayuan Ding, Sophia Busacca, Cezanne Bianco, Swapnil Sharma, Rae Lasko, Michelle Voisard, Sanchay Harneja, Darya Filippova, Gerry Meixiong, Kevin Cha, Amir Youssefi, Meyhaa Buvanesh, Howard Weingram, Sebastian Bierman-Lytle, Harpreet Singh Mangat, Kim Parikh, Saad Godil, and Alex Miller. Polaris: A Safety-focused LLM Constellation Architecture for Healthcare, 2024.
175. Ziwei Ji, Delong Chen, Etsuko Ishii, Samuel Cahyawijaya, Yejin Bang, Bryan Wilie, and Pascale Fung. LLM Internal States Reveal Hallucination Risk Faced With a Query,2024.
176. Cynthia Rudin. Stop explaining black box machine learning models for high stakes decisions and use interpretable models instead. *Nature Machine Intelligence*, 1(5):206–215, 2019.
177. Scott M Lundberg and Su-In Lee. A unified approach to interpreting model predictions. *Advances in Neural Information Processing Systems*, 30:4765–4774, 2017.
178. Mukund Sundararajan, Ankur Taly, and Qiqi Yan. Axiomatic attribution for deep networks. *Proceedings of the 34th International Conference on Machine Learning*, 70:3319–3328, 2017.
179. Jeffrey K Mak, Florian Störtz, and Peter Minary. Comprehensive computational analysis of epigenetic descriptors affecting crispr–cas9 off-target activity. *BMC Genomics*, 23(1):805, 2022.
180. Anupriya Jha, Matthew R Gazzara, and Yoseph Barash. Interpretable deep learning for chromatin-informed inference of transcriptional programs driven by somatic alterations across cancers. *Nature Communications*, 11(1):1–15, 2020.
181. Yu-Kai Lin and Arun Rai. The Scope of Software Patent Protection in the Digital Age: Evidence from Alice. *Information Systems Research*, 35(2):657–672, 2024.

182. Mildred Cho. Patently unpatentable: Implications of the Myriad court decision on genetic diagnostics. *Trends in Biotechnology*, 28(11):548–551, 2010.
183. Haroon Hasan Akhtar. Undiagnosed: The Failure to Determine Patentability Standards for Diagnostic Tests. *Journal of the Patent and Trademark Office Society*, 104(3):377–396, 2024.
184. Anusuya Das. Patentability Challenges in Personalized Medicine: A Fork in the Road Notes. *Indiana Law Review*, 57(2):455–478, 2023/2024.
185. Isabelle Huys, Geertrui Van Overwalle, and Gert Matthijs. Gene and genetic diagnostic method patent claims: A comparison under current European and US patent law. *European Journal of Human Genetics*, 19(10):1104–1107, 2011.
186. Jane C. Ginsburg. Copyright, Common Law, and Sui Generis Protection of Databases in the United States and Abroad Corporate Law Symposium: Intellectual Property Law for the Twenty-First Century. *University of Cincinnati Law Review*, 66(1):151–176, 1997/1998.
187. Romain Meys. Data Mining Under the Directive on Copyright and Related Rights in the Digital Single Market: Are European Database Protection Rules Still Threatening the Development of Artificial Intelligence? *GRUR International*, 69(5):457–473, 2020.
188. Masanori Arita. Open Access and Data Sharing of Nucleotide Sequence Data. *Data Science Journal*, 20:28–28, 2021.
189. AI firms must play fair when they use academic data in training. *Nature*, 632(8027):953–953, 2024.
190. Christi J. Guerrini, Margaret A. Curnutte, Jacob S. Sherkow, and Christopher T. Scott. The rise of the ethical license. *Nature Biotechnology*, 35(1):22–24, 2017.
191. Kevin Klyman. Acceptable Use Policies for Foundation Models. *Proceedings of the AAAI/ACM Conference on AI, Ethics, and Society*, 7(1):752–767, 2024.
192. Daniel McDuff, Tim Korjakow, Scott Cambo, Jesse Josua Benjamin, Jenny Lee, Yacine Jernite, Carlos Muñoz Ferrandis, Aaron Gokaslan, Alek Tarkowski, Joseph Lindley, A. Feder Cooper, and Danish Contractor. On the Standardization of Behavioral Use Clauses and Their Adoption for Responsible Licensing of AI, 2024.
193. Sulayman Sowe, Yongli Mou, Du Cheng, Lingxiao Kong, Alexander Tobias Neumann, and Stefan Decker. Understanding Open Source Large Language Models: An Exploratory Study. In *2024 2nd International Conference on Foundation and Large Language Models (FLLM)*, pages 132–140, 2024.
194. Jonathan Cui and David A Araujo. Rethinking use-restricted open-source licenses for regulating abuse of generative models. *Big Data & Society*, 11(1):20539517241229699, 2024.
195. Greg Gibson and Gregory P. Copenhaver. Consent and Internet-Enabled Human Genomics. *PLOS Genetics*, 6(6):e1000965, 2010.
196. Rachel Horton and Anneke Lucassen. Consent and Autonomy in the Genomics Era. *Current Genetic Medicine Reports*, 7(2):85–91, 2019.
197. A. Khan, B.j. Capps, M.y. Sum, C.n. Kuswanto, and K. Sim. Informed consent for human genetic and genomic studies: A systematic review. *Clinical Genetics*, 86(3):199–206, 2014.
198. Charles N. Rotimi and Patricia A. Marshall. Tailoring the process of informed consent in genetic and genomic research. *Genome Medicine*, 2(3):20, 2010.
199. Isabelle Budin-Ljøsne, Harriet J. A. Teare, Jane Kaye, Stephan Beck, Heidi Beate Bentzen, Luciana Caenazzo, Clive Collett, Flavio D'Abramo, Heike Felzmann, Teresa Finlay, Muhammad Kassim Javaid, Erica Jones, Višnja Katić, Amy Simpson, and Deborah Mascalzoni. Dynamic Consent: A potential solution to some of the challenges of modern biomedical research. *BMC Medical Ethics*, 18(1):4, 2017.
200. Sam HA Muller, Ghislaine JMW van Thiel, Menno Mostert, and Johannes JM van Delden. Dynamic consent, communication and return of results in large-scale health data reuse: Survey of public preferences. *Digital Health*, 9:20552076231190997, 2023.
201. Gabrielle Natalie Samuel and Bobbie Farsides. Public trust and 'ethics review' as a commodity: The case of Genomics England Limited and the UK's 100,000 genomes project. *Medicine, Health Care and Philosophy*, 21(2):159–168, 2018.
202. Jantina de Vries, Paulina Tindana, Katherine Littler, Michèle Ramsay, Charles Rotimi, Akin Abayomi, Nicola Mulder, and Bongani M. Mayosi. The H3Africa policy framework: Negotiating fairness in genomics. *Trends in Genetics*, 31(3):117–119, 2015.

203. Stefan Elbe and Gemma Buckland-Merrett. Data, disease and diplomacy: GISAID's innovative contribution to global health. *Global Challenges*, 1(1):33–46, 2017.
204. Sabina Leonelli. Open Science and Epistemic Diversity: Friends or Foes? *Philosophy of Science*, 89(5):991–1001, 2022.
205. Nathanael Sheehan and Sabina Leonelli. Reconciling data actionability and accountability in global health research: The case of SARS-CoV-2. *Global Public Health*, 20(1):2436422, 2025.
206. Amna Batool, Didar Zowghi, and Muneera Bano. AI governance: A systematic literature review. *AI and Ethics*, 2025.
207. Patricia Gomes Rêgo de Almeida, Carlos Denner dos Santos, and Josivania Silva Farias. Artificial Intelligence Regulation: A framework for governance. *Ethics and Information Technology*, 23(3):505–525, 2021.
208. Araz Taeihagh. Governance of artificial intelligence. *Policy and Society*, 40(2):137–157, 2021.
209. Marta Cantero Gamito and Christopher T Marsden. Artificial intelligence co-regulation? The role of standards in the EU AI Act. *International Journal of Law and Information Technology*, 32(1):eaae011, 2024.
210. Pascale Fung and Hubert Etienne. Confucius, cyberpunk and Mr. Science: Comparing AI ethics principles between China and the EU. *AI and Ethics*, 3(2):505–511, 2023.
211. Oskar J. Gstrein, Noman Haleem, and Andrej Zwitter. General-purpose AI regulation and the European Union AI Act. *Internet Policy Review*, 13(3):1–26, 2024.
212. Guangyu Qiao-Franco and Rongsheng Zhu. China's Artificial Intelligence Ethics: Policy Development in an Emergent Community of Practice. *Journal of Contemporary China*, 33(146):189–205, 2024.
213. Junhua Zhu. AI ethics with Chinese characteristics? Concerns and preferred solutions in Chinese academia. *AI & SOCIETY*, 39(3):1261–1274, 2024.

Channel Efficiency in ECG-Based Multi-label Cardiac Diagnosis

Dmitry Alexandrov, Claudia Dumitrescu, Sadia Purba, Anna Chau, and Joseph Picone

Abstract Electrocardiogram (ECG) analysis relies on preprocessed data and derived channels, which are linear combinations of raw signals, to aid in cardiac diagnosis. Medical professionals have developed these methods over the years, incorporating both clinical experience and technical knowledge of their monitoring equipment. The widespread use of these preprocessing steps has become standard practice in cardiology. However, recent advances in deep learning necessitate a reevaluation of these preprocessing steps. Deep learning models have demonstrated the ability to analyze raw signal or image data without requiring traditional preprocessing or derived channels. For example, architectures such as ResNet18 have achieved high performance on ECG datasets such as the Telehealth Network of Minas Gerais (TNMG) CODE Corpus. These architectures challenge the clinical assumption that derived channels are essential for accurate diagnosis. This shift raises the question of whether preprocessing methods developed for visual interpretation by clinicians remain necessary in machine learning contexts. In this chapter, we investigate a variety of baseline models for multi-label ECG classification. Our findings reveal that the performance of complex architectures, such as ResNet18, consistently declines as the number of input channels increases. We demonstrate that the original 8 channels collected using a standard 10-lead ECG can provide optimal performance on the TNMG Corpus. We describe the experimental setup used to compare models trained on raw ECG signals versus those trained on derived or preprocessed inputs. The results demonstrate that using raw signals not only simplifies the data pipeline during training

D. Alexandrov · C. Dumitrescu · S. Purba · A. Chau · J. Picone (✉)
The Neural Engineering Data Consortium, Temple University, Philadelphia, PA, USA
e-mail: picone@temple.edu

D. Alexandrov
e-mail: dmytro.bolshak@temple.edu

C. Dumitrescu
e-mail: claudia-anca.dumitrescu@temple.edu

S. Purba
e-mail: sadia.afrin.purba@temple.edu

A. Chau
e-mail: anna.chau@temple.edu

A. Ahmed and J. Picone (eds.), *Signal Processing and Biomedical Engineering Research*, https://doi.org/10.1007/978-3-032-18509-9_4

but also preserves the native morphology of waveforms and outperforms diagnostic models relying on engineered features. Overall, learning directly from raw ECG data streamlines model development, reduces computational overhead, improves robustness across varying conditions, and enhances interpretability by eliminating opaque preprocessing stages.

Keywords Cardiac diagnosis · ECG · Multi-label classification · Channel reduction · Deep learning

1 Introduction

The electrocardiogram (ECG) machine was invented by Willem Einthoven in 1903 [1]. For this groundbreaking work, Einthoven was awarded the Nobel Prize in Physiology or Medicine in 1924. Automated clinical ECG interpretation emerged in the early 1970s, marked by Ray Bonner's development of the IBM 5880 system, which utilized third-generation DSP boards to enable real time, on-site QRS complex detection [2]. Major device manufacturers quickly followed suit, and by the mid-1970s [3], manufacturers such as Philips and IBM had integrated proprietary QRS detection firmware into bedside monitors. These systems featured dual-filter pipelines to improve complex signal classification across multiple leads. In the early 1990s, the adoption of discrete wavelet transform (DWT) analysis [4] gained traction, offering multiresolution feature extraction that significantly enhanced performance—raising QRS detection accuracy above 99.8% on the MIT/BIH database. By the 2020s, deep learning (DL) approaches using big data resources had become commonplace in many medical applications, suggesting that it makes sense to revisit how ECG signals are preprocessed [5, 6].

The progression of ECG technology from Holter's early prototypes [7] to modern high-resolution recorders reflects a long-standing effort to increase data richness through added leads and higher sampling rates. By the 1960s, commercial multi-lead Holter systems recorded 3 to 12 leads continuously. While these improvements aimed to preserve signal fidelity, studies have shown that only a limited number of principal components are needed to capture most of the diagnostic information in multi-lead ECGs [8], while the addition of other leads adds noise. This implies that beyond a certain threshold, additional channels and higher sample rates contribute little new information, instead inflating storage and processing costs without proportional diagnostic gain.

In fact, today, there is great interest in wearable technology that does continuous monitoring from a single-lead [9] and low-cost systems that use a reduced number of leads [10]. Though the focus of this work is not an exploration of minimizing the number of leads for low-cost consumer devices or simplification of clinical equipment, our findings obviously have clinical and consumer implications. However, our focus here is more about how deep learning systems now offer the potential to simplify the interpretation of standard clinical ECG systems.

Access to state-of-the-art cardiovascular healthcare remains a critical issue in developing countries, primarily due to the severe shortage of cardiologists. For instance, in Brazil in 2022, there were only 8.40 cardiologists for every 100,000 individuals [11], while in Africa, a continent with a population of approximately 1.2 billion people, there are only about 2,000 recorded cardiologists, or 0.17 cardiologists per 100,000 individuals [12]. This alarming scarcity of official cardiologists requires the urgent need for innovative solutions, and although low-cost ECG machines are available in these regions, they fall short in providing comprehensive diagnoses due to their reliance on classical and naïve algorithms. These machines often lack the capacity to comprehend a wider range of present cardiovascular disease indicators, reducing their effectiveness in detecting indicators of cardiovascular diseases.

Hence, there is great interest in developing low-cost automated technology that can diagnose a broad range of conditions. Fortunately, DL technology has progressed to the point where this is now feasible provided adequate data exists to train such models. While there are several large publicly available datasets for medical applications [13–15], Ribeiro et al. [16, 17] have released a very significant corpus of ECGs that makes it possible to explore many dimensions of this problem space. This unique resource, which is well annotated and curated, is the focus of our work. In this chapter, we investigate the problem of signal preprocessing on the TNMG Corpus. However, there are many other interesting problems yet to be explored, such as the localization of critical events in the signal, often referred to as the micro-segmentation problem.

This chapter begins with a basic review of the signal processing steps used in ECG analysis in Sect. 2. We then discuss the problem of visual interpretation of an ECG signal in Sect. 3. Next, we review a variety of approaches to detect anomalies in ECGs based on DL principles in Sect. 4. In Sect. 5, we discuss the TNMG Corpus that we have used in our studies and review some of the work done on this impressive database. In Sect. 6, we present the results of our channel reduction experiments. Finally, in Sect. 7, we present evidence demonstrating that substantial reductions in channel dimensionality can be achieved without compromising diagnostic fidelity, thereby challenging the continued relevance of traditional ECG processing pipelines. We conclude with some comments on future research directions.

2 Signal Processing in ECG Analysis

Standard clinical ECG recordings [18] employ ten electrodes, as shown in Fig. 1, to collect eight signal channels. The postprocessing steps for these channels are shown in Fig. 2. Six precordial leads (V1–V6) are positioned on the chest, while four limb leads attach to the right arm (RA), left arm (LA), left leg (LL), and right leg (RL). The RL lead serves as ground. This configuration records eight raw waveforms: six from the precordial leads, plus DI (LA-RA potential difference) and DII (LL-RA potential difference).

Four additional leads are derived mathematically using simple differences and averages:

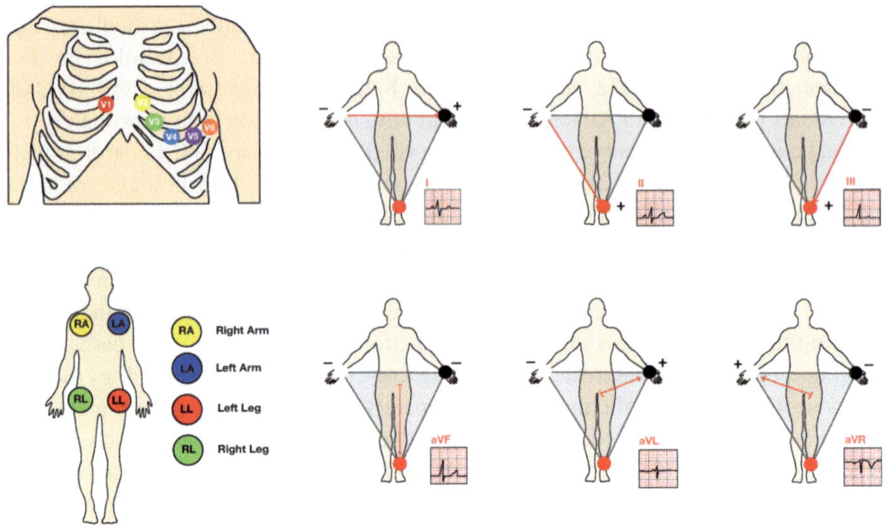

Fig. 1 The physical locations of the 10 electrodes used to record a 12-lead ECG

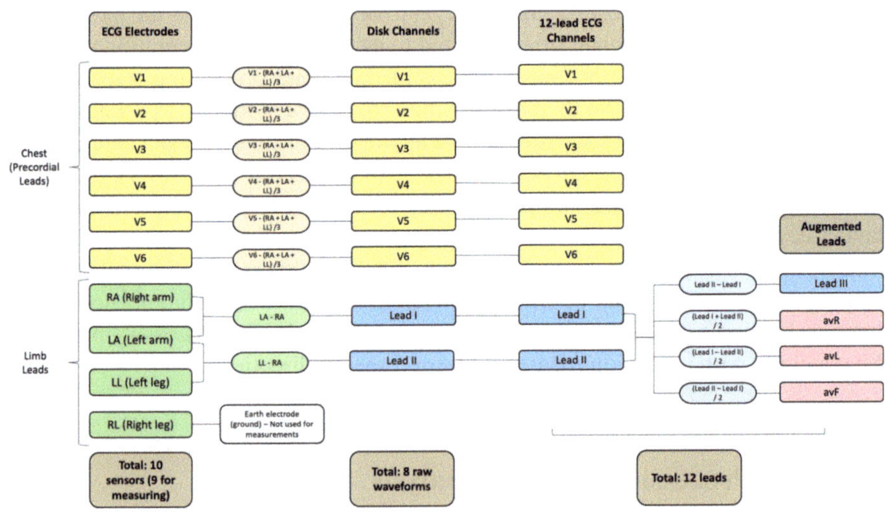

Fig. 2 Conversion of an ECG collected with 10 leads to 8 and 12-channel signals

$$\text{DIII} = \text{DII} - \text{DI}, \tag{1}$$

$$\text{aVR} = \frac{(\text{DI} + \text{DII})}{2}, \tag{2}$$

$$\text{aVL} = \frac{(\text{DI} - \text{DII})}{2}, \tag{3}$$

$$\text{aVF} = \frac{(\text{DII} - \text{DI})}{2}. \tag{4}$$

These computations are summarized in Fig. 2 and have been part of the art of interpreting an ECG for many decades. In this chapter, we explore whether these additional channels are needed for modern DL systems.

The derived ECG leads are based on established electrophysiological principles that map the heart's electrical activity onto anatomical planes using potential differences measured at specific body locations. Leads DI, DII, and DIII form the basis of Einthoven's triangle [19], a conceptual model that approximates the frontal plane of the body as an equilateral triangle with vertices at the left arm (LA), right arm (RA), and left leg (LL). In this configuration, DI measures the potential difference between LA and RA, DII measures LL to RA, and DIII derives from the vector sum relationship in Eq. (1) reflecting Kirchhoff's voltage law around the triangle. This is depicted in Fig. 1.

Goldberger's augmented leads—aVR, aVL, and aVF—extend Einthoven's model by creating unipolar measurements with reference to a virtual ground. These leads synthesize additional frontal plane perspectives: aVR views the heart from the right shoulder, aVL from the left shoulder, and aVF from the foot. Equations (2)–(4) are derived from averaging limb lead signals to create reference points, enabling the visualization of cardiac depolarization along across multiple axes without adding physical electrodes.

These derived leads provide clinicians with complementary spatial views of cardiac activity, improving diagnostic resolution. However, they are linear transformations of the original limb leads and introduce redundancy in the signal set. In DL applications, where models can extract latent spatial and temporal patterns automatically, the inclusion of derived leads may be unnecessary or counterproductive. The anatomical rationale for their construction remains valid in clinical interpretation, but their utility in data-driven models must be empirically justified, particularly in the context of input redundancy, multicollinearity, and computational efficiency.

3 Visual Interpretation of Cardiology Signals

An electrocardiogram (ECG) captures the electrical activity of the heart via surface electrodes, providing a multichannel signal used for clinical diagnoses. A classical ECG curve is shown in Fig. 3. This is an illustration of a typical normal sinus rhythm [20] in which critical points on the waveform are labeled as P, Q, R, S, J, T, and U [18]. Each normal cardiac cycle produces a characteristic waveform consisting of a P wave (atrial depolarization), a QRS complex (ventricular depolarization), and a T-wave (ventricular repolarization). In a normal sinus rhythm, impulses originate

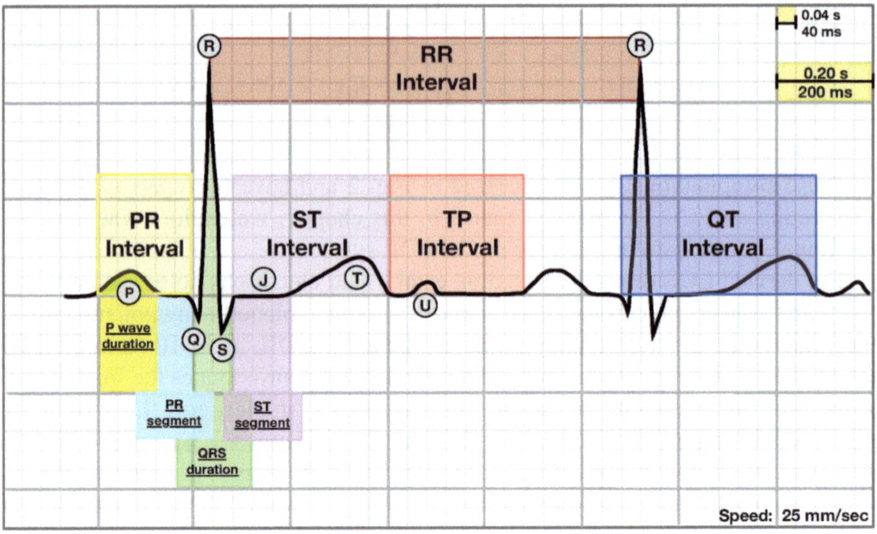

Fig. 3 A standard normal sinus rhythm with labeled P, QRS, and T waves

in the sinoatrial node (a small cluster of specialized cardiac muscle cells located in the right atrium, near the opening of the superior vena cava), producing a regular narrow-complex ECG at roughly 60–100 beats per minute. In this pattern, each QRS complex is preceded by a single upright P wave, and the PR interval (onset of the P wave to onset of the QRS) remains constant. This configuration reflects optimal atrioventricular conduction (refers to the way electrical impulses travel from the atria to the ventricles of the heart through the atrioventricular node and related conduction pathways) and is often considered the benchmark for normal cardiac electrical activity. However, such pristine patterns are relatively uncommon in practice.

Even among healthy individuals, deviations from this ideal are frequent. For instance, in a study [21] involving 26 healthy elderly men, 76.9% exhibited atrial ectopic beats (extra heartbeats that originate in the atria), and 76.9% had ventricular ectopic beats (extra heartbeats that originate in the ventricles instead of the sinoatrial node). Episodes of nonsustained ventricular tachycardia (abnormal heart rhythm that arises from the ventricles) were present in 11.5% of participants [21]. These findings suggest that transient arrhythmias (abnormal heart rhythms that occur for a short period of time) are common even in populations without known cardiac disease.

In a standard 12-lead ECG [22], shown in Fig. 2, different ECG leads offer complementary spatial views. For example, septal depolarization (the first phase of ventricular depolarization in which an electrical impulse spreads through the interventricular septum—the wall separating the right and left ventricles) produces small Q waves in lateral leads (I, aVL, V5–V6) and the initial R in lead V1, whereas ventricular depolarization toward or away from each lead causes the characteristic positive or negative deflections. These normal features provide the baseline for identifying abnormalities.

A recent ML study by Ribeiro et al. [16, 17] targeted exactly six common ECG abnormalities using DL: (1) first-degree AV block (1dAVb), (2) right bundle branch block (RBBB), (3) left bundle branch block (LBBB), (4) sinus bradycardia, (5) atrial fibrillation, and (6) sinus tachycardia. The following paragraphs briefly describe each condition's definition, physiology, ECG manifestations (with primary leads), and clinical significance. A more extended description can be found in [17].

(1) *First-degree AV block (1dAVb)* [23], described in Fig. 4, is defined by uniformly prolonged conduction through the AV node. All atrial impulses are conducted to the ventricles, but with a lengthened delay. On the ECG, this appears as a PR interval exceeding 200 ms while still maintaining one P wave before each QRS complex. This prolongation may be mild or marked (occasionally > 300 ms) and is most readily measured in lead II. In terms of physiology, it often reflects increased AV nodal delay due to enhanced vagal tone or aging fibrosis of the conduction system. It is frequently a benign finding seen in healthy young or athletic individuals and rarely causes symptoms [24]. Only when very long (> 0.3 s) or associated with other disease or medications (e.g., beta-blockers, digoxin, or Lyme carditis) does it warrant further evaluation.
(2) *Right bundle branch block (RBBB)* [25], described in Fig. 5. Right bundle branch block (RBBB) with rsR′ pattern in lead V1 and wide S in V6. It arises when the right-sided His-Purkinje conduction is delayed or interrupted. The left ventricle depolarizes normally, but the right ventricle must depolarize secondarily via slower cell-to-cell conduction. This yields a widened QRS (duration > 120 ms) and a characteristic pattern on the ECG. In leads V1–V3, one sees an rsR′ ("M-shaped") complex: a small initial R, a small S, then a secondary R′ from delayed right ventricular depolarization. In lateral leads (I, aVL, V5–V6), a broad, slurred S wave appears. For example, lead V1 shows an M-shaped QRS with a tall terminal R′, while lead V6 shows a wide terminal S wave [26]. These changes reflect the normal septal (LV) activation followed by delayed RV activation. The cardiac axis typically remains normal in isolated RBBBs. Clinically, RBBBs may be a benign finding (even in athletes) or indicate right ventricular strain (e.g., pulmonary pathology) or underlying conduction disease. ML-based algorithms detecting RBBB focus on the prolonged QRS and the distinctive V1–V6 morphologies. In practice, lead V1 (and V2) is examined for the "M" pattern and lead V6 for the wide S wave.
(3) *Left bundle branch block (LBBB)* [27], described in Fig. 6, involves delayed conduction in the left bundle. The impulse travels down the right bundle to activate the right ventricle first, then spreads slowly to the left ventricle. This produces a widened QRS (>120 ms) with a markedly different morphology than RBBB [28]. Key features include a broad, often notched (M-shaped) R-wave in lateral leads I, aVL, V5–V6, and a deep, wide S wave in right precordial leads V1–V3. The normal initial septal Q wave (left-to-right septal depolarization) is absent, and the overall QRS axis may shift leftward. Additional criteria are absent Q waves in V5–V6 and a prolonged R-wave peak time (>60 ms)

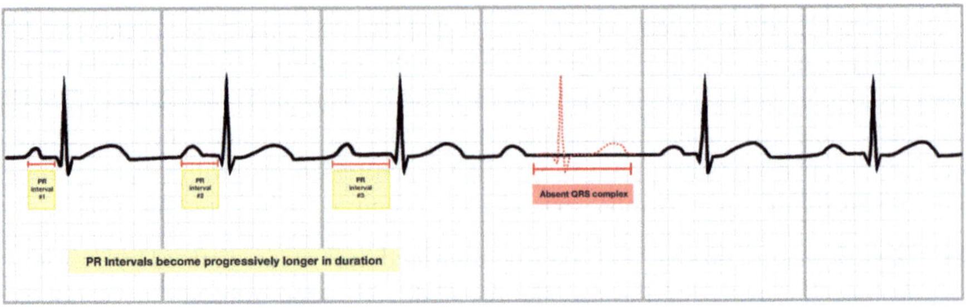

(a) description of a 1dAVb abnormality

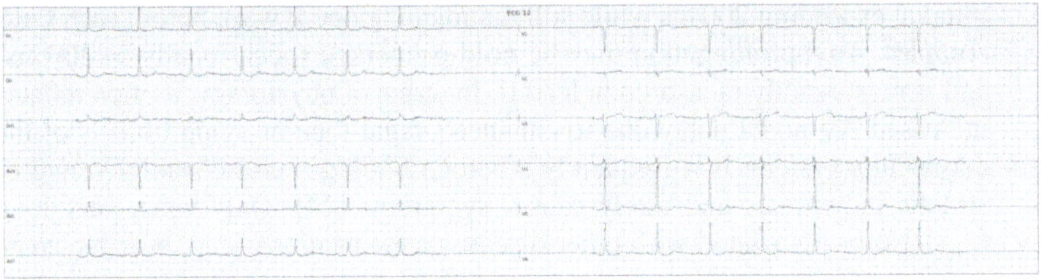

(b) abnormality present

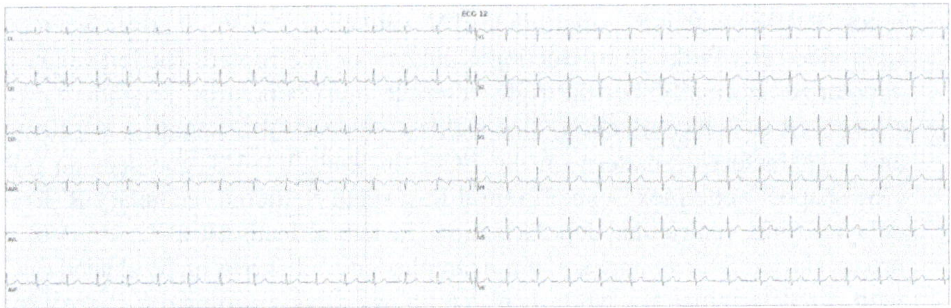

(c) no abnormality present

Fig. 4 First-degree AV block (1adAVb) showing a prolonged PR interval

in lateral leads. These ECG changes occur because the left ventricular depolarization is delayed. The result is tall R waves in V5–V6 and broad S waves in V1–V3. LBBB often signifies underlying cardiac disease (e.g., cardiomyopathy or ischemia) and complicates interpretation. An ML model must thus recognize LBBB as a distinct class of wide-QRS morphology. Clinically, leads V1 (showing a dominant S wave) and V6 (broad, often notched R) are inspected to confirm LBBB. A comparison of RBBB and LBBB is given in Fig. 5a.

(4) *Sinus bradycardia (SB)* [29], described in Fig. 7, is simply a sinus rhythm with a slow rate. It is defined as a resting ventricular rate below 60 beats per minute (bpm) in adults [30]. Importantly, the ECG features of sinus rhythm remain intact: there is a P wave before every QRS, the P wave morphology is normal (upright in leads I and II), and the PR interval is unchanged. The only

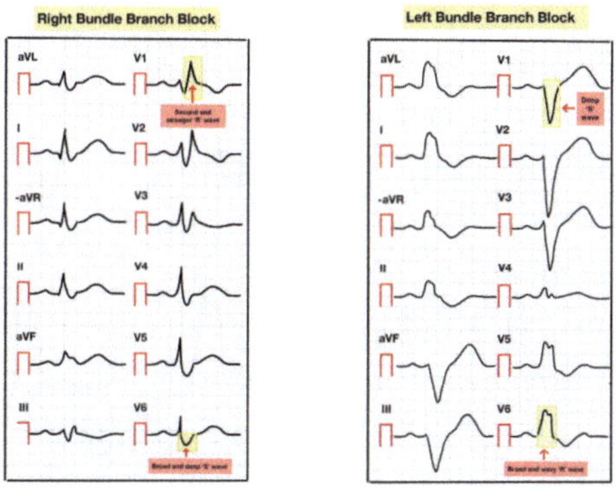

(a) description of a RBBB abnormality

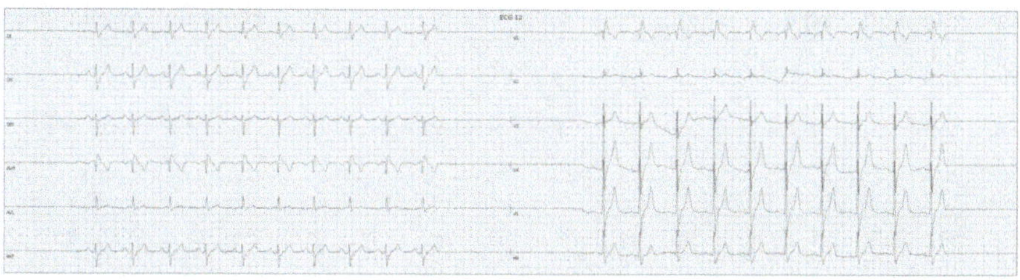

(b) abnormality present

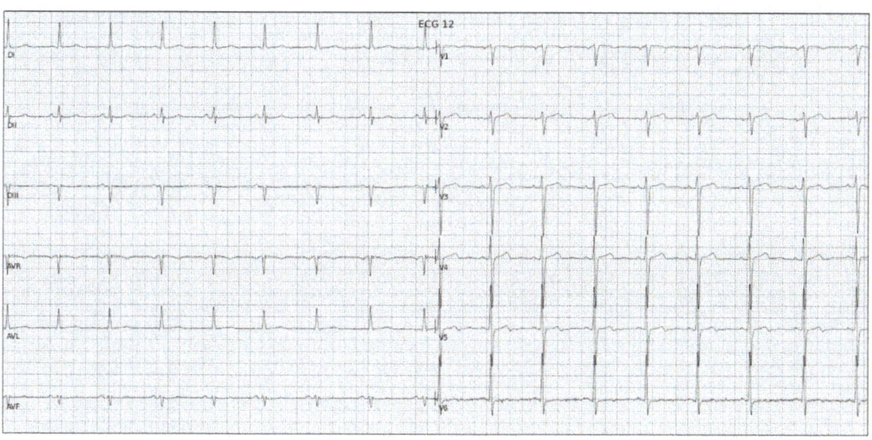

(c) no abnormality present

Fig. 5 Right bundle branch block (RBBB) with rsR′ pattern in lead V1 and wide S in V6

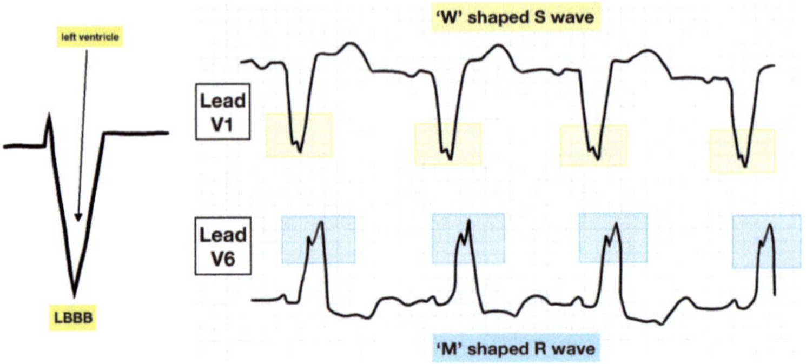

(a) description of a LBBB abnormality

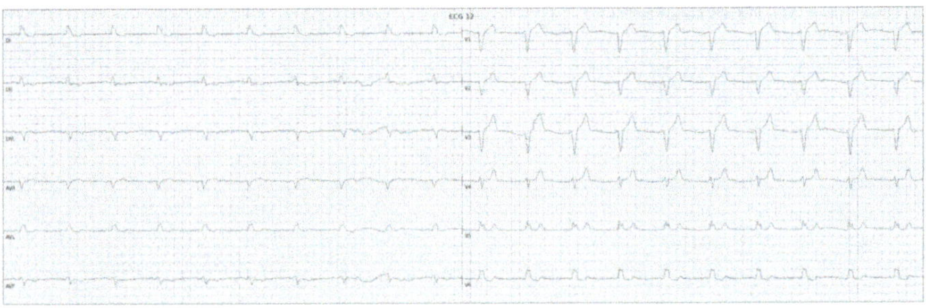

(b) abnormality present

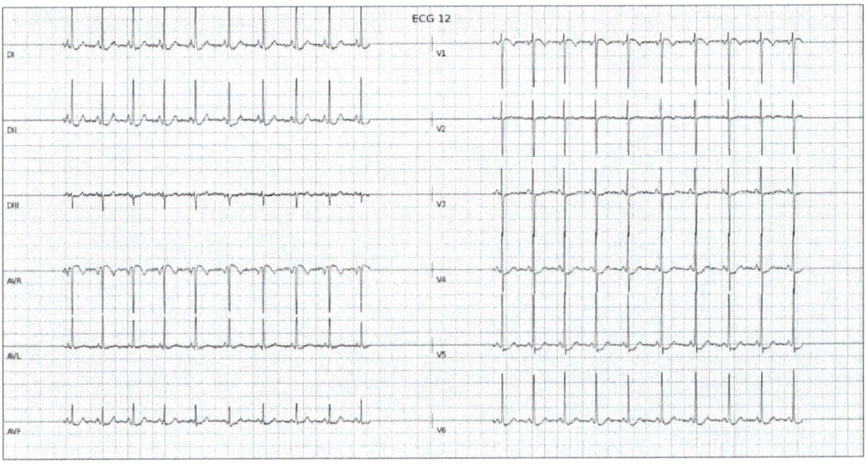

(c) no abnormality present

Fig. 6 Left bundle branch block (LBBB) with broad notched R in V6 and deep S in V1

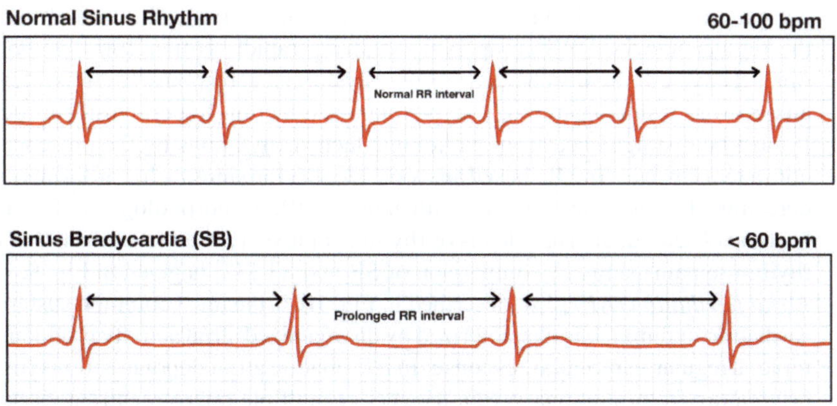

(a) a description of a SB abnormality

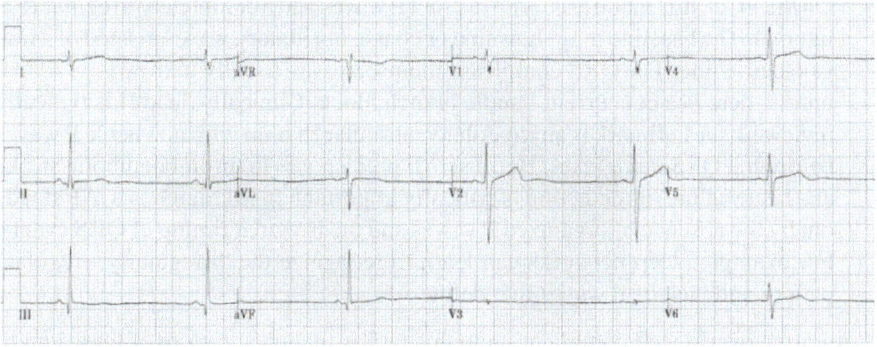

(b) abnormality present

Fig. 7 Sinus bradycardia (SB) in which a normal upright P wave precedes every QRS complex

difference is the longer R–R interval reflecting the slower rate. Often, prominent U waves (small deflections following the T-wave) may appear in precordial leads during the marked bradycardia. Sinus bradycardia commonly occurs physiologically during sleep or in well-trained athletes (reflecting high vagal tone) and typically causes no symptoms. It can also result from medications (beta-blockers, calcium-channel blockers) or conditions like hypothyroidism. In terms of diagnosis, lead II usually displays the clearest rhythm strip: a slow but regular rhythm (< 60 bpm) with one P-QRS-T sequence per cycle. ML systems classify sinus bradycardia by recognizing the slow but otherwise normal-appearing PQRST sequence.

(5) *Sinus tachycardia (ST)* [31], described in Fig. 8, is the converse: a sinus rhythm with an elevated rate (resting heart rate > 100 bpm in adults. The ECG again preserves normal sinus features (one P wave before each QRS, normal P wave axis), but the R–R intervals are shortened [32]. At very high rates, P waves may begin to merge with the preceding T wave or become difficult to distinguish,

but each complex still has a discrete P wave onset. Typically, lead II shows the regular narrow QRS complexes occurring rapidly (often 100–140 bpm). Sinus tachycardia is most often a normal physiological response (exercise, fever, anxiety) or a result of pathology (hypovolemia, infection, hyperthyroidism, etc.). It must be distinguished from pathologic tachyarrhythmias by the presence of consistent P waves. In ML-based analysis, identifying sinus tachycardia involves detecting the fast rate together with intact PQRST morphology. If P waves become obscured, alternate leads or rhythm context are used to confirm that the rhythm remains sinus. A comparison of SB and ST [32] is given in Fig. 9.

(6) *Atrial fibrillation (AF)* [33], described in Fig. 10, is the most common sustained cardiac arrhythmia. It is characterized by disorganized atrial activity producing a rapid, irregular ventricular response. In AF, chaotic atrial depolarizations (often originating from pulmonary vein triggers or multiple micro-reentrant circuits) bombard the AV node, causing inconsistent conduction to the ventricles [34, 35]. On the ECG, this manifests as no distinct P waves and a completely irregular R-R interval. The baseline may show fine or coarse fibrillatory waves instead of a true isoelectric line. The QRS complexes themselves are usually narrow (< 120 ms) unless there is a concurrent bundle branch block. Clinically, lead II is typically reviewed: the hallmark is an irregular ventricular rhythm with no visible P waves before the QRS complexes. Because AF predisposes to thromboembolic stroke and heart failure, its detection is critically important. In ML diagnosis, AF stands out by the absence of organized P waves and the irregular timing of QRS events. For example, deep networks can learn to recognize the "irregularly irregular" pattern and lack of P waves that define AF.

An often overlooked complication to this task is that one or more of these conditions can occur in an ECG. This kind of multi-label classification is more difficult because the training data usually does not support all combinations and permutations of the labels, making the data inherently sparse. We will discuss this in greater detail in Sect. 5.

The recent surge in artificial intelligence (AI) popularity, which stems from its success in applications such as multichannel signal and image processing, is due in part to its ability to circumvent the feature engineering problem and automatically learn complex features. Having overcome challenges like vanishing gradients and overfitting, DL in signal processing, once limited by high computational demands, delivered breakthroughs in performance using well-known networks such as Residual Network (ResNet) [36] and self-attention-based transformers [37]. Introduced as a solution to core limitations such as vanishing gradients, ResNet pioneered the use of residual connections, which allowed very deep networks to be trained efficiently, and resulted in significant gains in performance.

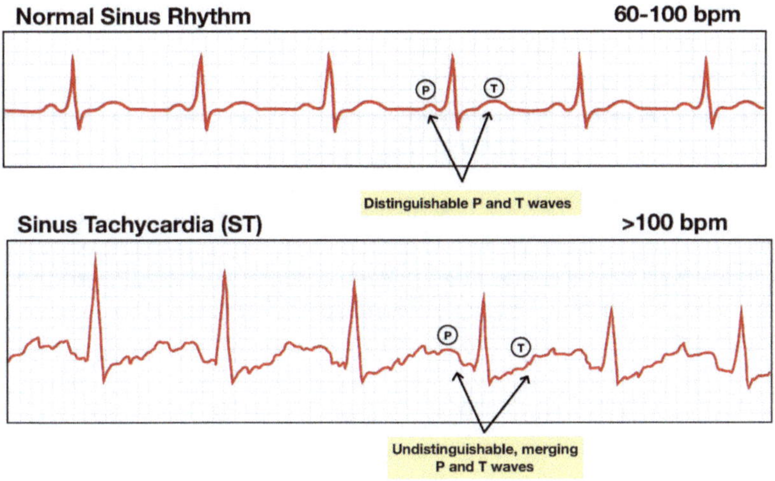

(a) description of a ST abnormality

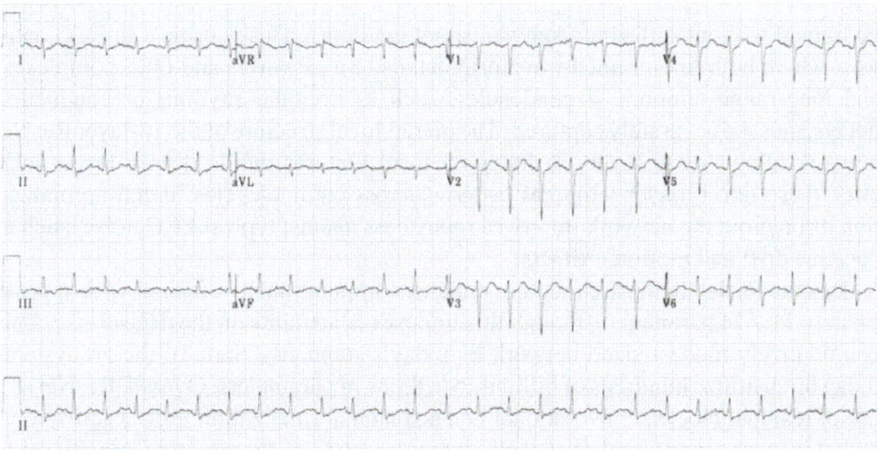

(b) abnormality present

Fig. 8 Sinus tachycardia in which P waves are hidden with each preceding T-wave

4 An Overview of State of the Art

The ResNet18 architecture, summarized in Figs. 11 and 12, and described in detail in [36], offers significant advantages for ECG signal processing by leveraging identity-based skip connections to learn residual functions. These connections preserve earlier feature representations and ensure stable gradient flow, enabling deeper networks to

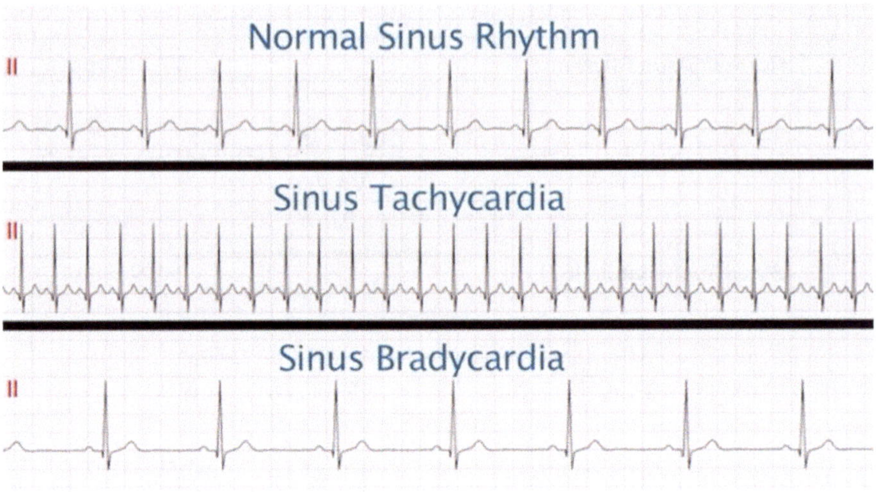

Fig. 9 A comparison of sinus bradycardia and sinus tachycardia

be trained without suffering from vanishing gradients. This is critical in ECG analysis, where both fine-grained waveform details (e.g., P waves and QRS complexes) and long-range temporal dependencies, such as irregular rhythms or conduction delays, must be accurately captured. The hierarchical structure of the 18-layer ResNet network allows early layers to focus on local features, while deeper layers integrate these into broader temporal patterns across cardiac cycles. Batch normalization throughout the network improves robustness against typical ECG noise, such as baseline drift and motion artifacts.

Despite its depth, the architecture remains computationally efficient with approximately 11.7 M parameters, though this imposes constraints on the dataset size. This is a relatively modest-sized network by today's standards. State-of-the-art systems today often utilize hundreds of billions to trillions of parameters. Overall, ResNet18's use of residual blocks, convolutional layers, batch normalization, ReLU activations, and skip connections makes it well suited for ECG tasks requiring sensitivity to both local morphology and global rhythm patterns. A major goal of this chapter is to demonstrate that modern deep learning systems are capable of automatically learning the optimal relationships between ECG channels, and hence, there is no longer a need to do model-based feature generation.

Recent studies by Ribeiro et al. [17], Pastika et al. [38], Sau et al. [39], and von Bachmann et al. [40] demonstrate that deep learning models can process raw ECG signals effectively without relying on derived channels or extensive preprocessing. These results raise the question of whether all recorded channels are necessary for accurate automated diagnosis. Rather than focusing solely on reducing the number of input channels, this chapter investigates whether models trained on reduced channel configurations can outperform or match those using the full 12-channel set. Experimental results are presented to evaluate performance across varying dataset sizes

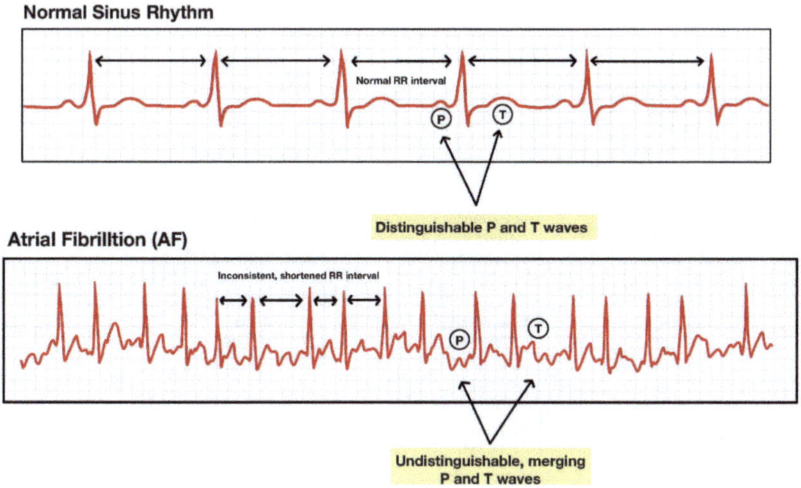

(a) description of an AF abnormality

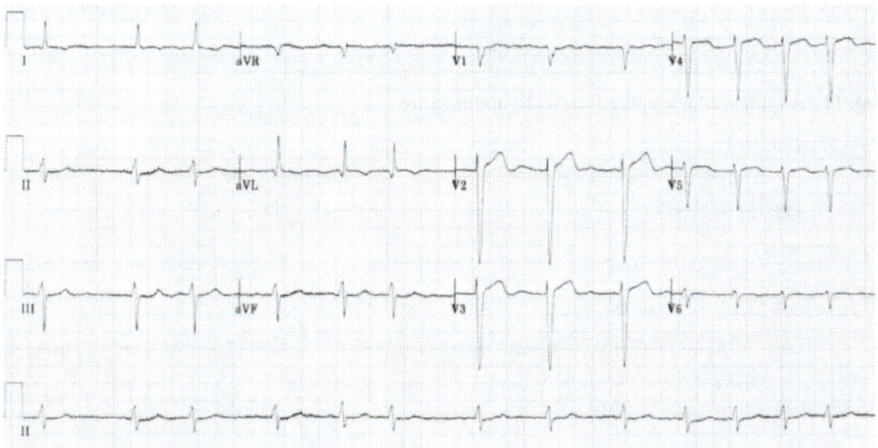

(b) abnormality present

Fig. 10 Atrial fibrillation with irregular R–R intervals and absent P waves in lead II

and lead configurations, isolating the diagnostic value of each channel group. The analysis emphasizes model accuracy, signal redundancy, and computational efficiency, establishing whether reduced configurations provide practical and statistical advantages in multi-label cardiac classification tasks.

The k-Nearest Neighbors (KNN) algorithm proved effective for initial attempts at automated diagnosis, achieving accuracy rates between 95 and 97% [41] in identifying basic arrhythmias. Support Vector Machines (SVM) demonstrated particular utility in handling the complex, non-linear nature of ECG signals, while Random

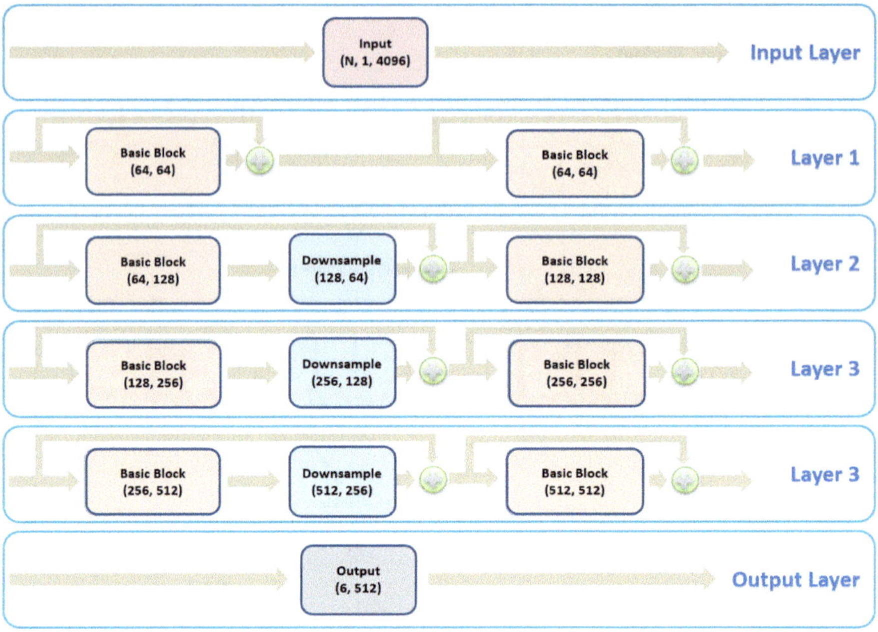

Fig. 11 A typical block in the ResNet18 architecture

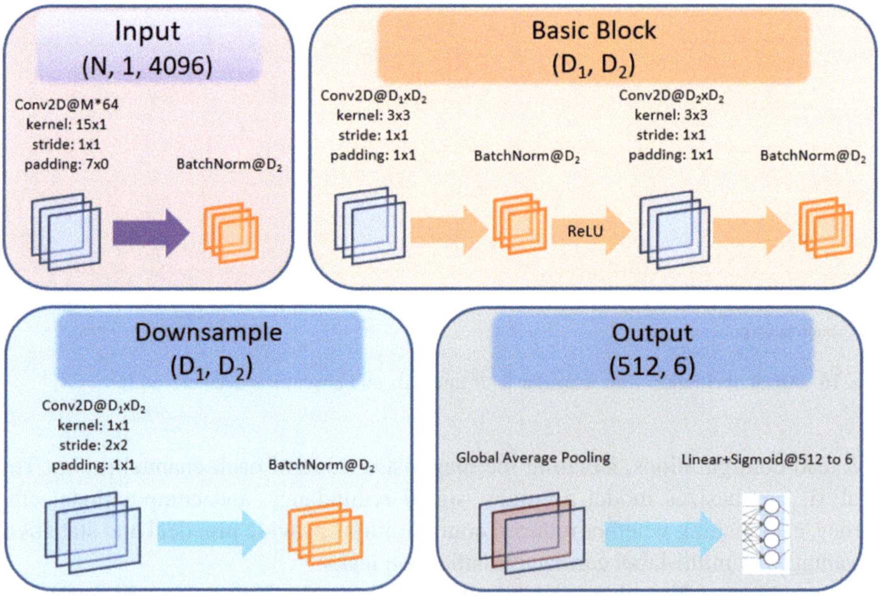

Fig. 12 The composite ResNet18 architecture uses four internal layers

Forest approaches offered robust performance in dealing with noise and signal variations [42, 43]. Early applications focused primarily on single-lead ECG interpretation, with researchers using feature extraction techniques to identify key signal characteristics. These features included QRS complex morphology, RR intervals (the length of a ventricular cardiac cycle), and various time-domain and frequency-domain measurements. The process of feature selection proved crucial, as the high-dimensional nature of ECG signals required careful consideration of which characteristics would provide the most diagnostic value.

A significant advancement came with the development of ensemble methods. By combining multiple algorithms, researchers achieved improved robustness and accuracy. Gradient boosting and AdaBoost techniques demonstrated success in handling the inherent variability of ECG signals, with accuracy rates between 85 and 95% for common cardiac abnormalities [44]. The challenge of preprocessing remained significant throughout this period. Traditional machine learning approaches required careful signal conditioning, including noise removal, baseline wander correction, and signal normalization. The preprocessing step proved both critical and limiting, as the quality of the analysis depended heavily on the effectiveness of these preliminary transformations.

Traditional models relied heavily on hand-crafted features, requiring significant domain expertise. Many early approaches struggled with large datasets and multiple lead configurations. The need for extensive signal preprocessing introduced potential points of failure and reduced generalizability. By the mid-2010s, researchers began exploring more sophisticated approaches. The introduction of neural networks marked a transition point, though early implementations still relied heavily on feature engineering. These initial neural network applications demonstrated promise in handling the temporal nature of ECG signals but were limited by available computational resources and dataset sizes [45]. The evolution toward deep learning approaches began to address many of these limitations. Convolutional Neural Networks (CNNs) showed promising results in automatically learning relevant features from raw ECG signals [46]. This capability reduced the dependence on manual feature engineering and improved the scalability of automated analysis systems.

Several key studies demonstrated the potential of neural network approaches during this transition period. Kiranyaz et al. [47] demonstrated the effectiveness of 1D CNNs for patient-specific ECG classification. Acharya et al. [46] showed that deep learning could effectively identify different types of arrhythmias. Hannun et al. [48] established that deep learning could match or exceed human performance in single-lead ECG interpretation. Long Short-Term Memory (LSTM) [49] networks became a powerful tool for classifying electrocardiogram (ECG) and cardiovascular data. Several studies [50–52] have explored their application in ECG classification, as these recurrent neural networks have demonstrated good performance in capturing temporal dependencies in sequential data. Despite their success, LSTM networks face certain challenges in ECG classification. One challenge is the need for large amounts of labeled data for training, which can be difficult to get in medical applications. Also, the computational requirements of LSTM networks may limit their use in resource-limited environments, although efforts have been made to develop lightweight versions for wearable devices [50–53].

Transformer-based models [54] have shown remarkable success in various domains, including natural language processing, computer vision, and time series forecasting. However, their application to ECG classification presents unique challenges, particularly when dealing with long-term sequential data [55]. Unlike discrete and highly semantic language tokens, ECG data consists of continuous numeric points with temporal redundancy and weak semantics. This characteristic makes it challenging for transformers to effectively depict the overall properties of time series, such as trends and periodic variations [56]. Such systems often require large amounts of training data and are extremely sensitive to imbalanced data. Despite the enormous size of the TNMG Corpus, the underrepresentation of certain combinations of labels poses significant challenges for transformer-based models. Hence, we selected a ResNet architecture that has performed fairly reliably across a wide range of applications.

The groundwork laid by these early machine learning applications proved essential for later developments in deep learning approaches. The challenges encountered and solutions developed during this period informed the design of more sophisticated systems, leading ultimately to comprehensive approaches such as those developed by Ribeiro et al. [17]. One significant development was the recognition that different types of cardiac abnormalities might require different analytical approaches. While some conditions could be identified through relatively simple feature analysis, others required more sophisticated pattern recognition capabilities. This understanding influenced the development of hybrid approaches that combined multiple analytical techniques.

The field continued to evolve with the introduction of transfer learning techniques, which allowed models trained on large datasets to be effectively adapted for more specific applications. This approach proved particularly valuable in addressing the challenge of limited data availability for rare cardiac conditions, such as Brugada syndrome, Wolff-Parkinson-White syndrome, and certain types of congenital long QT syndromes. These conditions are underrepresented in most clinical datasets, making it difficult to train accurate models directly. Transfer learning enables the reuse of generalized low-level ECG features learned from common conditions and adapts them to identify rare patterns through fine-tuning on small, condition-specific datasets [57, 58]. This strategy increases model robustness and diagnostic reach without requiring extensive new data collection.

The field has subsequently established several foundational principles for effective ECG analysis. The importance of end-to-end learning approaches that could process raw ECG signals directly became evident [59, 60]. An architecture capable of handling multiple lead configurations effectively proved crucial [17]. The need for robust validation approaches using clinically relevant metrics emerged as a key consideration [61]. Additionally, the importance of interpretability in automated analysis systems became increasingly recognized [62].

These developments set the stage for more comprehensive approaches to automated ECG analysis, leading to systems capable of handling multiple lead configurations and identifying a broader range of cardiac abnormalities. The progression from traditional machine learning to deep learning approaches marked a significant advancement in the field's capability to provide reliable, automated cardiac diagnosis support. Most systems today use transfer learning, fine-tuning, and reinforcement learning to achieve high performance. We are in the early stages of such work [63].

5 Baseline Performance on TNMG Code

The TNMG CODE Corpus (TNMG) [16] represents a major advancement in the field of cardiology and is the corpus we focus on in this study. TNMG is a dataset of ECG records collected by the Telehealth Network of Minas Gerais (TNMG) between 2010 and 2016 in 811 counties in the Brazilian state of Minas Gerais, organized by the Clinical Outcomes in Digital Electrocardiography (CODE) group. The dataset contains a total of 6,716,317 annotated records from 1,558,749 patients.

The dataset includes a curated "golden dataset" of 827 ECG recordings, which serves as a high-quality evaluation set. These recordings were independently annotated by two cardiologists. In cases of disagreement, a third specialist reviewed the annotations to establish a consensus. The dataset was labeled for six abnormalities discussed in Sect. 2 and shown in Table 1.

In Table 2, we show a distribution in the number of records and percentage of feature vectors in both datasets, where the presence or absence of each abnormality is marked as a binary vector in the same order as Table 1. It is clear that the majority of both datasets consists of healthy records. The imbalance in this data has a profound impact on our ability to train high-performance models. However, it is not our intention here to focus on techniques to deal with imbalance. Tokens with a single disease

Table 1 Annotations present in TNMG

Label	Description
1dAVb	First-degree atrioventricular block: A delay in the conduction of electrical impulses from the atria to the ventricles
RBBB	Right bundle branch block: A condition where the right side of the heart's electrical conduction system is impaired
LBBB	Left bundle branch block: A condition where the left side of the heart's electrical conduction system is impaired
SB	Sinus bradycardia: A slower-than-normal heart rhythm, defined as a heart rate below 60 beats per minute in adults
AF	Atrial fibrillation: An irregular heart rhythm, characterized by chaotic electrical activity
ST	Sinus tachycardia: A higher-than-normal heart rhythm, defined as a heart rate above 100 beats per minute in adults

Table 2 Distribution of classes in TNMG CODE

Feature vector	Train		Gold (eval)	
	#	%	#	%
000000	6,014,462	89.55000	681	82.34583
010000	145,208	2.16202	28	3.38573
000001	131,820	1.96268	35	4.23216
000010	100,865	1.50179	11	1.33011
000100	94,500	1.40702	15	1.81378
001000	86,487	1.28771	20	2.41838
100000	75,924	1.13044	25	3.02297
010010	11,910	0.17733	1	0.12092
110000	11,168	0.16628	0	0.00000
101000	7,580	0.11286	3	0.36276
001010	7,019	0.10451	0	0.00000
010100	5,713	0.08506	0	0.00000
100100	4,215	0.06276	0	0.00000
010001	3,408	0.05074	1	0.12092
001001	3,066	0.04565	0	0.00000
000011	2,860	0.04258	0	0.00000
100010	1,871	0.02786	1	0.12092
001100	1,625	0.02419	1	0.12092
011000	1,621	0.02414	4	0.48368
110100	1,165	0.01735	0	0.00000
100001	560	0.00834	1	0.12092
110010	515	0.00767	0	0.00000
000110	506	0.00753	0	0.00000
101100	331	0.00493	0	0.00000
011100	329	0.00490	0	0.00000
010011	292	0.00435	0	0.00000
101010	248	0.00369	0	0.00000
111000	220	0.00328	0	0.00000
011010	189	0.00281	0	0.00000
001011	143	0.00213	0	0.00000
100110	88	0.00131	0	0.00000
010110	82	0.00122	0	0.00000
110110	64	0.00095	0	0.00000
111100	56	0.00083	0	0.00000
011001	45	0.00067	0	0.00000
110001	43	0.00064	0	0.00000

(continued)

Table 2 (continued)

Feature vector	Train		Gold (eval)	
	#	%	#	%
100011	36	0.00054	0	0.00000
111010	19	0.00028	0	0.00000
101001	16	0.00024	0	0.00000
001110	14	0.00021	0	0.00000
000101	10	0.00015	0	0.00000
011011	5	0.00007	0	0.00000
010101	5	0.00007	0	0.00000
011110	5	0.00007	0	0.00000
111011	4	0.00006	0	0.00000
111110	3	0.00004	0	0.00000
110011	1	0.00001	0	0.00000
101110	1	0.00001	0	0.00000

occur in single-digit percentages. An even smaller fraction of tokens with multiple diseases appears in the corpus. Of equal concern is that tokens with multiple diseases are not well represented in the evaluation dataset, known as the gold standard dataset. Hence, despite the enormous amount of data present in TNMG, the sparsity of the data does pose challenges, especially for multi-label classification problems. These can be addressed using any of the popular techniques for dealing with imbalance [64–66].

The application of deep neural networks (DNNs) to ECG analysis represents a significant advancement in automated cardiac diagnosis. The study conducted by Ribeiro et al. [17] demonstrates both the potential and limitations of this approach on the TNMG CODE corpus. The researchers employed a residual network architecture (ResNet18) adapted for unidimensional signals. Their unique and careful approach to data labeling combined automatic analysis using the University of Glasgow ECG analysis program with expert cardiologist annotations, implementing a systematic procedure to resolve discrepancies. This dual-verification approach strengthens the reliability of their training data. We chose this study because it is one of the few large-scale, real-world uses of deep learning for ECG classification with open code and full dataset access. This made it a strong base for our work, with both solid methods and practical use.

For preprocessing, they resampled all ECGs to 400 Hz, zero-padded signals to 4096 samples per lead, used the derived channels, and applied z-score normalization. While their approach showed promising results, the impact of this preprocessing on model performance was not thoroughly investigated. Their decision to make all data the same length somewhat simplifies the sequential decoding nature of the problem but also makes issues such as normalization easier to resolve. Results are not significantly sensitive to issues such as the normalized sample frequency and methods of interpolation.

Newer studies by Pastika et al. [38] and von Bachmann et al. [40] have also adopted reduced lead configurations, utilizing 8-lead ECGs in their deep learning models for body mass index and electrolyte prediction, respectively. However, these studies did not extensively discuss the rationale for lead reduction. The choice of using raw leads stems from the fact that these additional leads are linear combinations of the raw leads, making them redundant in a deep learning context.

Following Ribeiro et al. [17], we utilize a ResNet18 architecture adapted for multi-label classification. To create input tensors for our model, we transform the time series ECG data into multichannel 3D tensors. Following [17], all signals are zero-padded to 4096 samples and undergo z-score normalization to standardize the data. The normalized signals are then reshaped into tensors of shape $(N, 1, 4096)$, where N represents the number of ECG channels. In this representation, each channel corresponds to a separate ECG lead, and the temporal samples span the width of the tensor. Although the tensor has three dimensions, the singleton height dimension (1) facilitates the processing of temporal data similarly to how CNNs handle spatial information in images.

A typical block in our architecture is shown in Fig. 11. The overall architecture is illustrated in Fig. 12. The "basic block" shown in Fig. 11 is the fundamental building unit of the system [36]. It has two convolutional layers, each followed by batch normalization and a ReLU nonlinearity, plus a shortcut (skip connection). We made minor modifications to the original architecture. The first convolutional layer of our model was modified to take either eight or twelve channels as input. The final layer was adapted to output probabilities for each of the six cardiac conditions by utilizing a sigmoid activation function. We employ the Adam optimization algorithm [67] with a learning rate of 0.001. Due to the multi-label nature of our task, we use Binary Cross-Entropy loss [68] as the objective function:

$$BCE(y, \hat{y}) = -\frac{1}{n}\sum\left[y_i \log(\hat{y}_i) + (1 - y_i)\log(1 - \hat{y}_i)\right], \tag{5}$$

where$n$ is the number of classes, y_i is the true label, and $\hat{y}_i$ is the predicted probability for class i.

In all experiments, the training process iterated over 10 epochs with a batch size of 32. We monitored training and validation losses, along with accuracy, micro-averaged F1 score, and macro-averaged F1 score to assess the model performance. This approach allows us to evaluate the effectiveness of our model in processing both raw and minimally preprocessed ECG signals.

Given that our problem is a multi-label classification task, we employ micro and macro F1 scores [69] as a key metric for evaluating model performance. These metrics provide a comprehensive assessment of the overall model's ability to identify several cardiac conditions at once. The micro F1 score calculates metrics by counting the true positives, false negatives, and false positives across all classes. It is computed as the harmonic mean of precision and recall:

$$\text{Micro F1} = 2 * \frac{(\text{Precision} * \text{Recall})}{(\text{Precision} + \text{Recall})}, \tag{6}$$

$$\text{Precision} = \frac{\text{TP}}{\text{TP} + \text{FP}}, \tag{7}$$

$$\text{Recall} = \frac{\text{TP}}{\text{TP} + \text{FN}}, \tag{8}$$

where TP, FP, and FN represent true positives, false positives, and false negatives, respectively.

In contrast, the macro F1 score calculates the F1 score for each class independently and then averages these scores. Micro F1 tends to give more weight to frequent classes, while macro F1 gives equal weight to all classes, regardless of their frequency in the dataset.

6 Channel Reduction

We conducted eight experiments to systematically evaluate the impact of ECG channel reduction and dataset size on deep learning performance. We used four different training dataset sizes: 2K (2,000), 20K (20,000), 200K (200,000), and 2, 000K (2,000,000) records, each tested with the 8-channel and 12-channel ECG configuration. For the 12-channel sets, we applied minimal preprocessing consisting of resampling to 400 Hz and $2x$ scaling of the signal amplitude. To address class imbalance, we chose the distribution for each training set that balances the frequency of occurrence of the class labels to the extent possible. However, as the dataset size increased, we had to include a higher proportion of healthy records due to their prevalence in the corpus.

In each experiment, we trained a separate ResNet18 model and evaluated its performance on the golden test set. We also utilized a fixed development set of 5,000 records in all experiments to monitor the training process and prevent over-fitting. The development set was balanced to represent a variety of combinations of cardiac conditions: approximately 4,000, examples were evenly split between single-condition cases and healthy records, 750 evenly split examples with two conditions, 249 examples containing three conditions, and one rare example with four concurrent conditions. The results for each experiment are shown in Table 3.

A breakdown of the errors by label is given in Table 4. Our results reveal a consistent pattern across all dataset sizes: the models trained on 8-channel ECG data outperformed those trained on 12-channel data. The performance difference was more pronounced in the smaller datasets and gradually diminished as the training dataset size increased.

Table 3 Micro F1 scores as a function of the training set size

Train size (K)	No. chans	Train	Dev	Eval
2	8	0.8810	0.7024	0.5029
2	12	0.8690	0.7050	0.2127
20	8	0.8870	0.8288	0.7022
20	12	0.8812	0.8366	0.5509
200	8	0.9310	0.8461	0.8421
200	12	0.9286	0.8545	0.7956
2,000	8	0.8809	0.7787	0.8649
2,000	12	0.8787	0.7708	0.8522

Table 4 A breakdown of errors on the eval set by label as a function of the amount of training data

Label	Freq	12 Channels (derived)			8 Channels (raw)		
		2K	200K	2M	2K	200K	2M
IdAVb	28	24	17	18	22	11	14
RBBB	34	27	6	5	64	5	6
LBBB	30	22	2	2	13	1	2
SB	16	51	26	8	49	22	8
AF	13	365	11	1	41	5	2
ST	37	44	13	9	70	13	8

We observed a significant decline in performance in both 8- and 12-channel models in experiments with 2,000K records. Performance on the training set, which constitutes closed-set testing, declined in going from 200K records to 2,000K records. Performance on the development set similarly declined. However, performance on the blind evaluation set improved slightly. We attribute this decrease to the inherent class imbalance in the larger dataset. As we expanded to a much higher number of records, the proportion of healthy ECG examples increased significantly. Although it reflects the prevalence of these records in the general population, this imbalance led to a bias in the model's predictions, favoring the majority class at the expense of less common combinations of cardiac conditions.

Another observation is that models trained on 2, 000K records performed poorly on the balanced development set but showed a noticeably higher performance on the evaluation set. This discrepancy is likely caused by the higher proportion of healthy records in the evaluation set, which more closely mirrors the distribution in the training data. These observations are an example of the importance of considering dataset composition and carefully balancing class distributions within datasets.

To assess the stability and reproducibility of our findings, we conducted several experiments to estimate the variance of the F1 scores on 8-channel data. For each of the three dataset sizes (2K, 20K and 200K) we performed five independent training

Table 5 Micro F1 scores as a function of the random seed

Data	Train	Dev	Eval
2K 8 Channels (1)	0.8898	0.7235	0.4596
2K 8 Channels (2)	0.8747	0.7047	0.4569
2K 8 Channels (3)	0.8766	0.7074	0.5111
2K 8 Channels (4)	0.8892	0.7163	0.4426
2K 8 Channels (5)	0.8720	0.6876	0.5251
StDev	0.0084	0.0136	0.0366
20K 8 Channels (1)	0.8887	0.8187	0.7335
20K 8 Channels (2)	0.8852	0.8132	0.7214
20K 8 Channels (3)	0.8880	0.8229	0.6897
20K 8 Channels (4)	0.8884	0.8241	0.6729
20K 8 Channels (5)	0.8882	0.8243	0.6905
StDev	0.0014	0.0047	0.0249
200K 8 Channels (1)	0.9312	0.8534	0.8251
200K 8 Channels (2)	0.9298	0.8523	0.8278
200K 8 Channels (3)	0.9307	0.8510	0.7831
200K 8 Channels (4)	0.9318	0.8451	0.8278
200K 8 Channels (5)	0.9298	0.8481	0.7752
StDev	0.0009	0.0034	0.0263

runs. Each run utilized a different random seed for data shuffling and model initialization. The results are shown in Table 5. As expected, there is a significant reduction in the standard deviation as the training set size increases.

This translates to an improvement in the statistical significance of these scores. For a sample size of 2K at 95% confidence, a difference in the F1 score of 0.0174 is statistically significant on the training data. For sample sizes of 20K and 200K, differences greater than 0.0054 and 0.0017, respectively, are significant. Hence, we see that the differences due to randomization in Table 4 are statistically significant, underscoring how sensitive these deep learning systems are to randomization (which makes reproducibility a challenge).

Nevertheless, the key point here is that performance for the 8-channel system is not statistically different from the 12-channel system, indicating that the deep learning system is able to implement whatever signal processing is necessary to extract meaningful information.

The research achieved impressive performance metrics, with F1 scores exceeding 80% and specificity above 99% across the six targeted conditions. The deep neural network matched or surpassed the diagnostic performance of cardiology residents. These results highlight the method's effectiveness for automated detection of common rhythm and conduction abnormalities. However, the study's scope remains narrow. It focuses on a limited set of conditions excluding many clinically relevant pathologies. The use of data exclusively from Brazilian healthcare settings also raises concerns about generalizability to broader populations. Additionally, the classification framework models each abnormality in isolation, without accounting for diagnostic interactions or comorbid patterns often encountered in real-world ECG interpretation. These limitations highlight the need to integrate such methods into the broader historical progression of ECG analysis and to evaluate their role alongside evolving modeling strategies.

A notable limitation emerges in the handling of borderline cases. The error analysis reveals that most DNN mistakes occurred in cases involving ECG interval measurements near diagnostic thresholds. This suggests potential limitations in the network's ability to handle cases that fall in diagnostic boundary zones—a crucial consideration for clinical implementation. The choice to use the ResNet18 architecture with a quarter of the layers used in previous studies warrants more detailed justification.

7 Feature Importance

While previous research established the effectiveness of deep learning for ECG analysis, the necessity of derived ECG channels in these modern approaches remained unexplored. We systematically investigated this question through a series of carefully designed experiments. We conducted an ablation analysis [70] with 250 randomized iterations for each channel. This process quantified feature importance by measuring the average decrease in micro F1 score when each channel was randomized. The analysis demonstrated that precordial leads (V1-V6) consistently showed higher importance scores, while derived channels (DIII, aVR, aVL, and aVF) often showed lower or negative scores, indicating redundancy rather than contribution.

Four systematic experiments, shown in Fig. 13, were conducted to evaluate the impact of ECG channel reduction and dataset size on deep learning performance. We used four different training dataset sizes: 2K, 20K, 200K, and 2,000K records, and 12-channel ECG configurations. For the 12-channel sets, minimal preprocessing was applied, consisting only of resampling to 400 Hz and $2x$ scaling of signal amplitude. A critical aspect of the experimental design was the careful balancing of feature vectors in each training set to address class imbalance. However, larger datasets necessarily included more healthy records due to their prevalence in the corpus.

Development and evaluation utilized three distinct datasets: the training set, a development set of 5,000 records for monitoring training progress and preventing overfitting, and a golden test set of 827 records. The development set was carefully balanced to represent various combinations of cardiac conditions: roughly 4,000 examples split evenly between single-condition cases and healthy records, 750 examples with two conditions, 249 examples containing three conditions, and one rare example with four concurrent conditions.

The experimental results revealed several key patterns. First, models trained on 8-channel ECG data consistently outperformed those using 12-channel configurations across all dataset sizes. This performance difference was most pronounced with smaller datasets and gradually diminished as training data increased. However, a

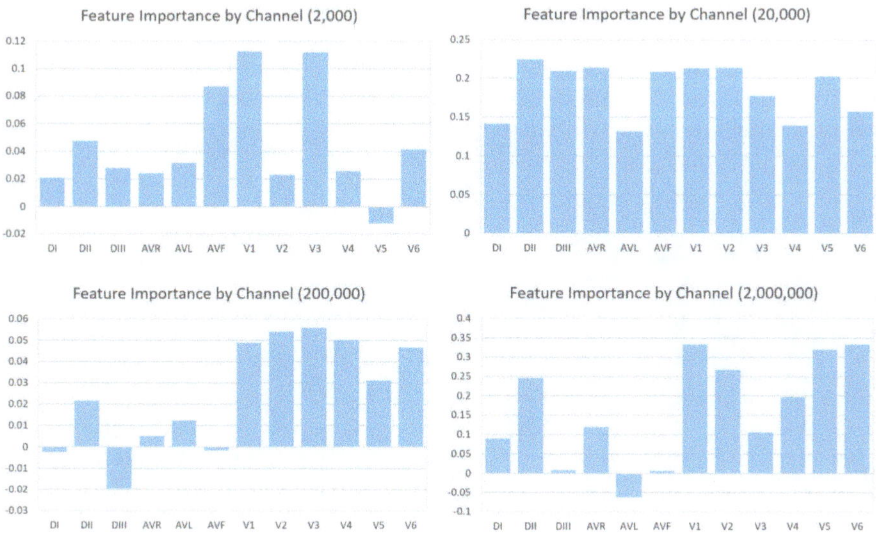

Fig. 13 Results of the ablation analysis

significant decline in performance occurred in both 8-channel and 12-channel models when using 2,000K records. All models were trained using the Adam optimizer with a learning rate of 0.001, batch size of 32, and binary cross-entropy loss. Each experiment used five training runs with varying random seeds. Models were trained for 50 epochs with input signals resampled to a fixed length of 2,048 samples and normalized per channel. Data loaders used four workers, and training was conducted on GPU hardware.

This unexpected performance decline with the largest dataset emerged from inherent class imbalance. As the dataset expanded to include a much higher number of records, the proportion of healthy ECG examples increased significantly. Although this reflects real-world prevalence, it created a bias in the model's predictions, favoring majority class identification at the expense of accurately detecting fewer common combinations of cardiac conditions (class imbalance).

An interesting observation emerged when comparing performance on different evaluation sets. Models trained on 2,000K records showed poor performance on the balanced development set but notably higher performance on the evaluation set. This discrepancy may arise from the evaluation set's distribution more closely aligning with the training data's bias toward healthy records.

We also addressed multicollinearity in the ECG signals, shown in Fig. 14. Multicollinearity [71] occurs when there is an approximately linear relationship between two or more independent variables in a regression model. While it is typically discussed in the context of regression models, it can also affect classification models, including our deep learning model for ECG classification. We employed the Variance Inflation Factor (VIF) analysis [72] on a balanced subset of 20,000 ECG recordings. VIF quantifies the severity of multicollinearity in regression analysis by measuring

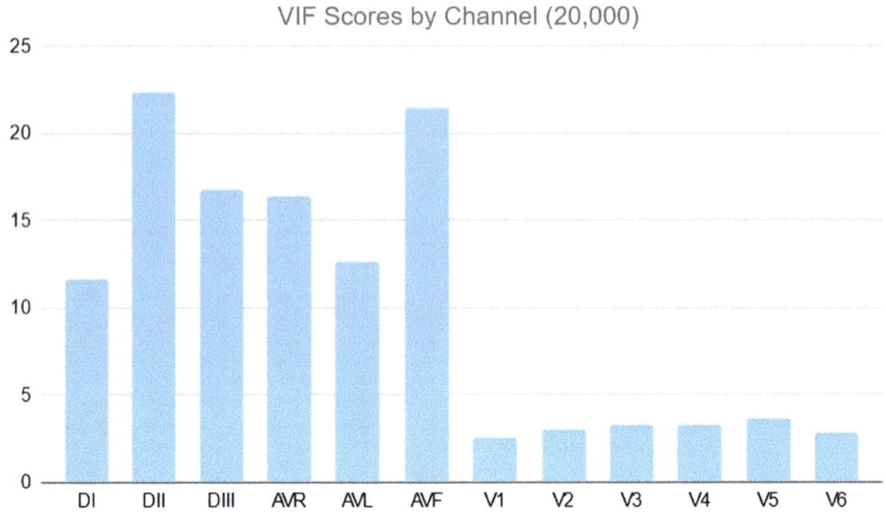

Fig. 14 VIF scores

how much the variance of a coefficient estimate is increased due to collinearity with other variables. Our analysis revealed substantial redundancy among limb leads and their derived channels, with DI and DII showing high VIF scores of 11.64 and 22.38, respectively. The derived channels (DIII, aVR, aVL, and aVF) similarly exhibited high VIF scores ranging from 12.69 to 21.48, indicating severe multicollinearity. In contrast, all precordial leads (V1-V6) demonstrated low VIF scores between 2.60 and 3.63, well below the standard multicollinearity threshold of 5.0. These results provide strong statistical evidence that limb leads, and their derivatives, contain redundant information, while precordial leads contribute unique and independent signal characteristics.

These results suggest that using only the eight independent leads effectively reduces this multicollinearity, potentially explaining the improved performance of the reduced channel configuration. This finding has important implications for signal processing in medical diagnostics more broadly. We validated the stability of these findings through additional experiments using various data splitting strategies: random $90-5-5$ splits, patient-stratified splits, and chronological splits. Results showed no statistically significant difference between these approaches, except for chronological splitting, in which changes in telehealth center operations over time affected performance.

8 Conclusions

Our findings challenge traditional ECG analysis approaches and suggest that modern deep learning methods can effectively learn from raw signals without derived features. In our study, by using only the eight independent leads and omitting the derived leads, we performed a form of variable selection that addresses the issue of multicollinearity in ECG data. Our findings in model training, ablation analysis, and VIF scores support the hypothesis, indicating that derived ECG channels introduce multicollinearity and do not provide any additional predictive power.

We point toward several future directions, including the investigation of optimal channel configurations for specific diagnostic tasks and the potential development of simplified ECG recording equipment. Our study also raises broader questions about feature engineering in medical signal processing, suggesting that traditional derived measurements may be unnecessary or detrimental when using modern machine learning approaches if there is adequate training data. Our study provides evidence supporting our hypothesis that extensive preprocessing of ECG data may not be beneficial for deep learning in cardiac diagnosis. Across various dataset sizes, models trained on raw signal outperformed those using derived channels and minor preprocessing. The ablation analysis further revealed that derived channels have little or a slightly negative impact on model performance.

These findings highlight the capability of deep learning algorithms to extract meaningful patterns from complex physiological data without relying on handcrafted features. This suggests a potential for a simpler, more direct approach for data input that may yield more accurate and robust models.

In future studies, we plan to further refine our approach by conducting additional experiments with 8-channel ECG data. We will focus on comparing the performance of models trained on raw 8-channel signals against those trained on extensively preprocessed 8-channel data, excluding the derived channels entirely and isolating the effects of preprocessing on the primary ECG leads.

Acknowledgements We would like to express our gratitude to Ribeiro et al. [16] for providing the TNMG CODE dataset used in this study, which was instrumental in our analysis. This material is based upon work supported in part by the Temple University Office of the Vice President for Research, and by the Temple University College of Science and Technology Research Experience for Undergraduates program. Any opinions, findings, and conclusions or recommendations expressed in this material are those of the author(s) and do not necessarily reflect the views of Temple University.

References

1. Barold, S. S. (2003). Willem Einthoven and the birth of clinical electrocardiography a hundred years ago. *Cardiac Electrophysiology Review*, *7*(1), 99–104. https://doi.org/10.1023/a:1023667812925.

2. Bonner, R. E., Crevasse, L., Ferrer, M. I., & Greenfield, J. C. J. (1972). A new computer program for analysis of scalar electrocardiograms. *Computers and Biomedical Research, an International Journal*, *5*(6), 629–653. https://doi.org/10.1016/0010-4809(72)90043-2.
3. Femtosim Clinical Inc. (n.d.). *Fifty Years of Physiologic Monitors*. Femtosim Clinical Inc. Retrieved May 21, 2025. url: http://www.femtosimclinical.com/History%20of%20Physiologic%20Monitors.htm.
4. Li, C., Zheng, C., & Tai, C. (1995). Detection of ECG characteristic points using wavelet transforms. *IEEE Transactions on Biomedical Engineering*, 42(1), 21–28. https://doi.org/10.1109/10.362922.
5. Quer, G., Arnaout, R., Henne, M., & Arnaout, R. (2021). Machine Learning and the Future of Cardiovascular Care: JACC State-of-the-Art Review. *Journal of the American College of Cardiology*, *77*(3), 300–313. https://doi.org/10.1016/j.jacc.2020.11.030.
6. Somani, S., Russak, A. J., Richter, F., Zhao, S., Vaid, A., Chaudhry, F., De Freitas, J. K., Naik, N., Miotto, R., Nadkarni, G. N., Narula, J., Argulian, E., & Glicksberg, B. S. (2021). Deep learning and the electrocardiogram: Review of the current state-of-the-art. *EP Europace*, *23*(8), 1179–1191. https://doi.org/10.1093/europace/euaa377.
7. Kennedy, H. L. (2006). The History, Science, and Innovation of Holter Technology. *Annals of Noninvasive Electrocardiology*, *11*(1), 85–94. https://doi.org/10.1111/j.1542-474X.2006.00067.x.
8. Castells, F., Laguna, P., Sörnmo, L., Bollmann, A., & Roig, J. M. (2007). Principal Component Analysis in ECG Signal Processing. *EURASIP Journal on Advances in Signal Processing*, 2007(1), 074580. https://doi.org/10.1155/2007/74580.
9. Abdou, A., & Krishnan, S. (2022). Horizons in Single-Lead ECG Analysis From Devices to Data. *Frontiers in Signal Processing*, *Volume 2-2022*. https://doi.org/10.3389/frsip.2022.866047.
10. Faruk, N., Abdulkarim, A., Emmanuel, I., Folawiyo, Y. Y., Adewole, K. S., Mojeed, H. A., Oloyede, A. A., Olawoyin, L. A., Sikiru, I. A., Nehemiah, M., Ya'u Gital, A., Chiroma, H., Ogunmodede, J. A., Almutairi, M., & Katibi, I. A. (2021). A comprehensive survey on low-cost ECG acquisition systems: Advances on design specifications, challenges and future direction. *Biocybernetics and Biomedical Engineering*, *41*(2), 474–502. https://doi.org/10.1016/j.bbe.2021.02.007.
11. Berwanger, O., & Santo, K. (2022). Cardiovascular Care in Brazil: Current Status, Challenges, and Opportunities. *Circulation*, *146*(6), 435–437. https://doi.org/10.1161/CIRCULATIONAHA.122.059320.
12. Gonçalves, M. A. A., Morais, H., Oliveira, G. M. M. de, & Mesquita, C. T. (2024). Challenges and Perspectives for Cardiology in the Developing World: Joint Views from Africa and Latin America. *International Journal of Cardiovascular Sciences*, *37*. https://doi.org/10.36660/ijcs.20240002
13. Wagner, P., Strodthoff, N., Bousseljot, R.-D., Kreiseler, D., Lunze, F. I., Samek, W., & Schaeffter, T. (2020). PTB-XL, a large publicly available electrocardiography dataset. *Scientific Data*, *7*(1), 154. https://doi.org/10.1038/s41597-020-0495-6.
14. Kim, Y.-G., Shin, D., Park, M. Y., Lee, S., Jeon, M. S., Yoon, D., & Park, R. W. (2017). ECG-ViEW II, a freely accessible electrocardiogram database. *PloS One*, *12*(4), e0176222. https://doi.org/10.1371/journal.pone.0176222.
15. Liu, H., Chen, D., Chen, D., Zhang, X., Li, H., Bian, L., Shu, M., & Wang, Y. (2022). A large-scale multi-label 12-lead electrocardiogram database with standardized diagnostic statements. *Scientific Data*, *9*(1), 272. https://doi.org/10.1038/s41597-022-01403-5.
16. Ribeiro, A. L. P., Paixão, G. M. M., Gomes, P. R., Ribeiro, M. H., Ribeiro, A. H., Canazart, J. A., Oliveira, D. M., Ferreira, M. P., Lima, E. M., Moraes, J. L. de, Castro, N., Ribeiro, L. B., & Macfarlane, P. W. (2019). Tele-electrocardiography and bigdata: The CODE (Clinical Outcomes in Digital Electrocardiography) study. *Journal of Electrocardiology*, 57S, S75–S78. https://doi.org/10.1016/j.jelectrocard.2019.09.008.
17. Ribeiro, A. H., Ribeiro, M. H., Paixão, G. M. M., Oliveira, D. M., Gomes, P. R., Canazart, J. A., Ferreira, M. P. S., Andersson, C. R., Macfarlane, P. W., Meira Jr., W., Schön, T. B., &

Ribeiro, A. L. P. (2020). Automatic diagnosis of the 12-lead ECG using a deep neural network. *Nature Communications*, 11(1), 1760. https://doi.org/10.1038/s41467-020-15432-4.

18. *-Lead ECG Placement: The Ultimate Guide*. (n.d.). Cables and Sensors. Retrieved May 25, 2025. url: https://www.cablesandsensors.com/pages/12-lead-ecg-placement-guide-with-illustrations#1.
19. Hampton, J., Hampton, J., & Adlam, D. (2024). *The ECG Made Easy* (10th ed.). Elsevier. url: https://shop.elsevier.com/books/the-ecg-made-easy/hampton/978-0-323-93766-5.
20. Burns, E., & Buttner, R. (2024, October 8). *Normal Sinus Rhythm*. Life in the Fastlane ECG Library. url: https://litfl.com/normal-sinus-rhythm-ecg-library/.
21. Wajngarten, M., Grupi, C., Bellotti, G. M., Lemos Da Luz, P., Gastào do Serro Azul, L., & Pileggi, F. (1990). Frequency and significance of cardiac rhythm disturbances in healthy elderly individuals. *Journal of Electrocardiology, 23*(2), 171–176. https://doi.org/10.1016/0022-0736(90)90138-R.
22. Kligfield, P., Gettes, L. S., Bailey, J. J., Childers, R., Deal, B. J., Hancock, E. W., van Herpen, G., Kors, J. A., Macfarlane, P., Mirvis, D. M., Pahlm, O., Rautaharju, P., & Wagner, G. S. (2007). Recommendations for the Standardization and Interpretation of the Electrocardiogram. *Circulation, 115*(10), 1306–1324. https://doi.org/10.1161/CIRCULATIONAHA.106.180200.
23. Parkes, C. (2025, March 11). *EKG Interpretation, part 9: Atrioventricular Blocks*. Level Up RN. url: https://leveluprn.com/blogs/ekg-interpretation/9-atrioventricular-blocks.
24. Kusumoto, F. M., Schoenfeld, M. H., Barrett, C., Edgerton, J. R., Ellenbogen, K. A., Gold, M. R., Goldschlager, N. F., Hamilton, R. M., Joglar, J. A., Kim, R. J., Lee, R., Marine, J. E., McLeod, C. J., Oken, K. R., Patton, K. K., Pellegrini, C. N., Selzman, K. A., Thompson, A., & Varosy, P. D. (2019). 2018 ACC/AHA/HRS Guideline on the Evaluation and Management of Patients With Bradycardia and Cardiac Conduction Delay: Executive Summary: A Report of the American College of Cardiology/American Heart Association Task Force on Clinical Practice Guidelines, and the Heart Rhythm Society. *Circulation, 140*(8), e333–e381. https://doi.org/10.1161/CIR.0000000000000627.
25. Burns, E., & Buttner, R. (2024, October 8). *Right Bundle Branch Block (RBBB)*. Life in the Fastlane ECG Library. url: https://litfl.com/right-bundle-branch-block-rbbb-ecg-library/.
26. Kashou, A.H., Shams, P. and Chhabra, L. (2024). "Right bundle branch block," in *StatPearls*, Treasure Island, FL, USA: StatPearls Publishing. [Online]. url: https://www.ncbi.nlm.nih.gov/books/NBK507872/.
27. Burns, E., & Buttner, R. (2024, October 8). *LeftBundle Branch Block (LBBB)*. Life in the Fastlane ECG Library. url: https://litfl.com/left-bundle-branch-block-lbbb-ecg-library/.
28. Scherbak, D., Shams, P., & Hicks, G. (2024). *Left Bundle Branch Block (LBBB)*. StatPearls Publishing. url: https://www.ncbi.nlm.nih.gov/books/NBK482167/.
29. Burns, E. (2024, October 8). *Sinus bradycardia*. Life in the Fastlane ECG Library. url: https://litfl.com/sinus-bradycardia-ecg-library/.
30. Hafeez, Y., & Grossman, S. (2023). *Sinus bradycardia*. StatPearls Publishing. url: https://www.ncbi.nlm.nih.gov/books/NBK493201/.
31. Burns, E., & Buttner, R. (2024, October 8). *Sinus Tachycardia*. Life in the Fastlane ECG Library. url: https://litfl.com/sinus-tachycardia-ecg-library/.
32. Henning, A., & Krawiec, C. (2023). *Sinus tachycardia*. StatPearls Publishing. url: https://www.ncbi.nlm.nih.gov/books/NBK553128/.
33. Serhal, H., Abdallah, N., Marion, J.-M., Chauvet, P., Oueidat, M., & Humeau-Heurtier, A. (2022). Overview on prediction, detection, and classification of atrial fibrillation using wavelets and AI on ECG. *Computers in Biology and Medicine*, 142, 105168. https://doi.org/10.1016/j.compbiomed.2021.105168.
34. Hindricks, G., Potpara, T., Dagres, N., Arbelo, E., Bax, J. J., Blomström-Lundqvist, C., Boriani, G., Castella, M., Dan, G.-A., Dilaveris, P. E., Fauchier, L., Filippatos, G., Kalman, J. M., La Meir, M., Lane, D. A., Lebeau, J.-P., Lettino, M., Lip, G. Y. H., Pinto, F. J., … Watkins, C. L. (2021). 2020 ESC Guidelines for the diagnosis and management of atrial fibrillation developed in collaboration with the European Association for Cardio-Thoracic Surgery (EACTS): The Task Force for the diagnosis and management of atrial fibrillation of the European Society

of Cardiology (ESC) Developed with the special contribution of the European Heart Rhythm Association (EHRA) of the ESC. *European Heart Journal*, 42(5), 373–498. https://doi.org/10.1093/eurheartj/ehaa612.

35. Joglar, J. A., Chung, M. K., Armbruster, A. L., Benjamin, E. J., Chyou, J. Y., Cronin, E. M., Deswal, A., Eckhardt, L. L., Goldberger, Z. D., Gopinathannair, R., Gorenek, B., Hess, P. L., Hlatky, M., Hogan, G., Ibeh, C., Indik, J. H., Kido, K., Kusumoto, F., Link, M. S., … Van Wagoner, D. R. (2024). 2023 ACC/AHA/ACCP/HRS Guideline for the Diagnosis and Management of Atrial Fibrillation: A Report of the American College of Cardiology/American Heart Association Joint Committee on Clinical Practice Guidelines. *Circulation*, 149(1), e1–e156. url: https://www.ahajournals.org/doi/https://doi.org/10.1161/CIR.0000000000001193.
36. He, K., Zhang, X., Ren, S., & Sun, J. (2016). Deep Residual Learning for Image Recognition. *Proceedings of the IEEE/CVF Conference on Computer Vision and Pattern Recognition (CVPR)*, 770–778. https://doi.org/10.1109/CVPR.2016.90.
37. Thundiyil, S. C., & Picone, J. (2025). Time Series Analysis from Classical Methods to Transformer-Based Approaches: A Review. In *Signal Processing in Medicine and Biology: Applications of Artificial Intelligence in Medicine and Biology* (Vol. 1, p. 56). Springer. url: https://isip.piconepress.com/publications/book_sections/2024/springer/transformers/. (in publication).
38. Pastika, L., Sau, A., Patlatzoglou, K., Sieliwonczyk, E., Ribeiro, A. H., McGurk, K. A., Khan, S., Mandic, D., Scott, W. R., Ware, J. S., Peters, N. S., Ribeiro, A. L. P., Kramer, D. B., Waks, J. W., & Ng, F. S. (2024). Artificial intelligence-enhanced electrocardiography derived body mass index as a predictor of future cardiometabolic disease. *Npj Digital Medicine*, 7(1), 167. https://doi.org/10.1038/s41746-024-01170-0.
39. Sau, A., Sieliwonczyk, E., Patlatzoglou, K., Pastika, L., McGurk, K. A., Ribeiro, A. H., Ribeiro, A. L. P., Ho, J. E., Peters, N. S., Ware, J. S., Tayal, U., Kramer, D. B., Waks, J. W., & Ng, F. S. (2025). Artificial intelligence-enhanced electrocardiography for the identification of a sex-related cardiovascular risk continuum: A retrospective cohort study. *The Lancet Digital Health*, 7(3), e184–e194. https://doi.org/10.1016/j.landig.2024.12.003.
40. von Bachmann, P., Gedon, D., Gustafsson, F. K., Ribeiro, A. H., Lampa, E., Gustafsson, S., Sundström, J., & Schön, T. B. (2024). Evaluating regression and probabilistic methods for ECG-based electrolyte prediction. *Scientific Reports*, 14(1), 15273. https://doi.org/10.1038/s41598-024-65223-w.
41. Kung, B.-H., Hu, P.-Y., Huang, C.-C., Lee, C.-C., Yao, C.-Y., & Kuan, C.-H. (2021). An Efficient ECG Classification System Using Resource-Saving Architecture and Random Forest. *IEEE Journal of Biomedical and Health Informatics*, 25(6), 1904–1914. https://doi.org/10.1109/JBHI.2020.3035191.
42. Melgani, F., & Bazi, Y. (2008). Classification of Electrocardiogram Signals With Support Vector Machines and Particle Swarm Optimization. *IEEE Transactions on Information Technology in Biomedicine*, 12(5), 667–677. https://doi.org/10.1109/TITB.2008.923147.
43. Turnip, A., Ilham Rizqywan, M., Kusumandari, D. E., Turnip, M., & Sihombing, P. (2018). Classification of ECG signal with Support Vector Machine Method for Arrhythmia Detection. *Journal of Physics: Conference Series*, 970(1), 012012. https://doi.org/10.1088/1742-6596/970/1/012012.
44. Reddy, S. D., Murugan, R., Nandi, A., & Goel, T. (2023). Classification of arrhythmia disease through electrocardiogram signals using sampling vector random forest classifier. *Multimedia Tools and Applications*, 82(17), 26797–26827. https://doi.org/10.1007/s11042-022-14304-x
45. Y. Zhang, J. Li, S. Wei, F. Zhou, & D. Li. (2021). Heartbeats Classification Using Hybrid Time-Frequency Analysis and Transfer Learning Based on ResNet. *IEEE Journal of Biomedical and Health Informatics*, 25(11), 4175–4184. https://doi.org/10.1109/JBHI.2021.3085318.
46. Acharya, U. R., Oh, S. L., Hagiwara, Y., Tan, J. H., Adam, M., Gertych, A., & Tan, R. S. (2017). A deep convolutional neural network model to classify heartbeats. *Computers in Biology and Medicine*, 89, 389–396. https://doi.org/10.1016/j.compbiomed.2017.08.022
47. Kiranyaz, S., Ince, T., & Gabbouj, M. (2016). Real-Time Patient-Specific ECG Classification by 1-D Convolutional Neural Networks. *IEEE Transactions on Biomedical Engineering*, 63(3), 664–675. https://doi.org/10.1109/TBME.2015.2468589.

48. Hannun, A. Y., Rajpurkar, P., Haghpanahi, M., Tison, G. H., Bourn, C., Turakhia, M. P., & Ng, A. Y. (2019). Cardiologist-level arrhythmia detection and classification in ambulatory electrocardiograms using a deep neural network. *Nature Medicine*, 25(1), 65–69. https://doi.org/10.1038/s41591-018-0268-3.
49. Hochreiter, S., & Schmidhuber, J. (1997). Long short-term memory. *Neural Computation, 9*(8), 1735–1780. https://doi.org/10.1162/neco.1997.9.8.1735.
50. S. Saadatnejad, M. Oveisi, & M. Hashemi. (2020). LSTM-Based ECG Classification for Continuous Monitoring on Personal Wearable Devices. *IEEE Journal of Biomedical and Health Informatics, 24*(2), 515–523. https://doi.org/10.1109/JBHI.2019.2911367.
51. Cheng, J., Zou, Q., & Zhao, Y. (2021). ECG signal classification based on deep CNN and BiLSTM. *BMC Medical Informatics and Decision Making*, 21(1), 365. https://doi.org/10.1186/s12911-021-01736-y.
52. J. Hu, W. L. Goh, & Y. Gao. (2024). LSTM-based ECG Signal Classification with Multi-level One-hot Encoding for Wearable Applications. *2024 IEEE Biomedical Circuits and Systems Conference (BioCAS)*, 1–5. https://doi.org/10.1109/BioCAS61083.2024.10798273.
53. Gagliano, L., Bou Assi, E., Nguyen, D. K., & Sawan, M. (2019). Bispectrum and Recurrent Neural Networks: Improved Classification of Interictal and Preictal States. *Scientific Reports*, 9(1), 15649. https://doi.org/10.1038/s41598-019-52152-2.
54. Vaswani, A., Shazeer, N., Parmar, N., Uszkoreit, J., Jones, L., Gomez, A. N., Kaiser, \Lukasz, & Polosukhin, I. (2017). Attention is All You Need. *Proceedings of the 31st International Conference on Neural Information Processing Systems*, 6000–6010. url: https://dl.acm.org/doi/https://doi.org/10.5555/3295222.3295349.
55. Zeng, A., Chen, M., Zhang, L., & Xu, Q. (2023). Are Transformers Effective for Time Series Forecasting? *Proceedings of the AAAI Conference on Artificial Intelligence*, 37(9), 11121–11128. https://doi.org/10.1609/aaai.v37i9.26317.
56. Cao, H., Huang, Z., Yao, T., Wang, J., He, H., & Wang, Y. (2023). InParformer: Evolutionary Decomposition Transformers with Interactive Parallel Attention for Long-Term Time Series Forecasting. *Proceedings of the AAAI Conference on Artificial Intelligence, 37*(6), 6906–6915. https://doi.org/10.1609/aaai.v37i6.25845.
57. Weimann, K., & Conrad, T. O. F. (2021). Transfer learning for ECG classification. *Scientific Reports*, 11(1), 5251. https://doi.org/10.1038/s41598-021-84374-8.
58. Van Steenkiste, G., van Loon, G., & Crevecoeur, G. (2020). Transfer Learning in ECG Classification from Human to Horse Using a Novel Parallel Neural Network Architecture. *Scientific Reports*, 10(1), 186. https://doi.org/10.1038/s41598-019-57025-2.
59. Ben-Moshe, N., Tsutsui, K., Brimer, S. B., Zvuloni, E., Sörnmo, L., & Behar, J. A. (2024). RawECGNet: Deep Learning Generalization for Atrial Fibrillation Detection From the Raw ECG. *IEEE Journal of Biomedical and Health Informatics, 28*(9), 5180–5188. https://doi.org/10.1109/JBHI.2024.3404877.
60. Xu, S. S., Mak, M.-W., & Cheung, C.-C. (2019). Towards End-to-End ECG Classification With Raw Signal Extraction and Deep Neural Networks. *IEEE Journal of Biomedical and Health Informatics, 23*(4), 1574–1584. https://doi.org/10.1109/JBHI.2018.2871510.
61. Moons, K. G. M., Damen, J. A. A., Kaul, T., Hooft, L., Andaur Navarro, C., Dhiman, P., Beam, A. L., Van Calster, B., Celi, L. A., Denaxas, S., Denniston, A. K., Ghassemi, M., Heinze, G., Kengne, A. P., Maier-Hein, L., Liu, X., Logullo, P., McCradden, M. D., Liu, N., … van Smeden, M. (2025). PROBAST+AI: an updated quality, risk of bias, and applicability assessment tool for prediction models using regression or artificial intelligence methods. *BMJ, 388*. https://doi.org/10.1136/bmj-2024-082505.
62. Salahuddin, Z., Woodruff, H. C., Chatterjee, A., & Lambin, P. (2022). Transparency of deep neural networks for medical image analysis: A review of interpretability methods. *Computers in Biology and Medicine, 140*, 105111. https://doi.org/10.1016/j.compbiomed.2021.105111.
63. Purba, S. A., Melles, A.-M., Hackel, D., Obeid, I., & Picone, J. (2025). Assessing the Visual Reasoning of Multimodal Language Models in Biomedical Applications. *Proceedings of the IEEE Signal Processing in Medicine and Biology Symposium (SPMB), 1*, 1–5. url: www.isip.piconepress.com/publications/unpublished/conferences/2025/ieee_spmb/segmentation/. (in review).

64. Singh, H. (2025, May 31). *10 Techniques to Solve Imbalanced Classes in Machine Learning (Updated 2025)*. Analytics Vidhya. url: https://www.analyticsvidhya.com/blog/2020/07/10-techniques-to-deal-with-class-imbalance-in-machine-learning/.
65. Seiffert, C., Khoshgoftaar, T. M., Van Hulse, J., & Napolitano, A. (2010). RUSBoost: A Hybrid Approach to Alleviating Class Imbalance. *IEEE Transactions on Systems, Man and Cybernetics. Part A, Systems and Humans, 2010, Vol.40 (1), p.185–197*. https://doi.org/10.1109/TSMCA.2009.2029559.
66. Chawla, N. V., Bowyer, K. W., Hall, L. O., & Kegelmeyer, W. P. (2002). SMOTE: synthetic minority over-sampling technique. *Journal of Artificial Intelligence Research, 16*, 321–357. https://doi.org/10.1109/TSMCA.2009.2029559.
67. Kingma, D. P., & Ba, J. L. (2015). Adam: A Method for Stochastic Optimization. *Proceedings of the International Conference on Learning Representations*, 1–15. url: https://arxiv.org/abs/1412.6980.
68. Mao, A., Mohri, M., & Zhong, Y. (2023). Cross-Entropy Loss Functions: Theoretical Analysis and Applications. In A. Krause, E. Brunskill, K. Cho, B. Engelhardt, S. Sabato, & J. Scarlett (Eds.), *Proceedings of the 40th International Conference on Machine Learning* (Vol. 202, pp. 23803–23828). PMLR. url: https://proceedings.mlr.press/v202/mao23b.html.
69. Manning, C. D., Raghavan, P., & Schütze, H. (2008). *Introduction to Information Retrieval*. Cambridge University Press. url: http://www.amazon.com/Introduction-Information-Retrieval-Christopher-Manning/dp/0521865719.
70. Meyes, R., Lu, M., Puiseau, C. W. de, & Meisen, T. (2019). *Ablation Studies in Artificial Neural Networks*. url: https://arxiv.org/abs/1901.08644.
71. Graham, M. H. (2003). Confronting Multicollinearity in Ecological Multiple Regression. *Ecology, 84*(11), 2809–2815. https://doi.org/10.1890/02-3114.
72. Belsley, D. A. (n.d.). *Regression diagnostics: Identifying influential data and sources of collinearity/David A. Belsley, Edwin Kuh, Roy E. Welsch*. New York : Wiley, c2004. https://doi.org/10.1002/0471725153.

Transformer or Not? A Review of Recent Progress in Time Series Analysis

Maohua Liu, De Yang, and Liqiang Zhao

Abstract While the Transformer has revolutionized natural language processing, its impact on time series analysis remains inconsistent, marked by both breakthroughs and setbacks. This paper provides a comprehensive review of the evolving role of the Transformer in time series analysis, examining its advantages, limitations, and recent advancements. We first explore the fundamental differences between text and time series data that challenge the direct adaptation of the Transformer to time series tasks. We then analyze why the Transformer has struggled in certain applications and how alternative models—such as CNNs, RNNs, Diffusion Models, and Mamba—have outperformed it in various scenarios. Additionally, we review key modifications and refinements designed to enhance the effectiveness of Transformer-based approaches for time series data. Beyond evaluating Transformer-based models, this paper also contextualizes their performance within the broader landscape of time series analysis by examining widely used non-Transformer methods. By synthesizing recent developments and emerging trends, this review offers valuable insights into the ongoing debate over whether to adopt Transformer architectures or alternative models. Ultimately, it serves as a guide for researchers and practitioners seeking the most effective methodologies for time series analysis.

Keywords Transformer · Diffusion model · Mamba · Foundation model · CNN · RNN · Time series

M. Liu (✉)
School of Electrical & Computer Engineering, University of Georgia, Athens, GA, USA
e-mail: maohua.liu@uga.edu

D. Yang
Institute for Artificial Intelligence, University of Georgia, Athens, GA, USA
e-mail: dy77706@uga.edu

L. Zhao
School of Math. and Info. S&T, Hebei Normal Univ. of S&T, Qinhuangdao, Hebei, China
e-mail: zhao_liqiang@126.com

A. Ahmed and J. Picone (eds.), *Signal Processing and Biomedical Engineering Research*, https://doi.org/10.1007/978-3-032-18509-9_5

1 Introduction

1.1 Time Series Analysis

Time series analysis (TSA) is crucial for understanding patterns, trends, and behaviors in data collected over time. It enables forecasting, anomaly detection, and decision-making in fields like finance, economics, weather, and healthcare. By analyzing temporal dependencies, it helps predict future values, optimize processes, and identify underlying structures, making it essential for planning and strategy in dynamic environments.

Many different TSA techniques have been developed. Traditional TSA techniques include correlation-based methods, frequency-domain and multi-timescale methods, non-linear methods, and regression-based methods. Modern TSA approaches include reinforcement learning, nonparametric methods, neural networks, and deep neural networks [1]. As an outstanding representative of deep neural networks, Transformer, which once brought great changes to natural language processing (NLP), is also widely used in TSA. Transformer is expected to play a huge role in TSA due to its effectiveness in handling long-term dependencies [1].

1.2 Transformer in Time Series Analysis

The Transformer is a deep learning architecture introduced in 2017, primarily designed for NLP tasks [2]. Unlike traditional models that rely on recurrent or convolutional layers, Transformer uses a mechanism called self-attention to process input data. This allows them to capture relationships between distant elements in a sequence, making them highly effective for handling long-term dependencies. The architecture consists of an encoder-decoder structure, where both parts are composed of multiple layers of self-attention and feedforward neural networks. Transformer has revolutionized NLP in tasks like translation, text generation, and sentiment analysis. Beyond NLP, their ability to model sequential data has led to widespread adoption in TSA.

Transformer functions like a vast knowledge graph, storing relationships between tokens within their attention weights, whereas RNNs and LSTMs act more like a short-term memory notebook, continually overwriting information as new inputs arrive. Unlike RNNs and LSTMs, which rely on hidden states to compress and retain information, Transformer eliminates the need for compression altogether, using a less biased attention mechanism that can handle long-term dependencies without compression. No matter how advanced a compression technique may be, it is fundamentally impossible to preserve all information within the limited capacity of hidden states. This inevitable loss of information weakens the ability of RNNs and LSTMs to capture long-term dependencies effectively. In contrast, Transformer sacrifices space for more complete information and richer dependency modeling, ensuring that no relationship is lost in the process. Rather than compressing past data, they store and compute relationships dynamically through self-attention, allowing them to retain all relevant information without degradation. Additionally, because each token can simultaneously attend to all others in the sequence, Transformer processes data in parallel rather than a recurrent way, dramatically improving efficiency compared to the sequential nature of RNNs. This trade-off—using more space to achieve better information retention and faster computation—is what makes Transformer a groundbreaking advancement over its predecessors. By maintaining full contextual awareness while enabling high-speed processing, they have revolutionized deep learning across multiple domains.

Inspired by the remarkable success of Transformer in NLP and its superior ability to capture long-term dependencies, researchers have sought to replicate this success in TSA. Given that many time series task—such as forecasting, anomaly detection, and classification—require understanding long-term dependencies, Transformer appears to be a promising solution.

Transformer has demonstrated success in several time series applications. In time series forecasting, the Informer model introduced probabilistic sparse attention to reduce the quadratic complexity of self-attention and improve forecasting efficiency [3], while Autoformer utilized seasonal-trend decomposition to enhance interpretability and performance [4]. In anomaly detection, Transformer applied attention mechanisms to detect anomalies in industrial sensor data [5]. In financial and healthcare applications, Transformer-based models have been used for stock market prediction [6] and ECG-based arrhythmia classification [7]. These successes proved the feasibility of Transformer in TSA, but their journey has been far from smooth.

Despite their potential, Transformers have not achieved the same breakthrough dominance in TSA due to data differences like features, continuity, temporal property, and distribution shift. On the contrary, traditional models such as ARIMA, RNNs, and LSTMs have built-in temporal structures, making them naturally suited for time series data. Furthermore, RNNs and LSTMs have strong inductive biases for sequential data, meaning they naturally process information in a stepwise manner or with local feature extraction. Transformer, however, treats sequences as a set of tokens with no inherent temporal constraints, making them less efficient at capturing local dependencies. Recently, new architectures have emerged, challenging the dominance of Transformer in TSA. These models have outperformed Transformer in some cases, leading researchers to reconsider the best architectures for time series tasks. Despite these challenges, Transformer is still evolving for time series applications. These innovations aim to overcome the weaknesses of Transformer in TSA while retaining their strengths.

Transformer has undeniably made an impact in TSA, enabling breakthroughs in forecasting, anomaly detection, and financial modeling. However, they have not completely displaced traditional models as they did in NLP. Instead, their journey has been marked by challenges, adaptations, and emerging. As research continues, it is likely that some more innovative models are on the way for defining the future of TSA.

1.3 Outline

This review examines the suitability of Transformer for TSA and explores the ongoing search for the most effective model in this domain. Below is an overview of the content covered in each section. Section 2 delves into the characteristics of time series data and compares them to text. Section 3 addresses the limitations of Transformer and discusses modifications designed to better handle time series data. Section 4 evaluates alternative deep learning models applied to TSA. Section 5 investigates the potential of Mamba for TSA, while Sect. 6 explores the application of Diffusion Models in this context. Section 7 talks about hybrid models. Finally, Sect. 8 introduces foundation models in TSA.

2 Time Series

Transformer has revolutionized NLP by replacing sequential models with self-attention mechanisms, enabling the capture of long-term dependencies and parallel computation. However, their application to time series data has yielded inconsistent results, falling short of the dominance they exhibit in NLP [3]. This disparity stems from fundamental differences between time series data and text, which influence how effectively Transformer can process each type of data.

As shown in the following table, time series data possesses unique characteristics that set it apart from text. A key distinction lies in its temporal continuity and strict sequential dependency [4]. Unlike text, where word order can often be rearranged without losing meaning, time series data relies heavily on chronological order. Each value is deeply correlated with past and future values, and even minor disruptions, such as a sudden drop in stock prices, can have immediate and far-reaching effects. This strict causality makes self-attention mechanisms, which treat all tokens as equally related, less suitable for modeling the precise temporal relationships inherent in time series data.

Time series data differ fundamentally from textual data in several key aspects that complicate the direct application of Transformer models. Unlike text, where discrete tokens carry semantic meaning, and order can often be flexible, time series are composed of continuous numerical values that exhibit strict temporal dependencies and causality. Each data point is highly correlated with its neighbors, and small variations may carry important signals, especially in domains like finance or healthcare. These characteristics often necessitate preprocessing techniques such as trend extraction, seasonal decomposition, and frequency-based analysis to highlight meaningful structures. Transformers, originally designed for NLP, do not inherently model these temporal dynamics. Their reliance on token embeddings and equal treatment of all positions through self-attention makes them less efficient in capturing localized, sequential dependencies without additional architectural modifications. Consequently, when applied directly to raw time series data, Transformers often require external feature engineering and decomposition methods to perform competitively with models that possess stronger temporal inductive biases.

These challenges explain why Transformer has not yet achieved the same level of success in TSA as they have in NLP. Their quadratic computational complexity makes them impractical for long sequences, their lack of strong inductive biases for sequential dependencies hinders their ability to learn causal relationships, and their reliance on large datasets makes them prone to overfitting [8]. Moreover, their attention mechanisms fail to naturally enforce temporal constraints, as they treat all time steps equally.

In summary, while Transformer has fundamentally transformed NLP, their application to time series data requires significant adaptations to address the unique properties of sequential numerical data. Text is contextual, sparse, and somewhat flexible in order, whereas time series is continuous, structured, and strictly sequential. These differences mean that self-attention alone is insufficient for time series tasks. Instead, hybrid approaches—combining Transformer with state-space models, CNNs, or recurrent structures—are emerging as more effective solutions. Future research may likely focus more on enhancing Transformer's compatibility with time series by integrating temporal biases, improving computational efficiency, and developing more sophisticated position encodings (Table 1).

3 Transformer's Limitations and Adjustments

Since time series data possesses fundamentally different characteristics from text, Transformer originally designed for NLP faces significant limitations when applied to time series tasks. These challenges stem from differences in temporal dependencies, data continuity, and structural constraints. In response, Transformer has undergone various modifications to better align with the unique demands of TSA. While these adaptations have led to improvements, they have yet to achieve the same level of success as in NLP, highlighting the ongoing need for further optimization and hybrid approaches.

As shown in Table 2, a significant limitation of Transformer in this domain is its computational complexity. Self-attention mechanisms scale quadratically with sequence length, which becomes a critical issue when dealing with time series data that can span thousands or even millions of timesteps. In contrast, NLP typically involves shorter sequences, such as sentences or paragraphs, making Transformer more computationally feasible. However, time series forecasting often requires processing extensive datasets collected over years at fine-grained intervals, leading to prohibitive memory and processing demands.

Another critical challenge is the lack of inherent temporal biases in Transformer. Unlike recurrent models like LSTMs and GRUs, which process sequences step by step and naturally respect temporal order, Transformer treats all time steps equally. This makes it difficult to enforce causality, as the model cannot inherently distinguish between meaningful past influences and irrelevant noise. Traditional time series

Table 1 Differences between time series and text

Differences	Time series	Text
Nature of dependencies	Strict temporal order; past affects the future in a causal manner	Contextual dependencies; words relate based on meaning
Information density	Dense and highly redundant; small variations in values across time	Sparse and symbolic; each word carries distinct meaning
Handling long sequences	Sequences can be extremely long, requiring efficient memory usage	Sequences are generally shorter and contain meaningful tokenized words
Positional encoding	Absolute and relative time positions are critical	Word order is important but can be flexible
Feature representation	Trend extraction, seasonality modeling, and numerical feature engineering	Semantic embeddings and context-based word representations
Computational complexity	Self-attention requires quadratic complexity; difficult for long sequences	More efficient sequences are generally shorter in NLP tasks
Temporal constraints	Strong constraints on sequence order; future values must not influence past values	Less rigid structure; words can be rearranged with little change in meaning
Handling missing data	Irregular sampling and missing values are common, requiring imputation	Structured and complete; missing words can often be inferred
Interpretability	More interpretable due to explicit time dependencies and known domain models	More challenging to interpret due to contextual meanings and abstract concepts

models, such as ARIMA, are explicitly designed to capture temporal dependencies, whereas Transformer requires additional mechanisms, such as the integration of the other architectures, to approximate similar behavior.

Feature representation further complicates the application of Transformer to time series data. In NLP, pre-trained embeddings effectively capture word semantics, enabling models to generalize across diverse contexts. Time series data, however, lacks such structured representations. Instead, it often requires explicit feature engineering—such as extracting trends, seasonal components, and anomalies—before meaningful patterns can be learned. Without these preprocessing steps, Transformer may struggle to differentiate between significant signals and recurring fluctuations, limiting their effectiveness.

Additionally, Transformer faces difficulties in handling missing data and irregular sampling intervals, which are common in real-world time series datasets. Gaps caused by sensor failures or inconsistent data collection can distort trends and relationships, making accurate modeling more challenging. In NLP, missing words can sometimes be inferred from context, but in time series, missing values can disrupt the continuity

Table 2 Transformer's limitations in TSA

Transformer's limitation in TSA	Description
High computational complexity	Self-attention has quadratic complexity, making it computationally expensive for long time series sequences
Lack of temporal bias	Unlike RNNs, Transformer does not have an inherent sequential processing mechanism, making them struggle with causal dependencies
Inefficiency in capturing local dependencies	Transformer processes all tokens equally, failing to focus on immediate past values that are often more relevant in time series forecasting
Handling missing data and irregular sampling	Many real-world time series datasets contain missing values or are sampled at irregular intervals, which Transformer is not well equipped to handle
Need for explicit feature engineering	Unlike NLP, where pre-trained word embeddings exist, time series models require handcrafted feature extraction such as trend decomposition and frequency-based analysis
Sensitive to data distribution shifts	Traditional Transformer assumes stationarity, meaning they struggle with time series data that undergo structural shifts over time
Difficulty in capturing long-term trends	Transformer may not efficiently capture seasonal trends and periodic fluctuations, requiring additional mechanisms such as decomposition-based modeling
Position encoding challenges	Standard positional encoding is not well suited for absolute time relationships, which are crucial in time series forecasting

of temporal dependencies, requiring specialized imputation techniques or model adaptations to address these issues effectively.

In summary, the rigid temporal structure of time series data, combined with its computational demands, lack of inherent temporal biases, and need for explicit feature engineering, poses significant challenges for Transformer.

Aforementioned limitations have spurred innovations such as optimized attention mechanisms, causal modeling techniques, and hybrid approaches that integrate Transformer with traditional TSA methods. By addressing these challenges, researchers aim to bridge the gap between the flexibility of Transformer and the structured nature of time series data, enabling more effective and efficient analysis.

Recent advancements in TSA mainly explored hybrid models that combine the strengths of different architectures to address the limitations of Transformer. These innovations aim to enhance their ability to handle the unique challenges of sequential numerical data, such as temporal dependencies, nonstationary patterns, and missing value. Below, we discuss several notable hybrid models and their contributions to the field.

PatchTST applies patch-based tokenization, inspired by Vision Transformers (ViTs), to better capture local temporal dependencies while leveraging attention for long-term forecasting [9]. By dividing time series into smaller patches, this approach improves the model's ability to detect localized patterns, making it more effective for tasks requiring both short- and long-term insights.

iTransformer improves time series forecasting by restructuring how input data is processed [10]. Instead of treating each time step as a token, it treats each variate as a token and embeds its values over time. This shift allows the model to focus on correlations between variates rather than direct temporal dependencies. Unlike other adaptations that modify the Transformer architecture, iTransformer keeps the original structure but changes the way data is represented. This design helps it handle long lookback windows effectively and avoids the loss of information caused by mixing all variates into a single token per timestamp. By optimizing input representation rather than altering the model itself, iTransformer improves forecasting accuracy while maintaining efficiency and interpretability.

The Nonstationary Transformer addresses the assumption of stationarity in time series, which is often violated in real-world data [11]. Traditional Transformer assumes that the underlying distribution of time series data remains stable, but real-world data often exhibit shifting patterns. Nonstationary Transformer, with the help of de-stationary attention mechanism, dynamically adapts its attention weights based on evolving trends, making it particularly useful for financial forecasting and climate modeling, where underlying dynamics frequently change.

TimesNet introduces a novel approach to TSA by transforming traditional 1D-dimensional time series data into a 2D-dimensional representation [12]. Instead of directly modeling complex temporal variations from raw one-dimensional sequences, which can be highly challenging, TimesNet leverages the multi-periodicity inherent in time series data. It decomposes temporal patterns into intraperiod and interperiod variations and maps them onto a two-dimensional tensor, where intraperiod variations align along columns and interperiod variations along rows. This transformation allows for more effective pattern extraction using two-dimensional convolutional kernels, which are well suited for capturing structural dependencies.

ETSformer uses novel attention mechanism for TSA by combining self-attention with exponential smoothing (ETS) models [13]. Traditional Transformers struggle with time series due to their lack of decomposition capability, interpretability, and efficiency for long-term forecasting. To address these issues, ETSformer introduces two novel attention mechanisms: exponential smoothing attention and frequency attention, replacing the standard self-attention mechanism. These modifications enhance both accuracy and computational efficiency by incorporating time series-specific properties. Additionally, ETSformer employs modular decomposition blocks that enable the model to break down time series data into interpretable components such as level, growth, and seasonality. This structured decomposition improves forecasting performance while maintaining interpretability.

Despite these advancements, Transformer has yet to fully dominate TSA. Their quadratic complexity, lack of inherent temporal bias, and inefficiencies in capturing structured dependencies continue to pose challenges. However, ongoing research is refining Transformer-based models for time series applications. Future work is likely to focus on efficient self-attention mechanisms, hybrid architectures, and domain-specific feature engineering to further bridge the gap between Transformer and time series modeling.

4 Non-transformer Architectures with Competitive Performance

As Transformer has yet to fully adapt to the unique characteristics of time series data, researchers have increasingly explored alternative models that may be better suited for these tasks. This shift has led to a renewed interest in architectures such as CNNs, RNNs, and LSTMs—models that Transformer previously outperformed in NLP. Unlike in NLP, where Transformer revolutionized sequence modeling, these traditional models still offer advantages in TSA, particularly in capturing local patterns, maintaining computational efficiency, and handling sequential dependencies. This growing exploration reflects the ongoing search for the most effective approach to modeling complex temporal data.

Just as Transformer actively incorporates elements from other models to compensate for their inherent limitations, other architectures also integrate Transformer-based innovations to enhance their own capabilities. Many alternative models adopt specific strategies to address their weaknesses, such as developing specialized mechanisms for capturing long-term dependencies, implementing encoder-decoder frameworks for structured representation learning, refining attention mechanisms to improve efficiency, and incorporating frequency-domain analysis to extract meaningful temporal patterns and long-term dependencies.

MICN (Multi-Scale Isometric Convolution Network) overcomes key limitations of Transformer-based methods in long-term time series forecasting [14]. While Transformer captures global correlations well, its self-attention mechanism is computationally costly and weak in modeling local dependencies. MICN addresses this by combining CNN-based local feature extraction with global correlation modeling, balancing the capture of both fluctuations and trends in time series data. MICN uses a multi-scale branch structure: down-sampled convolution extracts local features, while isometric convolution preserves global correlations. This design efficiently captures diverse temporal patterns at reduced computational cost. Unlike Transformer, which has quadratic complexity due to self-attention, MICN achieves linear complexity with optimized convolutional kernels, ensuring scalability. MICN excels in handling irregularly sampled data and outperforms Transformer-based models in forecasting accuracy, offering a highly efficient and accurate alternative. By integrating CNNs' ability to model fine-grained local dependencies with an efficient global correlation mechanism, MICN emerges as a powerful solution for time series analysis. Its balance of scalability, efficiency, and accuracy positions it as a leading architecture for long-term forecasting, highlighting its potential in surpassing Transformer in TSA.

LSTM-Attention-LSTM employs an encoder-decoder architecture enhanced by a custom attention mechanism to improve time series forecasting [15]. While it draws inspiration from the Transformer's structure and self-attention principles, it employs a uniquely designed attention mechanism tailored for sequential data. By integrating LSTMs, the model also excels at capturing local temporal features, enabling more efficient and structured learning of temporal patterns. The model consists of two LSTMs—one as the encoder and another as the decoder—linked by a custom attention mechanism. This attention mechanism plays a crucial role in overcoming a common limitation of encoder-decoder models: the inability of the decoder to retain information from very long input sequences. By calculating the interrelationship between different points in the time series, the attention layer enables the decoder to focus on the most relevant past observations, ensuring that important temporal dependencies are not lost. As a result, the model achieves better predictive performance than traditional LSTM-based architectures, particularly in cases where long-term trends and fluctuations need to be accounted for.

Graph Neural Networks for Time Series (GNN4TS) have recently gained prominence as a powerful tool for learning representations of non-Euclidean data, making them particularly effective for modeling complex multivariate time series [16]. Unlike traditional sequence-based models, GNNs are adept at capturing intricate spatial–temporal dependencies, enabling them to model both inter-variable relationships (connections between different variables in a multivariate time series) and intertemporal dependencies (patterns across different time points). Real-world time series data often involve complex, non-linear interactions among multiple variables, and GNNs offer a structurally aware approach that traditional temporal models cannot easily replicate. By representing multivariate time series as dynamic graphs, GNN-based methods can exploit spatial dependencies while incorporating temporal modeling techniques. This makes them highly effective in applications where interactions between variables—such as in sensor networks, financial markets, and traffic prediction—are critical for accurate forecasting. Recognizing these strengths, recent research has combined GNNs with various temporal modeling frameworks, including recurrent networks, state-space models, and attention mechanisms, to better capture both spatial and temporal patterns. These hybrid approaches have demonstrated significant potential and achieved promising results.

RNN-based model for time series tasks (RWKV-TS) introduces a revolutionary RNN-based approach that preserves the strengths of traditional RNNs while overcoming their historical weaknesses [17]. RWKV-TS represents a next-generation RNN-based model designed to address the shortcomings of traditional recurrent architectures in time series forecasting. While LSTMs and GRUs have long been staples for sequential modeling, their limitations in long-term dependencies have been increasingly exposed by the rise of Transformer-based models. Traditional

RNNs often suffer from vanishing gradient issues and computational inefficiencies, especially when processing long sequences. RWKV-TS tackles these challenges by introducing an optimized RNN architecture that can efficiently capture long-term dependencies like Transformer. Unlike conventional RNNs, which struggle with long-term dependencies, RWKV-TS integrates a novel attention-inspired mechanism. This mechanism selectively retains critical information across long sequences without incurring the heavy computational cost of full self-attention, striking a balance between efficiency and effectiveness. As a result, RWKV-TS achieves long-term sequence modeling capabilities comparable to Transformer while preserving the inherent advantages of RNNs. To illustrate its approach, consider the analogy of passing a message along a chain of individuals: traditional RNNs rely on each person remembering only the message from the previous person, leading to information loss over time. In contrast, Transformer allows everyone to listen to everyone else simultaneously, ensuring full retention but at a high computational cost. RWKV-TS finds a middle ground, enabling selective memory retention that ensures both computational efficiency and robust long-range dependency modeling. Extensive experiments have shown that RWKV-TS performs on par with or even surpasses state-of-the-art Transformer-based and CNN-based models, significantly reducing latency and memory usage. This efficiency makes RWKV-TS particularly appealing for real-world time series applications, especially in resource-constrained environments. The success of RWKV-TS underscores the renewed potential of RNN-based approaches for TSA. By demonstrating that RNNs can remain highly competitive with appropriate architectural refinements, RWKV-TS paves the way for further innovation in sequential data analysis. Its combination of scalability, efficiency, and forecasting accuracy positions it as a compelling candidate for both research and practical deployment in TSA, offering a promising alternative to Transformer for TSA.

Meta Prophet is a recent extension of Facebook's Prophet model, designed to improve forecasting accuracy across a diverse range of time series. While Prophet is known for its interpretable decomposition of time series into trend, seasonality, and holiday effects [18], it does not adapt well to data with varying structures or characteristics. Meta Prophet addresses this by incorporating a meta-learning strategy that enables the model to learn from prior forecasting tasks. This meta-learning layer allows it to identify and select the most appropriate forecasting configuration, such as Prophet, ARIMA, or neural network-based models, for a new dataset based on patterns learned from previous tasks [19]. As a result, Meta Prophet offers improved generalization, particularly in settings where datasets differ widely in their temporal behavior. This makes it especially useful for applications in retail demand forecasting, finance, and operations, where flexibility and reliability across different time series are essential [20].

5 Mamba

A growing number of recent models have abandoned self-attention mechanism altogether in favor of more efficient alternatives. Among them, Mamba stands out for its use of a selective state-space mechanism—a method that models sequence dynamics by continuously updating hidden states through structured linear recurrence. Unlike hybrid approaches that blend Transformer with other architectures, it is not a compromise but an entirely distinct framework with its own unique design philosophy. It retains the ability to capture long-term dependencies, much like a Transformer, but does so with significantly greater efficiency. Moreover, its inherent concurrency makes it particularly well suited for processing continuous time series data, overcoming some of the fundamental limitations of self-attention mechanisms.

At its core, Mamba introduces a novel approach to sequence modeling by leveraging selective state-space mechanisms, which efficiently encode temporal information. Unlike traditional self-attention mechanisms, which scale quadratically with sequence length, Mamba adopts a linear-time complexity computation. This design significantly reduces both memory usage and computational demands, enabling the model to handle long sequences and continuous data streams with remarkable ease [21]. As a result, Mamba is particularly well suited for applications that demand rapid and accurate processing of large-scale time series data, such as financial forecasting, healthcare monitoring, and climate modeling. Its architecture strikes a careful balance between capturing detailed long-term dependencies and addressing the practical constraints of real-world time series data.

Another key advantage of Mamba lies in its inherent flexibility. While Transformer relies on explicit positional encodings to maintain sequence order, Mamba's RNN-like recurrency naturally preserves the sequential structure of time series data. This eliminates the need for additional mechanisms to encode positional information, streamlining the model's design and enhancing its performance in real-world time series data. Moreover, Mamba's adaptability has spurred the development of numerous specialized models built on its foundational principles, underscoring its growing influence in the field of TSA.

Mamba's emergence marks a significant departure from the long-held belief that self-attention is the most effective mechanism for sequence modeling. While attention-based models have dominated areas like NLP and structured prediction tasks, Mamba's state-space approach demonstrates that alternative architectures can achieve comparable or even superior performance while drastically reducing computational costs. This paradigm shift has encouraged researchers to explore new architectures that prioritize efficiency, memory optimization, and the preservation of temporal structure. Such innovations are critical as datasets continue to grow in size and complexity, ensuring that sequence models remain scalable and practical.

The unique advantages of Mamba position it as a groundbreaking paradigm in sequence modeling. Its ability to combine computational efficiency with robust performance has inspired a wave of recent models in TSA that build upon its principles.

Simple Mamba (S-Mamba) is a novel time series analysis model that harnesses the efficiency and pattern-learning strengths of Mamba, a selective state-space model [22]. While Transformer-based models have shown strong performance in TSA by capturing complex dependencies in historical data, their quadratic computational complexity often leads to high processing costs, limiting their practicality for real-world applications that demand efficiency and scalability. Mamba, on the other hand, achieves near-linear complexity while effectively modeling long-term dependencies, offering a more computationally efficient solution for TSA tasks. S-Mamba capitalizes on these benefits by introducing a structured pipeline tailored for TSA. The process begins with autonomous tokenization of time points using a linear layer, enabling independent processing of each time variate. A bidirectional Mamba layer is then applied to extract inter-variate correlations, ensuring that relationships between different time series variables are accurately captured. To further enhance temporal modeling, a Feed-Forward Network is integrated to learn sequential dependencies. Finally, a linear mapping layer generates the forecasts, resulting in an efficient and streamlined prediction workflow. Extensive experiments have demonstrated S-Mamba's ability to achieve state-of-the-art forecasting performance with minimal computational overhead, solidifying its effectiveness as a scalable time series analysis model. Additional testing has further explored Mamba's potential in TSA, highlighting its versatility. By striking a balance between accuracy and efficiency, S-Mamba emerges as a compelling alternative to traditional Transformer-based methods, making it highly suitable for large-scale, real-world TSA applications.

SiMBA is designed by incorporating Einstein FFT (EinFFT), an advanced alternative to the traditional FFT used in standard Transformer [23]. EinFFT enhances stability and efficiency by computing specific eigenvalues. Furthermore, SiMBA integrates the Mamba block for sequence modeling, enabling robust long-term dependency capture while avoiding the computational inefficiencies of traditional attention mechanisms. Comprehensive performance evaluations across image and time series benchmarks reveal that SiMBA surpasses existing state-space machines and narrows the performance gap with Transformer in numerous tasks. Notably, SiMBA achieves state-of-the-art results on ImageNet, key transfer learning benchmarks like Stanford Cars and Flowers, and seven time series forecasting benchmarks. By combining EinFFT-based channel modeling with Mamba's efficient sequence processing, SiMBA delivers a scalable and high-performing alternative to attention-based architectures. This makes SiMBA a promising solution for both vision and TSA applications, offering a compelling balance of efficiency and accuracy.

TimeMachine is an advanced TSA model designed to address core challenges such as capturing long-term dependencies, achieving linear scalability, and maintaining computational efficiency [24]. Traditional TSA models often suffer from information loss over extended sequences, resulting in diminished performance as the prediction horizon grows. To overcome this, TimeMachine utilizes Mamba's efficient sequence modeling in processing long sequences with minimal memory overhead. At its core, TimeMachine features a novel quadruple-Mamba architecture, enabling it to effectively model both channel-mixing and channel-independent scenarios in multivariate TSA. This design allows the model to selectively extract key contextual information at multiple scales, balancing global trends with local fluctuations for more accurate long-term predictions. By integrating multi-scale representations, TimeMachine ensures that critical patterns—whether long-term trends or short-term variations—are efficiently captured, enhancing forecasting robustness. Extensive experiments on benchmark datasets demonstrate TimeMachine's superior prediction accuracy, scalability, and memory efficiency compared to existing models. Its ability to unify diverse contextual dependencies across multiple scales while maintaining linear complexity positions it as a state-of-the-art solution for large-scale TSA. As long-term forecasting remains a critical challenge across various domains, TimeMachine sets a new benchmark by combining Mamba's efficiency with a multi-scale predictive framework, delivering both precision and practicality for real-world TSA applications.

6 Diffusion Model

Diffusion models have established themselves as a highly effective class of generative models, excelling in producing high-quality outputs across diverse domains such as image synthesis, time series forecasting, and molecular generation. Their strength lies in their foundation in probabilistic modeling, where they generate data by iteratively transforming simple noise distributions into complex, structured representations through a series of learned transitions [25]. This step-by-step refinement process enables diffusion models to effectively capture high-dimensional data distributions, often outperforming traditional generative methods like Generative Adversarial Networks (GANs) [26] and Variational Autoencoders (VAEs) [27] in terms of both sample diversity and training stability.

The core principle of diffusion models revolves around learning to reverse a forward diffusion process. In this process, data is gradually corrupted by noise over multiple steps until it resembles pure Gaussian noise. The model then learns to invert this process, effectively denoising the data to reconstruct the original input. This design inherently enhances robustness, as diffusion models avoid issues like mode collapse, which commonly plague GANs. Additionally, the iterative nature of the diffusion process provides fine-grained control over the generation process, allowing for precise manipulation of outputs.

Recent innovations have broadened the applicability of diffusion models beyond their initial use in image generation, like TSA [28]. In time series forecasting, for instance, diffusion-based approaches have been developed to generate future sequences by iteratively refining predictions over multiple steps. This not only improves the accuracy of forecasts but also enhances uncertainty quantification and the handling of missing data.

TimeGrad uses diffusion models to iteratively refine predictions over multiple steps, improving uncertainty quantification and handling missing values [29]. It is particularly effective for probabilistic forecasting in multivariate time series. **Conditional Score-based Diffusion Model for Imputation** (CSDI) leverages diffusion models to impute missing values in time series data [30]. It uses a score-based approach to conditionally generate missing data points, making it robust for irregularly sampled time series. **Diffusion-TS** applies diffusion models to time series forecasting by learning the reverse process of noise addition [31]. It is designed to capture complex temporal dependencies and generate high-quality forecasts, especially for long-horizon predictions. **D3VAE** [32] integrates diffusion models with variational autoencoders (VAEs) for time series forecasting. It leverages the generative capabilities of diffusion models to improve the robustness and accuracy of forecasts in complex multivariate settings.

7 Hybrid Models

Due to Transformer, Mamba and Diffusion Model's advantages in their respective fields, there have been many attempts to combine them to bring out more comprehensive advantages. The following models are attempts in this regard.

MAT (integrating Mamba and Transformer) and **SST** (Structured State-space Transformer) are both hybrid models for long-short term time series forecasting that address challenges in capturing long-term dependencies and sparse semantic features in multivariate data [33, 34]. While Transformer excels at short-term correlations, its quadratic complexity limits efficiency and long-term dependency modeling compared to Mamba. MAT combines Mamba's long-term dependency modeling with Transformer's ability to capture short-term (relatively short-term) correlations and fine-grained features. This hybrid approach enables MAT to effectively model both global and local patterns, making it ideal for evolutionary trend forecasting in complex multivariate time series. Experiments on weather datasets show that MAT outperforms existing Time Series Forecasting models in accuracy, scalability, and memory efficiency. By merging Mamba's efficiency with Transformer's flexibility, MAT offers a scalable and adaptable solution for real-world forecasting, paving the way for further research in hybrid deep learning for sequential data.

Transformer-Modulated Diffusion Model (TMDM integrating Transformer and Diffusion Model) enhances multivariate time series (MTS) forecasting by combining Transformer and Diffusion Models into a unified framework [35]. While Transformer excels in MTS forecasting, they often lack uncertainty estimation, a critical factor for real-world decision-making. TMDM addresses this by using a conditional diffusion process to generate precise distribution forecasts instead of single-point predictions. TMDM employs Transformer to extract key insights from historical data, guiding the Diffusion Model's forward and reverse processes to ensure covariate-dependent and contextually relevant predictions. Unlike traditional Diffusion Models, TMDM integrates structured historical patterns, significantly improving forecasting accuracy. TMDM also introduces two novel metrics for evaluating uncertainty estimation. Experiments on six benchmark datasets show TMDM outperforms existing probabilistic forecasting methods, delivering reliable and interpretable predictions. By merging transformer-based forecasting with diffusion-based generative modeling, TMDM sets a new standard in probabilistic MTS forecasting, offering both accuracy and robust uncertainty quantification.

DiffImp (integrating Mamba and Diffusion Model) is an advanced probabilistic time series imputation model that improves denoising diffusion probabilistic models (DDPMs) by tackling challenges in sequence modeling and inter-variable dependencies [36]. While DDPMs excel at modeling complex distributions, they often struggle with high computational complexity and ineffective handling of bidirectional and inter-variable dependencies. DiffImp addresses these issues by integrating Mamba, a state-space model, as the denoising module for DDPMs. Mamba's near-linear complexity enables efficient scaling for long sequences, improving computational efficiency over Transformer or RNN-based methods. DiffImp also introduces custom SSM-based blocks to enhance bidirectional modeling and inter-variable dependency learning, ensuring robust handling of multivariate time series. Experiments across various datasets, missing scenarios, and ratios show DiffImp achieves state-of-the-art performance in probabilistic imputation. By combining DDPMs with Mamba's efficiency, DiffImp sets a new benchmark for uncertainty-aware time series imputation, offering a scalable and effective solution for missing data.

8 Foundation Models

Foundation models have revolutionized NLP, prompting researchers to explore similar breakthroughs in time series analysis. However, designing a foundational model for time series presents unique challenges due to the fundamental structural differences between text and time series data. Unlike language, where discrete tokens carry semantic meaning, time series data consists of continuous values with strong temporal dependencies. Besides, there are huge differences among time series data in different domains, just like two completely different languages. Researchers have approached those challenges from multiple perspectives, leading to the development of diverse foundation models.

Models derived from Large Language Model: Given the remarkable success of large language models (LLMs) and the fact that both text and time series share a sequential structure, researchers have explored adapting LLMs for time series analysis. This adaptation primarily involves two key modifications. The first is time series tokenization, which aims to assign semantic meaning to time series data by employing patching techniques and integrating additional features to enhance interpretability. The second adjustment addresses distribution drift across different domains, a common challenge in time series forecasting. Researchers have tackled this issue through various normalization strategies, ensuring that models generalize effectively across diverse datasets. These foundational adaptations have led to the development of specialized time series models based on LLM architectures, with notable examples including GPT4TS [37], Time-LLM [38], LLM4TS [39], Tempo [40], PromptCast [41].

Models pre-trained on time series datasets: The biggest challenge in developing such models lies not in the design and optimization of the model architecture, but in the time series dataset itself. Training large-scale foundation models requires vast amounts of high-quality data, similar to how large language models rely on extensive internet text and visual foundation models are trained on datasets like ImageNet. However, time series data is highly diverse and fragmented, spanning different domains, resolutions, and structures. This variability makes it difficult to establish a universal, high-quality time series dataset that effectively captures the full range of temporal patterns. A potential future direction could involve constructing a benchmark dataset, similar to ImageNet, tailored for specific time series domains, such as EEG signals or financial market data. Such foundation models for time series can generally be categorized into two groups. The first consists of models trained on heterogeneous datasets from multiple domains to achieve broad generalization, such as TimesFM [42], TimeGPT [43], and Chronos [44]. The second includes models trained on domain-specific datasets, designed to excel in specialized applications, such as LaBraM [45], BrainLM [46], and NeuroLM [47].

Non-transformer foundation models: Recently, several non-Transformer-based models have been developed to address the unique challenges of time series forecasting. One motivation behind these models is to better align with the intrinsic properties of time series data, such as temporal continuity and structured dependencies over time. Unlike traditional Transformer architectures, which rely on self-attention mechanisms that treat all time steps equally, these models are designed to more effectively capture sequential dynamics and preserve temporal relationships. Another key motivation is efficiency, as Transformer-based models suffer from quadratic complexity when handling long sequences. By adopting alternative architectures, these models aim to achieve near-linear computational complexity, making it possible to efficiently model long-term dependencies while significantly reducing memory and processing costs. Some of these foundation models are built upon Mamba, overcoming the scalability limitations of self-attention [48]. Others take inspiration from alternative methodologies, such as [49]. These emerging models represent a shift toward more specialized architectures that balance accuracy, scalability, and computational efficiency for time series forecasting.

Semi-transformer foundation models: In recent work, researchers have undertaken substantial modifications to the Transformer architecture to better align it with the inherent continuity and causality of time series data. Sundial is a notable example of this evolution [50]. While it retains the overall Transformer structure, its internal components have been redesigned using neural network elements tailored to time series dynamics. Additionally, Sundial incorporates an autoregressive mechanism for forecasting. As a new class of scalable and adaptable time series foundation models, Sundial advances generative forecasting by eliminating reliance on predefined distributions or discrete tokenization. It introduces a novel training objective called TimeFlow Loss, grounded in flow-matching, which enables native pre-training on continuous-valued data and supports the generation of multiple plausible future trajectories without depending on parametric distributional assumptions.

9 Conclusion

In this review, we examined the evolution of Transformer models in the context of time series analysis, highlighting both their potential and their limitations. Unlike in natural language processing, where Transformer architectures have become dominant, their application in time series has been more nuanced. This is due to inherent differences between textual and temporal data—such as continuity, strict sequential dependencies, and redundancy—which challenge the assumptions built into attention mechanisms.

We surveyed a wide range of adaptations to the Transformer framework, including efficient attention variants, hybrid models, and decomposition-based improvements. We also explored alternative architectures such as Mamba and diffusion models, which have shown considerable promise in modeling long-range dependencies with greater efficiency and domain adaptability. In addition, foundation models for time series are emerging, though challenges in data heterogeneity and domain alignment remain significant.

Through the survey, we revealed the critical role of architectural biases—such as recurrence and state-space modeling—in effectively capturing the structure of time series data. Equally important is the need for computationally efficient designs that can scale to long input sequences. We also highlight the growing trend toward pre-training models on domain-specific time series data, which can enhance adaptability and performance in specialized tasks. Looking ahead, we expect future research to focus on building cross-domain foundation models, improving methods for uncertainty quantification, and integrating time-aware mechanisms to better align with the unique properties of temporal data.

This review provides a foundation for researchers and practitioners to navigate the current landscape of models and make informed decisions about model selection, adaptation, and future exploration in time series analysis.

Conflict of Interest Statement All authors declare no conflict of interest.

References

1. Thundiyil, S., Picone, J., & McKenzie, S. (2024). Transformer architectures in time series analysis: A review. Lecture notes (Temple University, Philadelphia, PA, USA). Available: https://isip.piconepress.com/courses/temple/ece_8110/lectures/2024_00_spring/lecture_36a.pdf
2. Vaswani, A., Shazeer, N., Parmar, N., Uszkoreit, J., Jones, L., Gomez, A. N., ... & Polosukhin, I. (2017). Attention is all you need. Advances in neural information processing systems, 30.

3. Zhou, H., Zhang, S., Peng, J., Zhang, S., Li, J., Xiong, H., & Zhang, W. (2021, May). Informer: Beyond efficient transformer for long sequence time-series forecasting. In Proceedings of the AAAI conference on artificial intelligence (Vol. 35, No. 12, pp. 11106–11115).
4. Wu, H., Xu, J., Wang, J., & Long, M. (2021). Autoformer: Decomposition transformer with auto-correlation for long-term series forecasting. Advances in neural information processing systems, 34, 22419-22430.
5. Doshi, K., Abudalou, S., & Yilmaz, Y. (2022). Tisat: Time series anomaly transformer. arXiv preprint arXiv:2203.05167.
6. Wang, C., Chen, Y., Zhang, S., & Zhang, Q. (2022). Stock market index prediction using deep transformer model. Expert Systems with Applications, 208, 118128.
7. Hu, R., Chen, J., & Zhou, L. (2022). A transformer-based deep neural network for arrhythmia detection using continuous ECG signals. Computers in Biology and Medicine, 144, 105325.
8. Shin, Y., Choi, J., Wi, H., & Park, N. (2024, March). An attentive inductive bias for sequential recommendation beyond the self-attention. In Proceedings of the AAAI Conference on Artificial Intelligence (Vol. 38, No. 8, pp. 8984–8992).
9. Nie, Y., Nguyen, N. H., Sinthong, P., & Kalagnanam, J. (2022). A time series is worth 64 words: Long-term forecasting with transformers. arXiv preprint arXiv:2211.14730.
10. Liu, Y., Hu, T., Zhang, H., Wu, H., Wang, S., Ma, L., & Long, M. (2023). itransformer: Inverted transformers are effective for time series forecasting. arXiv preprint arXiv:2310.06625.
11. Liu, Y., Wu, H., Wang, J., & Long, M. (2022). Non-stationary transformers: Exploring the stationarity in time series forecasting. Advances in neural information processing systems, 35, 9881-9893.
12. Wu, H., Hu, T., Liu, Y., Zhou, H., Wang, J., & Long, M. (2022). Timesnet: Temporal 2d-variation modeling for general time series analysis. arXiv preprint arXiv:2210.02186.
13. Woo, G., Liu, C., Sahoo, D., Kumar, A., & Hoi, S. (2022). Etsformer: Exponential smoothing transformers for time-series forecasting. arXiv preprint arXiv:2202.01381.
14. Wang, H., Peng, J., Huang, F., Wang, J., Chen, J., & Xiao, Y. (2023). Micn: Multi-scale local and global context modeling for long-term series forecasting. In The eleventh international conference on learning representations.
15. Wen, X., & Li, W. (2023). Time series prediction based on LSTM-attention-LSTM model. IEEE access, 11, 48322-48331.
16. Jin, M., Koh, H. Y., Wen, Q., Zambon, D., Alippi, C., Webb, G. I., ... & Pan, S. (2024). A survey on graph neural networks for time series: Forecasting, classification, imputation, and anomaly detection. IEEE Transactions on Pattern Analysis and Machine Intelligence.
17. Hou, H., & Yu, F. R. (2024). Rwkv-ts: Beyond traditional recurrent neural network for time series tasks. arXiv preprint arXiv:2401.09093.
18. Taylor, S. J., & Letham, B. (2018). Forecasting at scale. *The American Statistician*, *72*(1), 37-45.
19. Montero-Manso, P., & Hyndman, R. J. (2021). Principles and algorithms for forecasting groups of time series: Locality and globality. *International Journal of Forecasting*, *37*(4), 1632-1653.
20. Salinas, D., Flunkert, V., Gasthaus, J., & Januschowski, T. (2020). DeepAR: Probabilistic forecasting with autoregressive recurrent networks. *International journal of forecasting*, *36*(3), 1181-1191.
21. Gu, A., & Dao, T. (2023). Mamba: Linear-time sequence modeling with selective state spaces. arXiv preprint arXiv:2312.00752.
22. Wang, Z., Kong, F., Feng, S., Wang, M., Yang, X., Zhao, H., Zhang, Y., et al. (2025). Is mamba effective for time series forecasting? Neurocomputing, 619, 129178
23. Patro, B. N., & Agneeswaran, V. S. (2024). Simba: Simplified mamba-based architecture for vision and multivariate time series. arXiv preprint arXiv:2403.15360.
24. Ahamed, M. A., & Cheng, Q. (2024). Timemachine: A time series is worth 4 mambas for long-term forecasting. In ECAI 2024: 27th European Conference on Artificial Intelligence, 19–24 October 2024, Santiago de Compostela, Spain-Including 13th Conference on Prestigious Applications of Intelligent Systems, European Conference on Artificial Intelli (Vol. 392, p. 1688).

25. Yang, L., Zhang, Z., Song, Y., Hong, S., Xu, R., Zhao, Y., Yang, M. H., et al. (2023). Diffusion models: A comprehensive survey of methods and applications. ACM Computing Surveys, 56(4), 1-39
26. Goodfellow, I. J., Pouget-Abadie, J., Mirza, M., Xu, B., Warde-Farley, D., Ozair, S., ... & Bengio, Y. (2014). Generative adversarial nets. *Advances in neural information processing systems, 27.*
27. Kingma, D. P., & Welling, M. (2019). An introduction to variational autoencoders. *Foundations and Trends® in Machine Learning, 12*(4), 307–392.
28. Lin, L., Li, Z., Li, R., Li, X., & Gao, J. (2024). Diffusion models for time-series applications: a survey. Frontiers of Information Technology & Electronic Engineering, 25(1), 19-41.
29. Rasul, K., Seward, C., Schuster, I., & Vollgraf, R. (2021, July). Autoregressive denoising diffusion models for multivariate probabilistic time series forecasting. In International conference on machine learning (pp. 8857–8868). PMLR.
30. Tashiro, Y., Song, J., Song, Y., & Ermon, S. (2021). Csdi: Conditional score-based diffusion models for probabilistic time series imputation. Advances in neural information processing systems, 34, 24804-24816.
31. Yuan, X., & Qiao, Y. (2024). Diffusion-ts: Interpretable diffusion for general time series generation. arXiv preprint arXiv:2403.01742.
32. Li, Y., Lu, X., Wang, Y., & Dou, D. (2022). Generative time series forecasting with diffusion, denoise, and disentanglement. Advances in Neural Information Processing Systems, 35, 23009-23022.
33. Zhang, W., Huang, J., Wang, R., Wei, C., Huang, W., & Qiao, Y. (2024, October). Integration of Mamba and Transformer-MAT for Long-Short Range Time Series Forecasting with Application to Weather Dynamics. In 2024 International Conference on Electrical, Communication and Computer Engineering (ICECCE) (pp. 1–6). IEEE.
34. Xu, X., Chen, C., Liang, Y., Huang, B., Bai, G., Zhao, L., & Shu, K. (2024). SST: Multi-Scale Hybrid Mamba-Transformer Experts for Long-Short Range Time Series Forecasting. arXiv preprint arXiv:2404.14757.
35. Li, Y., Chen, W., Hu, X., Chen, B., & Zhou, M. (2024). Transformer-modulated diffusion models for probabilistic multivariate time series forecasting. In The Twelfth International Conference on Learning Representations.
36. Gao, H., Shen, W., Qiu, X., Xu, R., Hu, J., & Yang, B. (2024). Diffimp: Efficient diffusion model for probabilistic time series imputation with bidirectional mamba backbone. arXiv preprint arXiv:2410.13338.
37. Zhou, T., Niu, P., Sun, L., & Jin, R. (2023). One fits all: Power general time series analysis by pretrained lm. Advances in neural information processing systems, 36, 43322-43355.
38. Jin, M., Wang, S., Ma, L., Chu, Z., Zhang, J. Y., Shi, X., ... & Wen, Q. (2023). Time-llm: Time series forecasting by reprogramming large language models. arXiv preprint arXiv:2310.01728.
39. Chang, C., Peng, W. C., & Chen, T. F. (2023). Llm4ts: Two-stage fine-tuning for time-series forecasting with pre-trained llms. CoRR.
40. Cao, D., Jia, F., Arik, S. O., Pfister, T., Zheng, Y., Ye, W., & Liu, Y. (2023). Tempo: Prompt-based generative pre-trained transformer for time series forecasting. arXiv preprint arXiv:2310. 04948.
41. Xue, H., & Salim, F. D. (2023). Promptcast: A new prompt-based learning paradigm for time series forecasting. IEEE Transactions on Knowledge and Data Engineering, 36(11), 6851-6864.
42. Das, A., Kong, W., Sen, R., & Zhou, Y. (2024, July). A decoder-only foundation model for time-series forecasting. In Forty-first International Conference on Machine Learning.
43. Garza, A., Challu, C., & Mergenthaler-Canseco, M. (2023). TimeGPT-1. arXiv preprint arXiv: 2310.03589.
44. Ansari, A. F., Stella, L., Turkmen, C., Zhang, X., Mercado, P., Shen, H., ... & Wang, Y. B. (2024). Chronos: Learning the language of time series.
45. Jiang, W. B., Zhao, L. M., & Lu, B. L. (2024). Large brain model for learning generic representations with tremendous EEG data in BCI. arXiv preprint arXiv:2405.18765.

46. Caro, J. O., Fonseca, A. H. D. O., Averill, C., Rizvi, S. A., Rosati, M., Cross, J. L., ... & van Dijk, D. (2023). BrainLM: A foundation model for brain activity recordings. bioRxiv, 2023–09.
47. Jiang, W. B., Wang, Y., Lu, B. L., & Li, D. (2024). NeuroLM: A Universal Multi-task Foundation Model for Bridging the Gap between Language and EEG Signals. arXiv preprint arXiv:2409.00101.
48. Bhethanabhotla, S. K., Swelam, O., Siems, J., Salinas, D., & Hutter, F. (2024). Mamba4Cast: Efficient Zero-Shot Time Series Forecasting with State Space Models. arXiv preprint arXiv:2410.09385.
49. Jollie, D., Sun, J., Zhang, Z., & Schaeffer, H. (2024). Time-series forecasting, knowledge distillation, and refinement within a multimodal pde foundation model. arXiv preprint arXiv:2409.11609.
50. Liu, Y., Qin, G., Shi, Z., Chen, Z., Yang, C., Huang, X., ... & Long, M. (2025). Sundial: A Family of Highly Capable Time Series Foundation Models. arXiv preprint arXiv:2502.00816.

Enabling Microsegmentation: Digital Pathology Corpora for Advanced Model Development

Dmitry Hackel, Maria Bagritsevich, Claudia Dumitrescu, Md. Abdullah Al Mamun, Sadia Purba, Dylan Heathcote, Iyad Obeid, and Joseph Picone

Abstract Pathology remains a foundational discipline in modern medicine, yet traditional microscopy workflows are constrained by manual slide preparation, subjective interpretation, and limited scalability. The advent of digital whole slide imaging has enabled seamless digitization of histological sections, though challenges such as inter-observer variability and workforce shortages persist. Recently, artificial intelligence-based decision-support tools—powered by deep-learning algorithms and high-performance computing—have demonstrated remarkable performance in pre-screening, case triage, and morphology segmentation. Artificial intelligence in diagnostic histopathology hinges on the availability of extensive, richly annotated image repositories. Therefore, in this chapter we introduce two important open source digital pathology corpora: the Temple University Hospital Digital Pathology Breast Tissue Subset (TUBR) and the Fox Chase Cancer Center Digital Pathology Breast Tissue Subset (FCBR). TUBR comprises 3,505 images from 296

D. Hackel · M. Bagritsevich · C. Dumitrescu · Md. A. Al Mamun · S. Purba · D. Heathcote · I. Obeid · J. Picone (✉)
The Neural Engineering Data Consortium, Temple University, Philadelphia, PA, USA
e-mail: picone@temple.edu

D. Hackel
e-mail: dmitry.hackel@temple.edu

M. Bagritsevich
e-mail: maria.bagritsevich@temple.edu

C. Dumitrescu
e-mail: claudia-anca.dumitrescu@temple.edu

Md. A. Al Mamun
e-mail: abdullah18@temple.edu

S. Purba
e-mail: sadia.afrin.purba@temple.edu

D. Heathcote
e-mail: dylan.heathcote@temple.edu

I. Obeid
e-mail: iobeid@temple.edu

A. Ahmed and J. Picone (eds.), *Signal Processing and Biomedical Engineering Research*, https://doi.org/10.1007/978-3-032-18509-9_6

patients (1.2 TB of data). FCBR contributes 1,463 oncology-focused slides from 1,397 patients (336 GB of data). Both corpora have been manually annotated and cover a spectrum of tissue states from normal and benign through pre-malignant intraepithelial lesions and invasive carcinomas. We also explore baseline performance of some well-established deep learning algorithms on an image classification task based on these corpora. Weighted DICE scores of 51.10% and 46.02% were achieved on TUBR and FCBR, respectively, using EfficientNet B0. We examine the challenges involved in training across both datasets to enhance overall system performance and assess the role accurate microsegmentations play in achieving clinically acceptable performance.

Keywords Histopathology · Whole-slide imaging · Artificial intelligence · Deep learning

1 Introduction

Pathology constitutes a cornerstone of modern medicine, providing essential support for accurate diagnosis, prognostication, and therapeutic planning across a wide range of tissue-based diseases. Traditionally, examination required a tedious workflow of preparing thin tissue sections on glass slides, followed by manual interpretation of stained tissue morphology under a microscope—an effort heavily reliant on the pathologist's expertise to discern subtle architectural, spatial, and cellular features. In support of enhancing this process, digital whole slide imaging, which emerged in the mid-1990s [1], enabled entire histological slides to be digitized into multi-resolution image pyramids. This allowed seamless, lossless zoom at arbitrary magnifications—unlike optical microscopes, which are limited to discrete objective lenses. This ensured consistent calibration, precise quantitative measurements, and preservation of cellular architecture across the specimen.

However, despite its potential, whole slide imaging did not mitigate inter-observer variability, a challenge that spans virtually all pathological subspecialties. A recent systematic review of 12 validation studies [2] reported that digital vs. light microscopy inter-observer κ values ranged from 0.45 ("moderate" agreement) to 0.75 ("substantial" agreement) across human pathology cases. More concerning was the fact that only 52% of pathologists reported being satisfied with their career choice [3], and the U.S. pathology workforce declined by 17.5% between 2007 and 2017 [4]. This decrease further worsens job satisfaction as remaining pathologists are being forced to increase workload, a difficult task to implement on a job requiring such precise analysis for each sample. To mitigate the negative impacts of the dwindling workforce, this chapter explores strategies for streamlining sample evaluation by equipping pathologists with advanced artificial intelligence (AI) based slide analysis tools.

Among the drivers of job dissatisfaction, it is worth mentioning that the declining headcounts, an aging workforce, and growing case complexity [5, 6], as well as insufficient exposure to pathology during undergraduate and medical education, have been shown to deter up to 40% of students from considering the specialty [7]. In response, AI-based decision-support tools have begun to fill critical gaps. In 2021, the FDA granted de novo authorization to Paige Prostate, the first AI-based in vitro diagnostic device for prostate cancer detection on digitized slides [8]. By leveraging advances in high-performance computing, gigapixel image storage, and deep-learning algorithms, such systems can pre-screen whole slide images (WSI), triage cases by risk, and highlight regions of interest—thereby reducing pathologist workload and enhancing diagnostic consistency beyond what whole slide imaging alone could achieve [9].

Conventional microscopy is a current standard within pathology. According to a recent study [10], 65% of pathologists remain reluctant to use the digital alternative, considering it unsuitable for routine diagnostic practice. Analog microscopy has a low to medium processing speed due to the need for manual transition of samples, which can lead to increased processing times that vary from a few minutes to hours [11]. In addition to eye strain, the need to be in a laboratory setting, especially when on call, and the increasing cost of manual labor, has fueled the push toward digitizing pathology. Digitized slides can be accessed and analyzed beyond the confines of the laboratory, offering significant advantages for pathologists on call, particularly in time-sensitive scenarios such as organ transplantation, where rapid remote assessment is critical.

The growing adoption of digital pathology within the field has provided a wealth of data that can be reviewed and utilized to further research into machine learning (ML) and deep learning (DL) technology. This technology has the potential to aid a pathologist's workflow. Such assistive resources enable automated slide triage (ordering cases by suspected severity), generation of annotated heatmaps highlighting regions of interest, and quantitative extraction of morphological features. Indeed, many studies report "near-perfect" performance: a recent meta-analysis of 48 AI algorithms across diverse disease types found a pooled sensitivity of 96.3% and specificity of 93.3% [12].

While the introduction of DL in digital pathology has demonstrated remarkable performance—achieving mean sensitivities of 95% and specificity of 92% for multiclass tasks—this often stems from evaluation of highly curated datasets with limited diversity. These datasets are typically collected from one or two institutions, which raises concerns about bias, lack of generalizability, and reproducibility. Slide preparation techniques vary by institution, posing serious challenges for DL technology that can easily overtrain. As the same review [12] reports, 99% of included studies had at least one domain at high or unclear risk of bias, often due to non-random sampling, inadequate reporting, and domain shifts across institutions or scanners. This makes it difficult to translate AI models into robust clinical tools across different healthcare settings. Hence, in this chapter, we focus on establishing a performance baseline using our recently released datasets from two distinct medical centers, explicitly addressing these limitations through cross-institutional evaluation and transparent reporting practices.

2 Digital Pathology Resources for Machine Learning

To develop robust deep learning models for digital pathology, a well-curated dataset encompassing a variety of normal, benign, and cancerous tissue architectures is essential. At Temple University's Neural Engineering Data Consortium (NEDC), two major open source digital pathology corpora have been established: the Temple University Hospital (TUH) Digital Pathology Corpus (TUDP) [13–15] and the Fox Chase Cancer Center (FCCC) Digital Pathology Corpus (FCDP) [16]. As the names suggest, TUDP was collected at Temple Hospital in Philadelphia—the major public hospital for the city. FCDP was collected at FCCC—a world-recognized leader in cancer research and treatment. TUDP contains over 100,000 high-resolution images, representing a diverse range of tissue types, including both normal and abnormal specimens, supporting a broad spectrum of research applications. In contrast, FCDP, with 14,276 images, is more specifically focused on oncological studies. These corpora have been further refined through extensive evaluation and annotation, particularly the breast pathology subsets, ensuring precise classification of cancerous, non-cancerous, and pre-cancerous structures. A comparison of these datasets to other publicly available datasets is given in Table 1. These enhanced datasets provide a critical foundation for training advanced AI models in digital pathology, facilitating improved diagnostic accuracy and research in cancer detection.

2.1 *Slide Preparation*

Traditional histological slide preparation [17] requires tissue sections to be thin enough to be mounted while maintaining the structural relationship between cells and extracellular components. This necessitates careful handling to ensure an accurate cancer diagnosis. To support tissue integrity during sectioning, specimens are either frozen and cut using a cryostat microtome for rapid diagnosis without chemical interference or infiltrated with a liquid agent such as epoxy resin or histological wax, which subsequently solidifies to encapsulate the sample. Before mounting, tissue specimens must be preserved using a fixative, commonly a formaldehyde suspension for 6–12 h, to prevent decomposition and eliminate microorganisms. Specimens are then dissected into 4 mm regions and processed with a hardening agent. Once prepared, paraffin sections of 3–5 μm thickness are cut to ensure only a single layer of cells is captured on a glass slide. These sections are then placed in a warm water bath for flattening and ease of handling before being mounted onto individual glass slides.

Table 1 An overview of publicly available breast histological H&E corpora

Dataset	Size (GB)	Subjects	Annotations	No samples	Data origin	Pixel size (μm/pixel)
ACROBAT	1164	1152	Invasive Cancer, Non-Malignant, Artefacts, Ductal Carcinoma, Lobular Carcinoma, Normal	4212 WSIs	Karolinska Institutet (SE)	0.91
BACH/ICIAR 2018	13	–	Normal, Benign, InSitu, Invasive	400 slides 30 WSIs	Ipatimup and INEB (PT)	0.50
BCNB	33	1058	Coordinates of Tumor	1058 WSIs	Beijing University of Posts and Telecommunications (CN)	–
BRACS	1100	189	Benign, Ductal Hyperplasia (2), Flat Epithelial Ductal Carcinoma, Invasive Carcinoma, Normal	547 WSIs	Institute for High Performance Computing and Networking (IT)	0.25
BreaKHist	2	82	4 Benign, 4 Malignant subtypes	7909 slides	P&D Lab, Federal University of Parana (Brazil)	multiple
CAMELYON16	1160	399	Tumor, Normal	400 WSI	Radboud University Medical Center, UMC Utrecht (NL)	0.24
CAMELYON17	2950	200	Negative, ITC, Micro-metastasis, Macro-metastasis	1399 WSIs	Radboud University Medical Center, UMC Utrecht (NL)	0.24
CPTAC-BRCA	113	134	Cancer, Normal	642 WSIs	Clinical Proteomic Tumor Analysis Consortium (USA)	0.25 0.50
GTEx-breast	80	894	Pathology (unlocated)	894 WSIs	Broad Institute of MIT and Harvard (USA)	0.50
HER2 / Warwick	20	86	HER2 Score	172 WSIs	University of Warwick, University of Nottingham, and AIDPATH consortium (UK)	0.23

(continued)

Table 1 (continued)

Dataset	Size (GB)	Subjects	Annotations	No samples	Data origin	Pixel size (µm/pixel)
HEROHE	820	360	Binary	500 WSIs	Institute of Research and Innovation in Health from Porto (PT)	0.24
SPIDER	85		18 morphologies	1925 slides	HistAI initiative	0.40
TCGA-BRCA	1640	1098	Nuclei segmentations	3111 WSIs	The Cancer Genome Atlas (USA)	0.25
TIGER	169	370	Tumor, Stroma, Lymphocytes, Necrosis	370 WSIs	Radboud University Medical Center of Nijmegen (NL), Jules Bordet Institut (BE) and TCGA-BRCA	0.50
TUBR	**1226**	**296**	**Artifact, Background, Ductal Carcinoma, Invasive Ductal Carcinoma, Inflammation, Non-neoplastic, Normal, Null, Suspicious**	**3505**	**Temple University Hospital (USA)**	**0.20**
FCBR	**336**	**1397**		**1463**	**Fox Chase Cancer Center (USA)**	**0.20**

2.2 *Immunostaining*

Immunostaining (e.g., staining) is essential for visibility during imaging [18]. Cells, apart from a few natural pigments like melanin, are mostly colorless. The routine hematoxylin and eosin (H&E) stain is used to provide general cellular structure information, staining nuclei blue and other cellular components in shades of pink. If further diagnostic information is required, special stains targeting specific cellular components are applied. For example, immunohistochemical (IHC) stains utilize antibodies to mark diagnostically relevant proteins, offering deeper insights into cellular processes. Once stained, the slides are covered with glass. Most slides can be archived for years without damage or discoloration.

Immunostaining plays a critical role in the annotation of digital pathology datasets for machine learning applications, as it can shift the distribution of the data, thereby making the model fail the generalization test. This is because different staining protocols highlight different cellular and molecular features while failing to reveal others, making cross-dataset transferability challenging. As a result, switching between datasets becomes more complicated due to variation in immunochemistry. In digital pathology, immunohistochemical (IHC) stains utilize antibodies to detect specific cellular markers, such as estrogen receptor (ER), progesterone receptor (PR), and human epidermal growth factor receptor 2 (HER2), which are essential in breast cancer classification [19]. These stains provide a detailed molecular profile of tissues, aiding in the differentiation of neoplastic from non-neoplastic regions, which is crucial for training deep learning models to recognize cancerous patterns with high specificity.

In contrast, Hematoxylin and Eosin (H&E) staining remains the gold standard for histological examination, offering a broad visualization of cellular and tissue structures. Hematoxylin stains cell nuclei blue-purple, while eosin highlights cytoplasm and extracellular matrix in pink, allowing machine learning models to extract key morphological features that distinguish normal, pre-cancerous, and malignant cells [20]. The integration of both staining techniques enhances model performance by expanding the diversity of input data, enabling algorithms to learn from both molecular and structural tissue characteristics. Moreover, computational pathology tools leveraging deep learning can process large-scale immunostained datasets efficiently, reducing interobserver variability and improving diagnostic consistency [20]. By incorporating a variety of immunostains, machine learning models gain a deeper contextual understanding of tissue pathology, facilitating advancements in automated cancer detection, classification, and prognostic assessment.

2.3 *Digitization*

Digital pathology involves converting physical glass slide samples into digital images stored using advanced compression techniques. This is achieved using a laser scanner that captures images at a resolution of 0.2 μm/pixel. Similar to a glass slide, a WSI contains one or more specimens along with the background space surrounding the histological sample. All scanning is performed using a Leica Biosystems Aperio AT2 high-volume scanner. The scanning process is described in detail in [13]. The scanner captures a low-resolution image and assigns multiple focus points on the tissue. A green rectangular box marks the scanning frame, and technicians can manually adjust the focus points and frame as needed. This manual review step is crucial in preventing scanning failures caused by inaccurate automatic focus placement.

There are cases where the scanner makes errors in its placement of the focus points and in determining the area of the scanning region. Both cases will cause a failure in the image processing of the specimen, thus causing a failure in scanning. Another error modality occurs when the scanning region is too large, which can cause unnecessary white space to be included in the WSI. In this case, the scanner might produce an image that is larger than necessary. Similarly, there is a risk of producing invalid images if the scanning region only partially encloses the specimen of interest. These events tend to occur with slides that are lightly stained or slides with

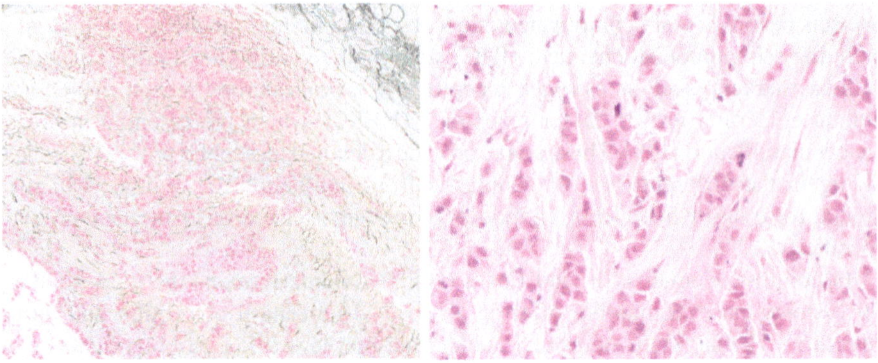

Fig. 1 Examples of a poor quality slide (left) and a high-quality slide (right) of the same structure

a significant amount of white space between tissue samples. In such cases, manual adjustment of focus points becomes required.

Errors caused by image blur frequently result in resolution issues within some pathology slides. Image blur is a result of the attenuation of high spatial frequencies which occur when the image is compressed or filtered. Blur can also be a result of image acquisition [21]. This makes the pre-scan snapshot phase of the scanning procedure labor intensive. We find that about 2% of the slides scanned are likely to experience a scanning failure. However, this number varies according to the type of stain applied to the slides. These failed slides were reviewed, readjusted, and rescanned. To date, we have scanned approximately 120,000 slides using these tools. An example of poor image quality is shown in Fig. 1.

The scanned image produced from the Aperio AT2 is stored using a ScanScope Virtual Slide (SVS) format [22]. An SVS file is a layered image representation that includes several thumbnails and the original source image. The source image is stored using JPEG compression with a quality factor of 70. (The specific image type in the Aperio ImageScope software is "SVS/JPEG 2". The parameter "Image Depth" is set to 1 and "Image Channels" is set to 3.) These parameters result in roughly an order of magnitude of compression over lossless compression with minimal image degradation.

A full-resolution image, or WSI, is stored as the baseline image using a tile size of 240×240 pixels (an image is represented as a series of adjacent tiles, or blocks). The following three layers contain downsampled versions of the image at resolutions of 4:1, 16:1, and 32:1. The final layer is a low-resolution thumbnail. Each of these layers is an image encoded using lossy JPEG encoding. An SVS file also contains a low-resolution picture of the slide's label as metadata and stores other information, such as the downsample and offset information. The number of layers generated depends on the size of the original image. Smaller images (e.g., the maximum dimension is less than 50K pixels) will generally have only two layers.

Our primary software for viewing SVS files is Aperio ImageScope [23]. This open source software features a wide variety of tools for image editing, adjusting, and annotation. The image adjustment tools include brightness and contrast controls, color balance, and color curve adjustment. These can be applied to all channels or to individual red, green, or blue (RGB) channels. These adjustments only apply to the viewed image and do not modify or overwrite the stored image. The settings applied to the current session can be saved and applied to other scanned images. The default presets for ImageScope are always applied to the scanned images. Image adjustment tools can then be utilized to calibrate the image features according to the pathologist's preferences. The stains applied on specimen images cause a specific structure to adopt a distinct color, and this color can be enhanced by adjusting the brightness, contrast, and the color channels of the image via the image adjustment tools. This is a beneficial tool for pathology diagnosis as it can be used to allow specific areas of investigation (such as cancerous tissue) to be more focused or to enhance the quality of lightly stained specimens.

As mentioned earlier, the Aperio AT2 scanner features a z-stacking option. The scanner can produce multiple images of a slide tissue that were scanned at different focal depths. This generates a 3D image that allows navigation of the image through different focal depths, which is analogous to the process pathologists use with an analog microscope. ImageScope features a tool that can adjust the focal depth that is similar to using the objective fine and coarse adjustments of a focus slider in a microscope [23]. This feature has yet to be explored in our work but is being used by other hospitals. The z-stacked images are very large in size, often several gigabytes, which poses additional challenges for machine learning research [13].

2.4 Data Organization and Anonymization

A detailed description of the file-naming conventions and organization of the data can be found in [15]. The file-naming convention is designed so that every file in the corpus has a unique filename, and simple UNIX commands can be used to locate data.

To ensure compliance with the Health Insurance Portability and Accountability Act (HIPAA), strict protocols are in place to maintain patient anonymity. Protocol No. 24943, approved by TU's Institutional Review Board, governs the handling of all data to prevent any exposure of patient-identifiable information. The deidentification process follows a methodology similar to the one used in the Temple University Hospital EEG Seizure Corpus [24]. Each patient is assigned a unique, randomized 8-character letter sequence.

SVS files initially contain low-resolution tiled images of slide labels, which may include patient initials and specimen IDs. These labels are manually removed before data release. Together, these measures ensure that all released data is completely anonymized while maintaining the integrity required for research.

2.5 Annotation Guidelines

Pathology is the primary method for the confirmation of breast cancer, assessing tumor margins, and evaluating cancer proliferation. Assessing these metrics can be challenging due to the diverse nature of pathology. The variety of different tissue architectures a pathologist needs to assess is vast, which makes our dataset extremely valuable among the existing alternatives, as shown in Table 1. Annotation refers to the manual delineation of both biological and non-biological structures by an annotator. We refer to these as patch-level annotations since they include a polygon defining the region and a label. Our annotation methodology is discussed in detail in [16, 25]. Here we summarize some key issues that make our annotations unique and extremely valuable for machine learning.

Annotation labels need to ensure precise margins to minimize extraneous structures such as background fibrous or adipose tissue. Proper margins or delineation of structures of interest are important for models that will be trained to do fine-grained segmentation image analysis, which we refer to as microsegmentation. The ability to delineate differences on a cellular level is incredibly important in pathology. Accurate delineation ensures that the model learns to distinguish the true boundaries of pathological structures, which is critical for reliable feature extraction, tissue classification, and overall model performance.

Annotation of an entire slide is extremely time-consuming and is prohibitively expensive. Therefore, images are *partially annotated*, focusing on a few instances of what we consider prominent or significant structures. These structures are either selected based on patient reports, which identify regions of diagnostic interest, or on an annotator's decision of what would be most valuable for ML training. Selective annotation ensures efficient use of time and resources while still capturing clinically relevant tissue architecture.

Areas of interest in pathological slides can vary significantly. Although it would be simple to label tissue as either normal or cancerous, the vast array of different structures poses a real issue for models. In addition, there are many non-cancerous or pre-cancerous structures that would inform a pathologist about making medical determinations. We use a system that includes five labels for relevant pathologies and four labels for non-cancerous or anomalous tissues. The full list of labels can be found in [15] and is repeated here in Table 2. We briefly describe these labels below.

Normal ductal terminal units and lobules are labeled with the symbol (NORM), which is crucial for training models to allow for the differentiation between normal ducts and pathological ones. One of the many things that makes this type of problem challenging is that a small percentage (less than 1%) of the overall slide area contains a cell or group of cells that represent areas of interest. Further, there can be more than one tissue on each slide, and each of these tissue samples can have multiple labels. Hence, the ability to segment images and localize where events of interest occur in the data is critical. Because of the huge imbalance in the frequency of occurrence of the labels in Table 2, care must be taken during training to avoid models from focusing on normal tissue or white backgrounds.

Table 2 Labels for the TUH and FCCC breast tissue corpora

Label	Description/features
Normal (NORM)	Normal ducts and lobules
Ductal Carcinoma in Situ (DCIS)	Ductal carcinoma in situ, and lobular carcinoma in situ
Invasive Ductal Carcinoma (INDC)	Invasive ductal carcinoma, invasive lobular carcinoma, and invasive mammary carcinoma
Non-Neoplastic (NNEO)	Fibrosis, hyperplasia, intraductal papilloma, adenosis, ectasia, etc
Inflammation (INFL)	Areas of inflammation
Artifact (ARTF)	Grease pen marks, stitches, foreign bodies, etc
Indistinguishable (NULL)	Indistinguishable tissue, normally due to issues with the cut/stain
Suspected (SUSP)	Regions that are at risk of developing into cancerous regions
Background (BCKG)	Stroma, no ducts or lobules

2.5.1 An Overview of Anatomical and Physiological Considerations

When we discuss how we annotate the data, it is important to understand the basis of the anatomical and physiological components of breast tissue and how cancerous cells develop [26, 27]. The mammary gland parenchyma consists primarily of adipose stroma within a connective tissue matrix enriched with interstitial fluid. It contains lobular alveoli responsible for milk synthesis, which is transported through lactiferous ducts that converge and open at the nipple. The terminal ductal-lobular units (TDLUs) are a primary interest for pathologists as most primary breast cancers originate in these regions. Lobules, derived from terminal duct lobular units (TDLUs), are hormonally responsive structures composed of glandular epithelial cells encircled by myoepithelial cells, a basement membrane, and fibrocollagenous stroma. These lobules are organized in clusters and play a crucial role in milk secretion.

The epithelial–stromal interfaces within the lobules are composed of an epithelial–myoepithelial layer supported by elastic fibers, fibroblasts, and capillaries, with fiber deposition varying according to hormonal status and age, thereby adding complexity to the interpretation process. The hormonal impacts on the ductal units due to menopause, adolescence, or pregnancy mean a far greater variability in lobular appearance. Standard acini and ducts are made up of three layers: the basement membrane, myoepithelial layer, and epithelial lining. These ducts carry milk and are found throughout the TDLUs. The epithelial layer is usually only one thick cell which lines the ducts. However, multiple layers of cells, hyperplasia, are a common occurrence. The degree or rate of growth of these epithelial cells determines if a patient has atypical ductal hyperplasia or ductal carcinoma in situ.

A significant contributor to the complexity of breast tissue analysis, particularly in pathological interpretation, is a dynamic and differing morphological variability of terminal ductal units driven by hormonal influences. Unlike other tissue architecture which may remain uniform, the breast undergoes structural changes depending on factors such as adolescence, menstrual cycling, pregnancy, lactation, and menopause. These states change the proliferative activity and differentiation status of epithelial and stromal components within the terminal duct units. For example, during pregnancy and lactation, lobules become hypertrophic and exhibit increased acinar formation with prominent secretory activity. This makes the ducts look atypical and could be mistaken for non-neoplastic. Women who are going through menopause have reduced lobular units, atrophic epithelial lining, and increased stromal fibrosis or adipose replacement. These changes could pose a challenge for the development of models that can accurately analyze images for women of all ages and hormone levels.

Morphological variability in pathology is also shaped by genetic and ancestral factors, contributing to disparities in breast cancer characteristics observed within different racial populations. It has been well established that the mutations in BRCA1/BRCA2 genes significantly elevate the risk of developing ductal and invasive carcinoma. Individuals of African American descent tend to develop higher-grade tumors, more frequently present with basal-like or triple-negative subtypes, and are often diagnosed at a younger age. Race can also play a role in the tissue architecture of the terminal ductal units, meaning that the structure and image analysis of breast tissue specimens can have the same diagnosis, but may look different depending on a multitude of factors. This highlights the importance of data curation and diversity of data when developing such models. It is therefore an ethical concern to design models that uphold fairness, transparency, robustness, and ensure a high degree of explainability and privacy.

2.5.2 Cancerous Structures

The two primary labels utilized for cancerous structures are ductal carcinoma in situ (DCIS) and invasive ductal carcinoma (INDC). Although there is some differentiation between what constitutes ductal carcinoma among pathologists, cross-referencing with patient reports and metadata files ensures consistency and accuracy in labeling. DCIS is defined by abnormal growth of epithelial cells throughout the duct. DCIS presents itself in a multitude of ways. Cribriform formations look like sponges as microcysts from within the tumor inside the duct. Micropapillary DCIS is defined by its microcalcification on mammograms and ranges in size. Other common formations of DCIS include apocrine, comedo, with comedonecrosis, or papillary DCIS. All forms are characterized by abnormal cells confined within the ductal units of the breast, as cancer tends to proliferate along the path of least resistance. Ductal carcinoma in situ (DCIS) is typically graded on a scale of 1–3, and in some cases 1–5, based on the mitotic activity (the rate at which cells are dividing and creating new cells).

Once DCIS breaches the basement membrane of the ductal epithelium and invades the surrounding stromal tissue, it is classified as INDC. The INDC label makes up 70% to 80% of breast cancer cases [28]. The various presentations of invasive ductal carcinoma patterns include nests, cords, or sheets of small squamous cells. The tumor cells often exhibit moderate to marked nuclear pleomorphism and increased mitotic activity [29]. Tubule formation is a key histological feature assessed during grading. Tumors with more than 75% tubule formation receive a lower grade, indicating better differentiation and prognosis. However, many INDC occurrences display less tubule formation, correlating with higher grades and more aggressive behavior. The heterogeneity in the histological presentation of INDC poses challenges for AI models in accurately classifying and predicting outcomes, especially when rare subtypes or poorly differentiated forms are underrepresented in training datasets.

Non-neoplastic (NNEO) structures are a broad term that, in a clinical context, is nonspecific and encompasses a wide range of structural types. In TUBR and FCBR, NNEO structures refer to a range of structures that may either present themselves as a benign formation, pre-cancerous, or associated with cancer. The term non-neoplastic can also encompass lesions with pre-cancerous potential, such as atypical ductal hyperplasia, as well as benign structures like cysts that may mimic pathological features.

A NNEO structure refers to the formation of calcium deposits, often detectable on mammograms. These deposits can result from abnormal cellular activity, including cancer proliferation, which increases calcium output that may accumulate and form solid structures [30]. These calcium microcalcifications are likely a byproduct of dysregulated intracellular calcium transport channels due to oncological transformations. The regulatory imbalance, induced by oncological cells, causes an overexpression of calcium pumps such as SPCA2 and PMCA2 [30], creating an environment that allows calcium to react with bone matrix proteins and phosphate transporters to develop a microenvironment conducive to mineralization. This results in microcalcification [31]. The loss of calcium homeostasis and resulting structures could be leveraged to improve model training. We are exploring this in ongoing research and have released a corpus to support this research [32].

NNEO labels include benign proliferative or reactive changes that could mimic cancers, improving model specificity. Inflammation labels (INFL) are given to areas of lymphocyte reactions to either non-neoplastic or cancerous structures. The distinction between these two labels allows for the distinction between an immune response and possible malignant progression.

2.5.3 Non-cancerous and Ambiguous Structures

Artifact (ARTF) refers to the identification of grease-labels or other non-biological artifacts found within slides and serves to prevent misclassification. These artifacts, if not properly labeled, could be misinterpreted by the model as meaningful biological signals, potentially leading to false positives.

Null (NULL) labels are given to areas that are considered undistinguishable tissue, either due to a blurry image or poor stain techniques. These regions are intentionally excluded from training to prevent introducing uncertainty or noise to the model.

The suspicious (SUSP) label is given to tissues that are suspected to be precancerous or are incredibly ambiguous. Background labels (BCKG) serve to prevent models from learning patterns in irrelevant regions. Background labels are given to any structure that does not follow in the prior labeling selection and primarily is made up of adipose and fibrous tissue. Since a large percentage of the area of a slide is non-tissue (white), we provide a BCKG label so ML systems can be trained on tissue patterns that are not one of our five significant labels. Without this label, an ML system will tend toward modeling a white image as background and will confuse images of irrelevant tissue as one of the meaningful five classes.

Maintaining balanced label distributions is essential to mitigate bias and improve model robustness. Although randomization and weighted labels can be utilized to mitigate label biases, the importance of ensuring a good label distribution is vital. Class imbalance poses a significant challenge in medical datasets, especially when tasked with identifying uncommon tissue architectures. In digital pathology databases, the abundance of normal and background tissue is significantly more common than carcinogenic or non-neoplastic structures. If the annotation process is not meticulous, a model would be disproportionately accurate on normal tissue architecture while severely underperforming on cancerous structures. This overfitting of dominant labels can lead to models effectively ignoring minority classes. To address this issue, curating these datasets with sufficient representation of all relevant classes has been an important annotation objective, in addition to adding weighted loss functions and randomization during training.

2.6 *Typical Examples of Annotated Slides*

In Fig. 2, we show normal ducts and normal lobules, components of a TDLU, surrounded by stromal background tissue. The stroma is the pink-stained tissue found throughout the slides, which functions as supportive tissue around the ducts and lobules. Figure 2 illustrates fibrous stroma surrounding the regions of interest, with the inclusion of a few fat cells, which appear as white round shapes. The NORM label is applied to regions where lobules remain circular with no instances of hyperplasia and where the ducts are clearly defined with both layers of the lining intact. The lobules around the duct are composed of small glands that are called acini, which are formed during puberty and represent secretory units of the breast. The acini and the ducts are lined by two different cell layers. The outer layer is made up of myoepithelial cells that are mostly just clear cytoplasm and function to contract and push milk through the breasts. The inner layer is made up of epithelial cells that produce breast milk and are usually where abnormal cell growth develops. The stroma consists of connective tissue, blood vessels, and fat cells. The difference between fatty stroma and fibrous stroma can be seen in Fig. 3.

NNEO annotations encompass a range of structures that aid in contextualizing the interpretation of specific tissue architecture. The NNEO label typically includes entities such as atypical hyperplasia, intraductal papilloma, calcifications, benign

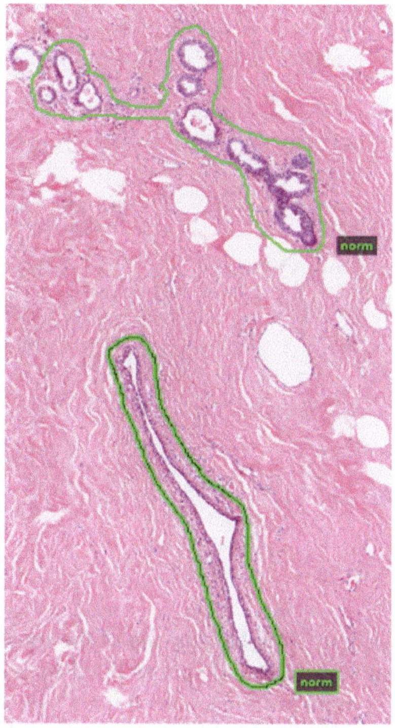

Fig. 2 A normal terminal duct unit surrounded by healthy fibrous stroma

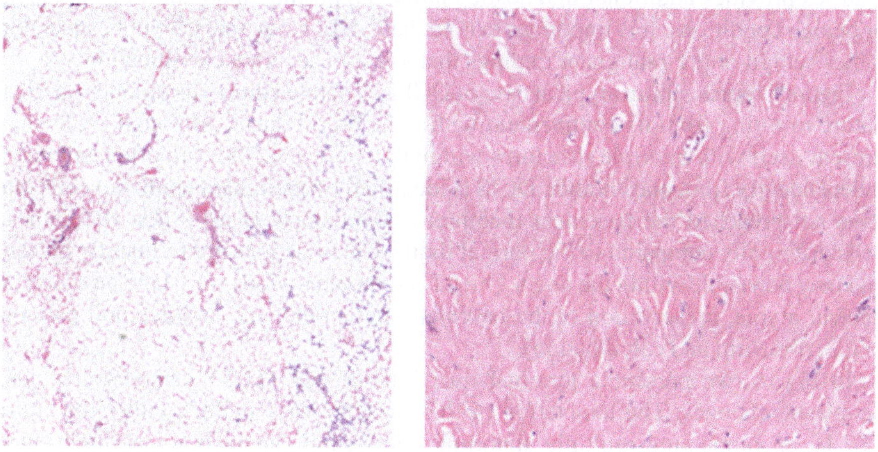

Fig. 3 Examples of fatty stroma (left) and fibrous stroma (right)

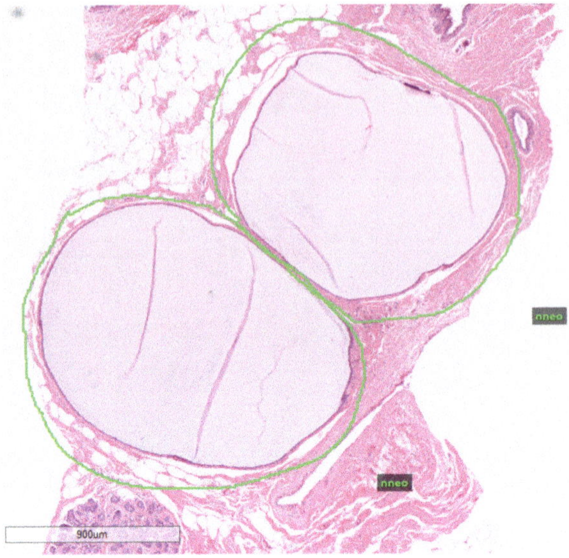

Fig. 4 Microcysts annotated as NNEO

structures, and adenosis. Figure 4 highlights one of the most common NNEO structures: cysts. Cysts present as pale round spaces lined by a single layer of cuboidal epithelium and typically contain proteinaceous fluid. Cysts and cancer often coexist, as tumor growth can obstruct normal ductal drainage pathways and is frequently influenced by hormonal activity. The presence of cysts and other non-neoplastic features provides additional diagnostic clarity, supporting annotation of malignant or benign conditions. It is important to note that NNEO structures are not cancer, but rather benign or pre-malignant findings that assist in accurate classification.

Figure 5 represents an atypical case of DCIS, annotated due to its unique spherical nature resulting from its confinement within the breast ducts. Although DCIS may present itself with multiple structures within the same area, they are often annotated independently from one another. DCIS is identified by malignant epithelial cells that have not breached the basement membrane. Immunochemistry often stains these cells differently compared to the normal tissue. Calcifications are frequently present and may be visible within the ducts. Ductal carcinoma regions were labeled based on nuclear atypia, preservation of the basement membrane, and architectural patterns (e.g., cribriform and calcification).

In Fig. 6, we present a terminal duct lobular unit (TDLU) undergoing non-neoplastic changes. The annotation of this structure may include either the entire TDLU, encompassing multiple abnormal ducts simultaneously, or a single duct. This annotation approach is consistent across many labels, contributing to a more diverse and representative dataset for model training.

To aid in this process, it is important to understand the characteristic non-neoplastic changes such as gynecomastia. Also known as gynecomastoid hyperplasia or gynecomastia, this condition is characterized by two cell-layer linings instead of

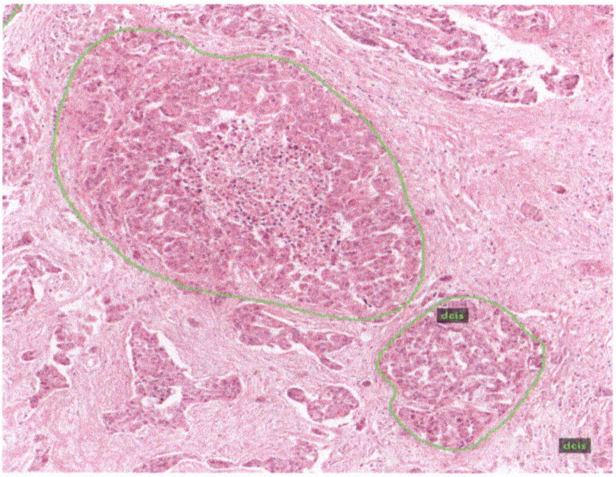

Fig. 5 A typical DCIS presentation of malignant epithelial cells confined to the ductal-lobular system

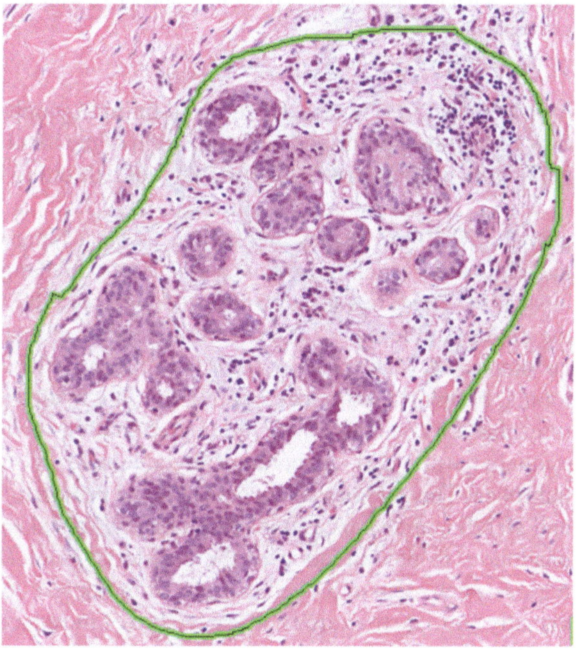

Fig. 6 A TDLU undergoing non-neoplastic changes (gynecomastoid hyperplasia and benign proliferation)

one within the epithelial cell wall of the lobule or duct. Those formations are harder to detect since they require high magnification down to the cell layer of the lobule. The cell can appear mushroom-like if the bi-layer juts into the lumen, where it can become narrower toward the center of the lumen. Hyperplastic changes arise when the epithelial lining proliferates, encroaching on the lumen and leaving minimal or no empty space within. An additional feature of hyperplasia is represented by mushroom-like, irregular expansions of the cell. It can be classified as elongated, ovoid-shaped nuclei

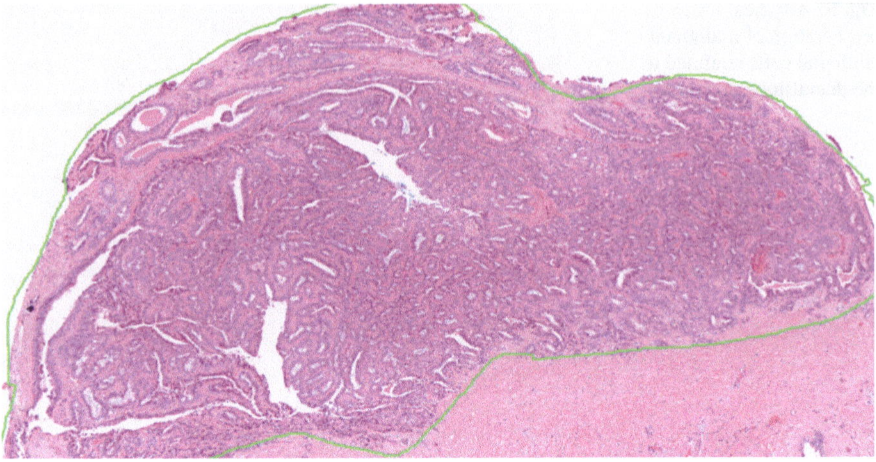

Fig. 7 A benign intraductal papilloma is annotated with an NNEO label that encompasses the entire papilloma

(dot-like particles) within the cell wall lining. Lymphocytes appearing as small dark dots can be seen, pointing to inflammation of the lobule. These visual markers guided annotators in consistently labeling regions exhibiting hyperplasia and inflammation.

In Fig. 7, we show an example of an intraductal papilloma, another non-neoplastic growth. Papillary lesions are known to be a heterogeneous group histologically and clinically, which represents a significant challenge in making an accurate diagnosis [33]. These lesions are annotated with the NNEO label, usually encompassing the entire papilloma or, at a minimum, ensuring that key architectural features are encompassed. This group includes both benign and cancerous formations, as well as lobular neoplasia. Intraductal papillomas are made up of well-kept, vein-like structures that are contained, with little inflammation surrounding the region. Their inner epithelial walls and surrounding myoepithelial tissue are highly proliferated and unbroken. This region may include atypical benign tumors within. Due to its massive and weblike appearance, this region can be mistaken for DCIS. It is important for annotators to pay attention to the classic signs of DCIS when annotating similar regions, such as an unbroken border and/or nuclear abnormalities. Lack of those in a visually similar structure is characteristic of intraductal papilloma and other benign papillary lesions.

2.7 Annotation Challenges

Producing high-quality digital pathology data for breast cancer is challenging, particularly in distinguishing non-neoplastic structures from early-stage carcinogenic lesions and ensuring consistency in pathology diagnoses. One of the most complex distinctions is between atypical ductal hyperplasia (ADH) and ductal carcinoma in situ (DCIS), as these lesions share overlapping morphological features, leading to high interobserver variability even among expert pathologists [34]. Studies have shown that agreement rates in diagnosing ADH versus DCIS can be as low as 48%, highlighting the subjective nature of classification [35]. This inconsistency extends to broader breast cancer research, where pathologists often disagree on borderline or pre-malignant conditions, complicating the creation of reliable training datasets for machine learning. Deep learning models require consistent ground truth labels, and if expert pathologists struggle with diagnostic uniformity, AI systems risk learning from misclassified images, reducing their accuracy and real-world applicability [36].

Rare biological structures are underrepresented and pose a significant challenge for models. The lack of sufficient examples of uncommon cases makes it difficult to train models. Both metaplastic breast carcinoma and tubular carcinoma are invasive ductal carcinomas that are representative of less than 2% of all breast cancers. For example, tubular carcinoma can be mistaken by models as normal ductal formation because they are defined by small, angulated tubules with minimal atypia. Metaplastic breast carcinoma is poorly differentiated. It exhibits highly variable morphology, often including spindle cells, squamous components, or even bone or cartilage-like tissue. The lack of frequency in these presentations of INDC can lead to significant misclassification of models. This variability in the morphological presentation of carcinoma further complicates the task of accurately classifying these structures.

Another issue with large datasets like this is the possibility of annotator bias. Due to the partially annotated nature of the data, annotation bias could pose a real issue and be translated into the model. This was mitigated by multiple cross-reviews. Furthermore, not all breast cancer diagnoses can be determined solely through digital pathology. While histological analysis of stained tissue samples is fundamental for diagnosing breast cancer, certain molecular and genetic subtypes require additional testing beyond digital imaging. For example, triple-negative breast cancer (TNBC), an aggressive subtype that lacks ER, PR, and HER2 expression, cannot be diagnosed based purely on histopathology [37]. Immunohistochemistry (IHC) and molecular profiling techniques such as fluorescence in situ hybridization (FISH) or next-generation sequencing (NGS) are necessary to determine receptor status and guide treatment decisions [38]. This limitation underscores the need for multimodal diagnostic approaches, where digital pathology is complemented by molecular testing to provide a comprehensive assessment of breast cancer cases.

3 Database Analysis

Both breast tissue subsets discussed in this chapter are subsets of much larger corpora. A brief comparison of the full corpora from which these were extracted is given in Table 3. Because these two corpora were collected at different hospitals, the data is coded differently, which makes direct comparisons difficult. Hence, one important goal of this work was to normalize the data so they could be used simultaneously for technology development. Each database contains an ample amount of breast tissue data. FCDP is a bit richer in terms of occurrences of cancer because that is a specialty for FCCC. TUDP better reflects general pathology studies at a major urban public hospital.

Annotations are created by undergraduate bioengineering or pre-health students who are trained by senior personnel and board-certified pathologists. These students go through extensive training and supervision by senior annotators before they are allowed to annotate production data. In the early stages of the project, they were trained directly by TUH pathologists. However, once we developed adequate experience in house, a pipeline was developed in which senior annotators would train and supervise junior annotators. A significant amount of data has been reviewed by at least three annotators. Informal annotator agreement experiments have shown over the years that annotators have performed well and compare favorably to experts with a Kappa statistic above 0.8. Clinically trained pathologists have reviewed some of the data as part of their supervision process and provided ample feedback on our annotation process.

In addition, to aid in the annotation process, for TUDP, anonymized patient reports are available for reference. The TUDP reports, which are fairly detailed, were manually anonymized and held on a secure server. Though there is some structure to these reports, they are difficult to parse automatically for specific information due to the complex nature of the language. Hence, we mainly used them to validate challenging cases. Data selection to create various subsets of the database is usually done based on manual review of the reports.

The FCDP data was completely anonymized before we were given access. FCDP includes metadata for each slide that has been aggregated in a spreadsheet [15]. This spreadsheet contains a wealth of information about the tissue samples and diagnoses. The metadata usually do not identify non-neoplastic or abnormal structures but do indicate the existence of ductal carcinoma or invasive ductal carcinoma.

A comparison of the statistics for the annotations for the two breast tissue subsets is given in Table 4. In both cases, we selected all the available data from the master corpora for annotation. The breast tissue subset within these corpora contains a larger number of WSI than most publicly available breast tissue datasets (see Table 1). These subsets contain 4,968 WSI annotated with the nine different labels shown in Table 2.

Table 3 A comparison of TUDP and FCDP

Attribute	TUDP (v1.0.0)	FCDP (v1.0.0)
No. files	99,123	14,276
Amount of data	21.123 GB	3.497 GB
Average file size	456 MB	237 MB
Approximate composition (top four categories)	C64 (kidney): 16% C61 (prostate): 13% C34 (lung): 10% C50 (breast): 10%	Urinary/prostate: 37% Breast: 21% Gastro: 18% Gyneco: 8%

The FCDP and TUDP datasets include more annotated instances than all other breast-focused datasets analyzed, with the exception of the SPIDER subset from HistAI initiative. While many of the alternative datasets rely on binary classification (e.g., cancer vs non-cancer), FCDP and TUDP adopt a more granular annotation framework. While binary classification may suffice for certain machine learning tasks, we argue that a more detailed annotation structure is essential to capture the complexities of tissue architecture and nuanced morphological heterogeneity in breast tissue, enabling more robust and biologically informed model development.

Table 4 A comparison of the annotations for TUBR and FCBR

Attribute	TUBR (v5.0.0)	FCBR (v3.0.1)
No. Files:	1,652	765
/train	932	373
/dev	921	325
/eval	3,505	1,463
Total		
Amount of data (Gbytes)	1.226	3.497
No. labels	22,283	20,085
Avg. no. labels per slide	6.36	13.73
Annotated area per slide (%)	1.96%	2.53%
No. Labels:	4,797 [21.53%]	325 [1.62%]
Norm	1,048 [4.70%]	1,209 [6.02%]
DCIS	1,512 [6.79%]	10,955 [54.54%]
INDC	7,162 [32.14%]	728 [3.62%]
NNEO	970 [4.35%]	1,215 [6.05%]
INFL	1,118 [5.02%]	914 [4.55%]
ARTF	618 [2.77%]	1,090 [5.43%]
NULL	115 [0.52%]	24 [0.12%]
SUSP	4,953 [22.18%]	3,625 [18.05%]
BCKG		
Area of labels (pixels^2):	2 [0.01%]	250 [1.24%]
64 × 64	130 [0.58%]	1,001 [4.98%]
128 × 128	1,132 [5.08%]	2,677 [13.33%]
256 × 256	4,346 [19.50%]	5,051 [25.15%]
512 × 512	6,555 [29.42%]	5,651 [28.14%]
1,024 × 1,024	4,892 [21.95%]	3,764 [18.74%]
2,048 × 2,048	3,372 [15.13%]	1,363 [6.79%]
4,096 × 4,096	1,433 [6.43%]	304 [1.51%]
8,192 × 8,192	386 [1.73%]	23 [0.11%]
16,384 × 16,384	35 [0.16%]	1 [0.00%]
32,768 × 32,768	0 [0.00%]	0 [0.00%]
65,536 × 65,536		

3.1 The TUDP Breast Tissue Subset (TUBR)

Temple University Hospital Breast Tissue Subset (TUBR) contains 3,505 high-resolution images with various degrees of carcinogenicity, ranging from samples with healthy background, ducts, and lobules to samples with definitive ductal and invasive carcinoma. This subset contains 8,035 non-cancerous, 6,222 carcinogenic, and 2,714 cancerous identified structures. This totals 22,241 labels and an average of 6.3 annotations per slide (see Table 4). The most common structures identified, aside from background, are non-neoplastic, normal ductal structures and invasive ductal carcinoma. Due to the general nature of pathologies Temple Hospital investigates, a wide range of non-neoplastic changes and normal structures can be found in this subset. The dataset provides a unique archive of various benign non-cancerous structures such as papilloma, fibroadenoma, ductal and lobular hyperplasia, and inflammatory lesions. The dataset is also rich in various samples of normal healthy ducts and lobules, with various degrees of branching and imaging clarity, which can be beneficial as a control for mistakenly marking healthy structures as non-neoplastic or carcinogenic. Each individual slide has at least one background label to equalize the difference in staining saturation and scanning discoloration.

We can see that a fairly small percentage of the overall slide area has been annotated (1.96%). Fully annotating each slide is simply not feasible or cost-effective, so our annotation team focused on events that would be most meaningful for ML technology development. In Table 5, we show the average dimensions of an annotated region sorted by label. DCIS and NNEO are the largest, and INFL are the smallest. Inflammation often accompanies the carcinogenic structure because it is an immune response from the body. Therefore, many times we see it as an "additional layer" surrounding abnormality as the body tries to respond to it. In addition, lymphocytes, the cell which we label as INFL, are composed of very small cells. The "structure" of inflammation is not really a structure at all but rather a clump of lymphocytes. Their small nature and their proximity to the structure of interest result in smaller-sized regions.

3.2 FCCC Breast Tissue Subset

Fox Chase Cancer Center Breast Tissue Subset (FCBR) contains 12,164 non-cancerous, 1,967 carcinogenic, and 5,954 cancerous identified structures. This totals 20,086 labels and an average of 13.7 annotations per slide, as shown in Table 4. Each individual slide has at least one background label to allow ML to equalize the difference in staining saturation and scanning discoloration. When compared to TUBR, FCBR is more heavily weighted toward malignant pathology, offering a greater proportion of invasive ductal carcinoma (INDC). This makes it valuable for training models and validating models focused on the identification of cancerous structures.

Table 5 A comparison of annotation dimensions

Label	TUBR	FCBR
dcis:	0 [0.00%]	3 [0.25%]
64 × 64	0 [0.00%]	91 [7.53%]
128 × 128	5 [0.48%]	147 [12.16%]
256 × 256	137 [13.07%]	311 [25.72%]
512 × 512	970 [4.35%]	427 [35.32%]
1,024 × 1,024	346 [33.02%]	210 [17.37%]
2,048 × 2,048	91 [8.68%]	19 [1.57%]
4,096 × 4,096	7 [0.67%]	1 [0.08%]
8,192 × 8,192	0 [0.00%]	0 [0.00%]
16,384 × 16,384	0 [0.00%]	0 [0.00%]
32,768 × 32,768	0 [0.00%]	0 [0.00%]
65,536 × 65,536		
indc:	0 [0.00%]	4 [0.04%]
64 × 64	4 [0.26%]	132 [1.20%]
128 × 128	47 [3.11%]	1,031 [9.41%]
256 × 256	179 [11.84%]	3,093 [28.23%]
512 × 512	450 [29.76%]	3,545 [32.26%]
1,024 × 1,024	551 [36.44%]	2,181 [19.91%]
2,048 × 2,048	254 [16.80%]	790 [7.21%]
4,096 × 4,096	26 [1.72%]	169 [1.54%]
8,192 × 8,192	1 [0.07%]	10 [0.09%]
16,384 × 16,384	0 [0.00%]	0 [0.00%]
32,768 × 32,768	0 [0.00%]	0 [0.00%]
65,536 × 65,536		
nneo:	0 [0.00%]	0 [0.00%]
64 × 64	2 [0.03%]	8 [1.10%]
128 × 128	126 [1.76%]	133 [18.27%]
256 × 256	1,398 [19.52%]	266 [36.54%]
512 × 512	3,010 [42.03%]	223 [30.63%]
1,024 × 1,024	1,933 [26.99%]	82 [11.26%]
2,048 × 2,048	597 [8.34%]	15 [2.06%]
4,096 × 4,096	84 [1.17%]	1 [0.14%]
8,192 × 8,192	12 [0.17%]	0 [0.00%]
16,384 × 16,384	0 [0.00%]	0 [0.00%]
32,768 × 32,768	0 [0.00%]	0 [0.00%]
65,536 × 65,536		
infl:	0 [0.00%]	66 [0.00%]
64 × 64	74 [7.63%]	310 [0.26%]
128 × 128	225 [23.20%]	453 [3.11%]
256 × 256	337 [34.74%]	282 [11.84%]
512 × 512	277 [28.56%]	9 [29.76%]
1,024 × 1,024	50 [5.15%]	1 [36.44%]
2,048 × 2,048	7 [0.72%]	0 [16.80%]
4,096 × 4,096	0 [0.00%]	0 [1.72%]
8,192 × 8,192	0 [0.00%]	0 [0.07%]
16,384 × 16,384	0 [0.00%]	0 [0.00%]
32,768 × 32,768	0 [0.00%]	0 [0.00%]
65,536 × 65,536		

The FCBR Corpus includes extensive metadata about each slide [15]. In Table 6, we show a histogram of the ICDO-10 codes [39] for both the tissue site and the tumor site. The breast tissue subset was the subset of FCDP that c50.* codes for both. A few entries did not contain tumor site codes and were assigned the value "cxx.x" as a placeholder. The most common combination of values for these were c50.9 ("Malignant neoplasm of breast of unspecified site") and c50.4 ("Malignant neoplasm of upper-outer quadrant of breast"). Over 90% of the samples are contained in the top-10 code pairs.

FCBR metadata also contains a wealth of metadata about the diagnosis. Diagnosis from a single pathology slide is, of course, a challenging problem (and the reason we are investing research in these resources). In Table 7, we show a histogram of several important metadata fields that can be used to interpret the slides and guide ML. The first column contains a clustering of the slides into four values [15]:

Table 6 A histogram of tissue site and tumor site codes for FCBR

Tissue site (ICDO code)	Tumor site (ICDO code)	Count
c50.9	c50.4	334
c50.9	c50.8	278
c50.9	c50.9	158
c50.4	c50.4	155
c50.8	c50.8	92
c50.9	c50.2	86
c50.9	c50.5	70
c50.9	c50.1	62
c50.2	c50.2	55
c50.9	c50.3	50
c50.5	c50.5	36
c50.3	c50.3	26
c50.1	c50.1	25
c50.9	cxx.x	20
c50.0	c50.0	3
c50.9	c50.6	2
c50.9	c50.0	2
c50.1	c50.4	2
c50.9	c63.9	1
c50.8	c50.4	1
c50.8	cxx.x	1
c50.6	c50.6	1
c50.5	c50.4	1
c50.4	cxx.x	1
c50.3	c50.2	1

high grade (hg), intermediate grade (ig), low grade (lg), and unknown (UNK). The second column shows the behavior of cells within the slide: invasive ductal carcinoma (MALIG-PRIMARY), ductal carcinoma in situ (CA IN SITU), or unknown tissue behavior. These judgements were made using all the available metadata and manual annotations. The next three columns represent selected metadata fields that were used to support these judgements. Note that only the top-10 most frequent combinations are shown. The top-10 values for "hg" covers 54% of the corpus, while the top-10 for "ig" covers 29% of the corpus. Together these two categories cover over 80% of the corpus. FCBR is much richer with respect to indications of cancer than TUBR.

These datasets are one-of-a-kind, meticulously curated, and multi-reviewed data corpora, distinguished by their accompanying unique annotations. Each annotation has undergone validation based on patient reports and has been cross-reviewed. This thorough review enhances their reliability for training deep models in digital pathology, particularly for breast tissue analysis. The combination of extensive image diversity and precision makes the corpora invaluable resources for advancing research in cancer detection and image analysis. Given that the dataset originates from two entirely different sources but shares the same format, they offer a practical framework for evaluating a model's capacity for generalization across separate data domains.

4 Baseline Experiments

There has been a rapid growth in interest in automated methods for the interpretation of digital pathology images in recent years. Though many current models have limited capabilities in terms of the type of diagnosis they can provide (e.g., only making whole slide classifications), the range of analysis techniques available has grown significantly over the years. A generation of models first debuted with only two prediction classes (cancerous and non-cancerous). However, models have since evolved to detect a broader range of pathologies, including multiple cancer types, pre-cancerous lesions, and various non-cancerous conditions (see Table 1). Advances in AI-driven digital pathology enable these models to identify complex biomarkers and subtle histopathological features, facilitating more precise and personalized treatment options. Furthermore, the AI systems can integrate molecular and genomic data, providing a deeper understanding of disease mechanisms, prognosis, and potential therapeutic responses, thus transforming the landscape of diagnostic pathology.

Accurate identification of the labeled regions described in Table 2 requires analysis of high-resolution images at full resolution. This is a problem we refer to as microsegmentation. Since only 2% of the image area is annotated, the problem is further complicated by many of the usual challenges in dealing with imbalance in data. Also, like many healthcare-related applications, false alarm rates are very important, since pathologists often need to report on an image even if there is only one or two areas indicated the potential for cancer. To avoid over-reporting, false alarm rates need to be vanishingly small, and that is often very difficult to achieve for modern ML systems. In this section, we provide some baseline experiments using

Table 7 The top-10 entries for derived grades for FCBR

Derived grade	Behavior	ICDO (Histology)	Cancer status	Count
hg	MALIG-PRIMARY	8500/3	No evidence of this cancer	399
hg	MALIG-PRIMARY	8500/3	Evidence of this cancer	170
hg	MALIG-PRIMARY	8500/3	Unknown, indeterminate whether this cancer present	91
hg	MALIG-PRIMARY	8522/3	No evidence of this cancer	34
hg	MALIG-PRIMARY	8520/3	No evidence of this cancer	31
hg	MALIG-PRIMARY	8523/3	No evidence of this cancer	19
hg	MALIG-PRIMARY	8520/3	Evidence of this cancer	18
hg	MALIG-PRIMARY	8522/3	Evidence of this cancer	14
hg	MALIG-PRIMARY	8520/3	Unknown, indeterminate whether this cancer present	8
hg	CA IN SITU	8501/2	No evidence of this cancer	7
ig	MALIG-PRIMARY	8500/3	No evidence of this cancer	220
ig	MALIG-PRIMARY	8520/3	No evidence of this cancer	65
ig	MALIG-PRIMARY	8522/3	No evidence of this cancer	35
ig	MALIG-PRIMARY	8500/3	Evidence of this cancer	33
ig	MALIG-PRIMARY	8500/3	Unknown, indeterminate whether this cancer present	30
ig	MALIG-PRIMARY	8522/3	Evidence of this cancer	14
ig	MALIG-PRIMARY	8520/3	Evidence of this cancer	13
ig	CA IN SITU	8523/2	No Evidence of this cancer	8
ig	MALIG-PRIMARY	8524/3	No evidence of this cancer	5
ig	MALIG-PRIMARY	8523/3	No evidence of this cancer	5
lg	MALIG-PRIMARY	8500/3	No evidence of this cancer	23
lg	MALIG-PRIMARY	8480/3	No evidence of this cancer	5

(continued)

Table 7 (continued)

Derived grade	Behavior	ICDO (Histology)	Cancer status	Count
lg	MALIG-PRIMARY	8500/3	Unknown, indeterminate whether this cancer present	4
lg	MALIG-PRIMARY	8523/3	No evidence of this cancer	3
lg	MALIG-PRIMARY	8520/3	No evidence of this cancer	3
lg	MALIG-PRIMARY	8522/3	No evidence of this cancer	2
lg	MALIG-PRIMARY	9020/3	Evidence of this cancer	1
lg	MALIG-PRIMARY	8523/3	Unknown, indeterminate whether this cancer present	1
lg	MALIG-PRIMARY	8522/3	Evidence of this cancer	1
lg	MALIG-PRIMARY	8522/3	Unknown, indeterminate whether this cancer present	1
UNK	UNKNOWN			29

common-off-the-shelf (COTS) technology to demonstrate the challenges faced in the automatic interpretation of digital pathology images.

The data was split into three subsets (/train, /dev, and /eval) following standard practices to make the sets mutually exclusive from a patient perspective, and to achieve approximately a 60% train, 20% dev, and 20% eval split. We also tried to balance the number of labels occurring in each subset, though in practice, there are limits to what can be achieved given the number of labeled regions available in the corpus. We prefer a 60/20/20 split, so there are more instances of the less frequently occurring labels in the evaluation set. This gives a much more realistic measure of performance in practice.

Evaluation of performance for this type of data is challenging and often an under-appreciated problem. Since the vast majority of an image is BCKG, common popular measures such as the DICE score [40] that score every pixel are not accurate since they over-emphasize BCKG. Instead, we must use weighted approaches [41] that focus only on annotated areas and assess the accuracy of the detected boundaries of an annotated patch. Factoring in the accuracy of a segmentation is extremely critical in this type of application.

However, since the images are partially annotated, we must align hypotheses with reference annotations and evaluate the extent to which the partially annotated data is correctly segmented and identified. We cannot really assess the accuracy of hypotheses that correspond to unannotated regions since we don't know the ground truth. Therefore, we have developed a modified version of the DICE measure that compares the degree of overlap between two comparable patches in the reference and hypothesized annotations. This scoring system is available as part of our open source digital pathology tools [42]. We refer to this modified DICE score as MDICE.

Mathematically, for each annotated reference polygon area R_i with a matched hypothesis area H_i, we define $MDICE_i = (R_i \cap H_i)/R_i$. The overall score is $MDICE = \sum_i MDICE_i = \left(\sum_i R_i \cap H_i\right)/\left(\sum_i R_i\right)$. This simple modification of a standard DICE score rewards coverage of annotated tissue without penalizing predictions in unannotated regions. It is one of two metrics we will use to analyze performance.

We will also analyze performance using a modified F1 (MF1) score. In this case, since there is a predominance of images that do not contain indications of cancerous morphologies, we will use an F1 score averaged across 5 classes of interest (DCIS, INDC, NNEO, INFL, and NORM), and one class indicating background (BCKG). We compute the F1 score for each of these two groups for an overall judgment about the image based on the annotated label, weight the averaged F1 score for the first group by 90% and the second group by 10%, and sum them. We ignore NULL, SUSP, and ARTF. MF1 is reported as a percentage similar to an F1 score—higher is better. This modified metric, in our experience, gives a much better estimate of the clinical relevance of a hypothesis.

In this section, we will present results using several popular COTS approaches. Though we have generated results for a much wider range of algorithms discussed in [43, 44], here we focus on three relatively robust and popular algorithms:

- *ResNet18*: a deep convolutional neural network designed to enable training of very deep architectures by addressing the vanishing gradient problem through the use of residual connections [45–47].
- *EB0/EB7*: two variants from the EfficientNet family of convolutional neural networks developed to optimize both accuracy and efficiency (speed, memory, and power usage) [48].
- *ViT-16/ViT-32*: an alternative to convolutional neural networks that analyzes images using 16×16 or 32×32 pixel patches, serializes patches into vectors, and uses a transformer-based architecture for classification [49].

These algorithms were selected based on our experiences that they give good performance across a wide range of signal and image processing applications. The first two are based on convolutional neural networks (CNNs), which have proven to be extremely powerful for signal and image applications. The third algorithm is based on a transformer architecture. Transformers, which use self-attention to model context, are widely considered to be a disruptive force in ML due to their ability to encode long-term sequential and spatial dependences [44]. We have adapted these to a variety of tasks involving EEG, digital pathology, and cardiology applications.

In our experiments, all models were fine-tuned from pretrained weights using the standard PyTorch implementation of ResNet18, EfficientNet (B0/B7), and Vision Transformer (ViT-16/32) architectures available in the Torchvision package [50]. To improve generalization and mitigate overfitting, we applied standard data augmentation techniques, including resized cropping, horizontal and vertical flipping, color jittering, and random erasing, followed by normalization.

4.1 Window Size and Sub-image Selection

Our first set of experiments was designed to optimize the process of selecting an analysis window. The goal with these experiments was to understand how much spatial context was needed to accurately classify a patch. We evaluated two approaches to selecting an image to represent an annotated region. In the first approach, we drew a rectangular bounding box around the image and then experimented with different resampling approaches to make the resulting box a uniform size and/or square (a common shape in image processing applications). We refer to this as the "BBOX" method. Since annotated regions are irregularly shaped, this method has the potential to include significant amounts of tissue that do not share the same label as the reference label. This can result in divergence of a model during training.

The second approach, which we ultimately preferred because of its simplicity and overall good performance, was to compute the center of mass of an annotated region, and then select a window of data centered around the center of mass of the region. We refer to this as the "COM" method. Since annotated regions can be elongated and irregularly shaped, this approach attempted to identify the most significant portion of the region. We included a guard band that allowed the area of the window to be expanded by a percentage (e.g., "g25" means we expand the area by 25%). This has the potential to account for imperfections in the annotations, but also can have an adverse effect of including too much out-of-class data adjacent to the patch of interest. Our hypothesis is that a good ML algorithm should be able to learn to differentiate between correctly labeled data and mislabeled data when it sees enough training data.

The results of several experiments with the COM method are given in Table 8 for TUBR using the MF1 metric. We see that a window of 1024 × 1024 pixels with no guard band (g00) gives the best performance. Results for FCBR are comparable. The results in Table 8 somewhat validate the data in Table 5, in which we see that the majority of the annotated regions for labels of interest are in the 1024 × 1024 range in size. This also somewhat validates the accuracy of the manual annotations since they seem to contain enough information to reliably classify the window. In fact, experiments with greedy algorithms such as Random Forests have confirmed this in that performance on the training data when overtraining can be quite good. Hence, for the remainder or our experiments, we focused on the com_w1024 × 1024_g00 data.

Using the com_w1024 × 1024_g00 data, we evaluated the three popular algorithms mentioned above. The results are shown in Table 9, which also uses the MF1 metric. In this case, the TUBR experiments involved training and evaluating on TUBR. Similarly, the FCBR results were generated by training and evaluating solely on FCBR. We see that EB7 performs slightly better than the other two algorithms. We also see that performance on the dev set generally correlates well with performance on the blind evaluation set. However, there is some evidence of overtraining since the error rates on the training data (closed-set testing) are surprisingly low. On one hand, since the datasets are reasonably large, this suggests there is ample information

Table 8 Patch-level classification results (MF1) as a function of the window dimensions for TUBR

Arch	Dataset	Train	Dev	Eval
eb0	com_w0128 × 0128_g00	22.42	39.91	43.01
eb0	com_w0128 × 0128_g25	13.89	35.42	39.87
eb0	com_w0128 × 0128_g50	9.09	35.31	40.45
eb0	com_w0256 × 0256_g00	4.13	30.45	34.56
eb0	com_w0256 × 0256_g25	15.51	25.87	32.03
eb0	com_w0256 × 0256_g50	8.16	28.69	35.02
eb0	com_w0512 × 0512_g00	4.95	25.53	31.43
eb0	com_w0512 × 0512_g25	15.77	28.22	33.94
eb0	com_w0512 × 0512_g50	7.87	26.31	32.68
eb0	com_w1024 × 1024_g00	8.12	22.43	28.21
eb0	com_w1024 × 1024_g25	13.15	26.92	30.11
eb0	com_w1024 × 1024_g50	13.03	25.33	32.56
eb7	com_w0128 × 0128_g00	8.98	34.97	38.32
eb7	com_w0128 × 0128_g25	13.41	34.20	39.59
eb7	com_w0128 × 0128_g50	11.42	31.48	36.46
eb7	com_w0256 × 0256_g00	8.28	29.36	32.52
eb7	com_w0256 × 0256_g25	8.57	25.24	30.03
eb7	com_w0256 × 0256_g50	6.81	28.85	33.56
eb7	com_w0512 × 0512_g00	8.44	22.06	27.77
eb7	com_w0512 × 0512_g25	5.71	22.23	26.57
eb7	com_w0512 × 0512_g50	9.03	21.82	29.15
eb7	com_w1024 × 1024_g00	4.46	21.86	25.95
eb7	com_w1024 × 1024_g25	3.63	23.66	27.36
eb7	com_w1024 × 1024_g50	3.48	20.58	25.31

in these large windows for good classification performance. On the other hand, it suggests that overall open-set testing performance could be improved if generalization were improved. Since ViT is a transformer-based architecture, we expect it to outperform the other algorithms given enough training data. Therefore, its slightly inferior performance might be an indication that the database is not sufficiently large.

4.2 Performance on a Sequential Decoding Task Using a Frame-Based Analysis

Whole-slide pathology images demand models that capture both fine-grained cell morphology and large-scale tissue architecture. Traditional CNNs, which are used

Table 9 A comparison (MF1) of selected popular algorithms on window-based classification

Arch	TUBR			FCBR		
	Train	Dev	Eval	Train	Dev	Eval
ResNet18	15.03	25.51	34.01	21.46	33.73	31.91
EB0	8.12	22.43	28.22	21.09	27.81	29.40
EB7	0.01	22.06	25.39	3.67	26.09	24.85
ViT-16	0.17	20.66	30.08	0.08	27.87	27.56
ViT-32	5.28	24.94	29.54	10.55	26.28	27.33

in ResNet18 and EB7, excel at capturing local texture patterns, but must rely on very deep stacks (or large filters) to model larger amounts of context. This is inefficient for these extremely high-resolution pathology images. In contrast, vision transformers use self-attention to flexibly integrate information across the entire image, encoding long-range dependencies that CNNs inherently miss. Indeed, systematic comparisons report that transformer-based models often outperform CNNs in cancer segmentation tasks precisely because they encode global context [51]. For example, virtually all pure transformer architectures exceeded CNNs and ResNet architectures in histopathology segmentation, thanks to their image-wide attention [52]. This ability to "see" the whole tissue at once is critical for breast cancer detection, where tumor regions may only be distinguishable by their relationship to distant anatomical landmarks or by subtle patterns that span many cells.

These limitations motivated our exploration of CNNs better suited for such requirements. EfficientNet offers a principled scaling of network depth, width, and resolution to balance representational power and computational demand. While larger variants like EB7, which use approximately 66M parameters, achieve excellent performance, their memory and FLOPS requirements render them impractical for gigapixel whole slide images (WSIs) without hardware acceleration or significant image downsampling. In contrast, EB0, which uses approximately 5.3M parameters, delivers competitive accuracy with significantly reduced computational overhead—an advantage confirmed in our experiments in Tables 8 and 9, where EB0 performance was competitive with EB7. These findings suggest that in data-scarce histopathology pipelines, smaller, well-scaled CNNs can generalize more reliably than their deeper, more resource-intensive counterpart.

In our next set of baseline experiments, we took the models from the previous experiment and integrated them into a frame-level classification scheme. Previously, the systems were only exposed to perfectly registrated images centered around the manually derived annotation. This is often what we call an oracle experiment, in which we ask the question how good can performance be if the segmentation is perfect. But, as is well known in sequential decoding tasks like speech recognition and EEG interpretation, segmentation is the hardest part of the problem. For this reason, state-of-the-art algorithms often do simultaneous segmentation and classification, so both aspects of the algorithm can be jointly optimized. Hence, in this section, we explore performance when these three algorithms are inserted into a system that

Table 10 A comparison of decoder performance (MDICE) using pretrained models

Arch	TUBR			FCBR		
	Train	Dev	Eval	Train	Dev	Eval
ResNet18	45.21	41.98	39.13	51.97	45.98	48.50
EB0	69.83	56.60	51.72	62.32	54.49	55.02
EB7	67.17	55.67	50.72	57.98	47.00	49.39
ViT-16	47.75	39.00	35.23	50.26	41.78	41.58
ViT-32	57.30	48.57	43.64	52.24	43.64	44.65

iterates over an image frame by frame, uses a 1024×1024 analysis window, and classifies each frame.

Of course, this adds a level of complexity to the task because these frame-level decisions must be aggregated into patches and assessed against the manual annotations by comparing the similarities of these patches. Converting frame-level hypotheses to patch-level hypotheses is a well-studied problem and normally involves using a heuristic algorithm to decide which frames should be aggregated and which frames should be discarded. Typical approaches to this problem include smoothing filters [53], majority filters [54], and morphological techniques based on edges and boundary detection [55]. Ideally, this is a problem best solved using another deep learning system as a postprocessor, but we often lack adequate data to train such systems. We have implemented a variant of these approaches known as flood fill that uses morphological operations to aggregate similarly labeled frames [56]. We use this in conjunction with MDICE to evaluate the overall performance of a segmentation and classification algorithm.

In Table 10, we provide results for frame-level decoders using a 256×256 frame and a 1024×1024 window, and using a pretrained model from Table 9. For this analysis, we use the MDICE metric. We see that EB0 is the most promising approach. Note that these results are not yet optimal because the model being used was only trained on the perfectly registrated images used in Table 9. The model has not been exposed to situations where an analysis window only partially overlaps with an annotated region. Hence, there is still a mismatch between training and decoding. We believe this contributed to the slight degradation in performance for EB7 compared to EB0.

4.3 A Comparison of Postprocessing Algorithms

In computational pathology, the initial output of a frame-level deep learning model often appears as a disordered distribution of isolated points, resembling a constellation of scattered stars. Each point represents a localized tissue frame confidently classified by the model. However, the larger histological context such as the broad expanses of stroma or the complex arrangements of tumor region remains obscured by

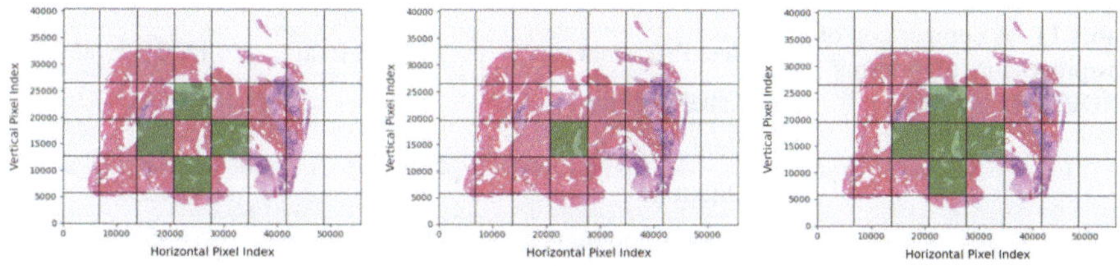

Fig. 8 An overview of frame-to-patch postprocessing

the gaps between these discrete predictions. To address this limitation, we introduce a framework composed of five complementary algorithms, each employing a distinct strategy to integrate the scattered outputs. Collectively, these methods aim to interpolate the intervening spaces and resolve boundary ambiguities, thereby transforming fragmented predictions into a coherent, polygonal representation of the underlying tissue architecture. An illustration of this process is provided in Fig. 8.

The first method employed is the Label Priority [56, 57] postprocessor, which operates according to a predetermined hierarchical ranking of tissue classes. Beginning with the highest priority label, the algorithm aggregates all predicted frames of that class and extends them spatially to occupy the intervening regions, thereby forming contiguous areas. It then proceeds sequentially through the remaining classes in descending order of priority, filling only those regions not already assigned. Once a frame has been allocated to a higher priority class, it is never reassigned, ensuring that the most critical biological structures are preserved irrespective of local model confidence.

The second method is the Confidence Priority [58] postprocessor, which resolves spatial conflicts based on prediction confidence rather than a predefined hierarchy. Each tissue class first expands from its predicted frames to establish potential regions. When overlap occurs between competing regions, the algorithm evaluates all frames within the contested area and computes the mean confidence score for each class. The class with the highest average confidence is assigned to the disputed region, overwriting lower-confidence predictions. In this way, regional boundaries are determined by the aggregate certainty of the model's predictions.

The third method is Dilation [59], which applies morphological expansion to reinforce the presence of high-priority labels. Using a maximum filter, this approach enlarges the boundaries of dominant regions iteratively, allowing them to incorporate adjacent unlabeled or lower priority cells. Unlike approaches aimed primarily at filling gaps, dilation emphasizes outward growth, producing contiguous and consolidated regions that emphasize the most critical tissue classes.

The final two methods act as smoothing mechanisms to refine local inconsistencies: Mode Smoothing [60] and k-Nearest Neighbors (kNN) Smoothing [60]. Both operate over localized neighborhoods of size 3×3 pixels. Mode Smoothing assigns the central cell the most frequently occurring label within its neighborhood, thereby eliminating isolated misclassifications and enforcing local homogeneity. In contrast, the kNN Smoothing incorporates similarity-based weighting: the central

Table 11 A comparison of postprocessing algorithms (MDICE) using EB0

Post Processor	TUBR Eval	FCBR Eval
Label priority	53.81	52.51
Confidence priority	50.86	47.93
Dilation	40.56	43.95
Mode smoothing	52.81	51.95
kNN	54.38	55.17

cell's label is updated according to the consensus of the k most similar neighbors within its window, enabling a more nuanced correction that accounts for feature similarity as well as spatial proximity. Together, these smoothing techniques suppress noise, remove spurious predictions, and improve the geometric coherence of regional boundaries.

The evaluation results, summarized in Table 11, indicate that the kNN Smoothing achieved the highest performance across both metrics, with MDICE scores of 54.38 (TUBR) and 55.17 (FCBR), suggesting its effectiveness in refining local inconsistencies through similarity-based smoothing. Label Priority also performed comparatively well (53.81, 52.51), outperforming the Mode Smoothing (52.81, 51.95) and Confidence Priority (50.86, 47.93). In contrast, Dilation yielded the lowest results (40.56, 43.95), indicating that aggressive morphological expansion may be less effective for producing coherent tissue maps in this context. Overall, smoothing-based strategies demonstrated the greatest improvement in predictive performance.

5 Joint Segmentation and Classification

The emergence of powerful statistical methods such as hidden Markov Models (HMMs) in the 1990s [60] demonstrated the value of jointly optimizing segmentation and classification by doing an exhaustive search within the decoding process. These systems share encoders for both localization and diagnostic labeling, facilitating shared feature learning, implicit attention, and reduced error propagation from separate preprocessors. Their ability to localize information in the data made them extremely valuable as diagnostic and discovery tools. Though systems such as SWIN [62, 63] and U-NET [64, 65] have emerged that emulate this approach in a deep learning context, adapting these systems to high-resolution pathology images has proven to be a challenge. However, our unit tests with these algorithms on smaller manageable datasets have shown that their performance is comparable to the algorithms previously discussed.

As an alternative, we have implemented frame-based versions of ResNet, EfficientNet, and ViT that are trained on large amounts of pathology data. To avoid data imbalance, we have selected approximately 100,000 windows from each corpus that best represent the diversity of the data and balance the ratio of labels to background. We evaluated these systems on WSIs and show the results in Table 12, MDICE scores

Table 12 The performance of optimized systems (MDICE)

Arch	TUBR			FCBR		
	Train (s)	Dev (s)	Eval (s)	Train (s)	Dev (s)	Eval (s)
ResNet18	40.87	37.42	33.28	44.40	44.34	42.11
EB0	62.32	55.31	51.10	66.83	49.49	46.02
EB7	56.03	47.10	43.81	31.53	31.31	32.41
ViT-16	62.68	48.32	44.72	26.11	24.44	25.07
ViT-32	57.12	46.34	41.39	29.58	26.11	28.06

Table 13 Complexity analysis for three popular ML algorithms

Arch	No. Parameters	Training Time	Decoding Time
ResNet18	11.69 M	2,356 (1.0x)	1,597 (1.0x)
EB0	5.30 M	3,900 (1.7x)	1,520 (1.0x)
EB7	66.35 M	4,883 (2.1x)	1,735 (1.1x)
ViT-16	86.60 M	3,713 (1.6x)	1,726 (1.1x)
ViT-32	88.20 M	3,869 (1.6x)	1,623 (1.0x)

are shown for the fully optimized systems. We again see that a simpler system, EB0, outperforms the other systems at significantly reduced complexity.

In Table 13, we present a complexity analysis of our three leading systems. The parameter counts inversely correlate with performance—the system with the least number of parameters, EB0, performs best. The computational complexity was compared using the same machine (AMD EPYC™ 7413 Processor 24-core 2.65GHz 128MB Cache) and a single Nvidia A40 GPU. All jobs were run using a single GPU so we could generate a fair comparison. As expected, EB0 is quite fast and produces competitive performance, and the ViT systems use the most resources but lag slightly in performance.

6 Cross-Modal Training Results

The makeup of training data strongly shapes the model architecture and training strategy. Consider two example datasets: a large, heterogeneous collection like TUDP (~100k images of diverse normal and pathological tissues) versus a focused set like FCDP (~14k images emphasizing oncological cases, including a breast cancer subset with detailed annotations). These differences lead to very different learning problems. For instance, FCDP's breast subset contains a high proportion of malignant tissue, whereas TUDP's slides include many normal biopsies and benign findings. A model trained on TUDP would see many more normal patterns, risking that it learns to under-detect cancer (a majority-class bias). By contrast, training on FCDP might make the model sensitive to cancerous features but possibly prone to false positives when deployed on general data.

The origin and diversity of images also influence generalization. Models trained on the narrow domain of FCDP's cancer cases might perform very well on similar breast images but poorly on new hospitals or on other tissue types. In contrast, a model trained on TUDP's wide-ranging tissues might be more robust across subtypes. Analogously, multi-center studies (e.g., separate lung or breast histology cohorts) often find that stain differences or scanner artifacts shift the distribution [65]. For example, in a breast histology benchmark the target dataset (BreakHis) contained thousands of patches from one source and had a very different class balance than the small ICIAR source set. In practice, one must watch how differing dataset composition affects the model's sensitivity and specificity. A model trained on high-prevalence cancer data might achieve high sensitivity but at the cost of specificity when applied to general screening images. Conversely, one trained mostly on normal tissue might under-predict disease and suffer low recall. Careful calibration, perhaps through threshold tuning or meta-learning of prevalence, is necessary to maintain an appropriate balance between false positives and false negatives, given each dataset's statistics.

In Table 14, we present a series of experiments in which we evaluate mismatched training conditions. In the first two groups of experiments, each system is trained on one corpus and evaluated on both. In the third set of experiments, the systems are trained on the pooled data. We see that the results in this table are mixed, indicating we still don't have a firm grasp on how to utilize the two datasets simultaneously. The overall best performance on TUBR is achieved by EB7 and by pooling the training data. However, that same system performs below EB0 on FCBR. EB0 surprisingly gives better performance on TUBR when trained on FCBR, which is some indication of the amount of experimental noise in this evaluation. EB0 and EB7 performed similarly when trained and evaluated on TUBR, but EB7 is superior on FCBR. This is a lesson learned all too often in ML—learning how to use more data is an important part of the process of improving performance by adding new data. These experiments suggest that more research is needed to understand how best to leverage the pooled dataset for better generalization.

To assess whether the performance differences observed between the EB0 and EB7 models were statistically significant, a two-tailed significance test was conducted on the results presented in Table 14. The analysis indicates that the difference in mean performance scores (50.24 versus 59.37) is statistically significant at the 95% confidence interval ($p = 0.000083$). This result confirms, with high confidence, that the superior performance of EB7 is not due to random variation. The finding supports the conclusion that scaling model capacity within the EfficientNet family provides measurable and statistically robust performance improvements in digital pathology tasks.

Table 14 The impact of mismatched training conditions (MDICE)

Arch	Training condition	TUBR			FCBR		
		Train	Dev	Eval	Train	Dev	Eval
ResNet18	TUBR	40.87	37.43	33.28	31.52	29.05	31.66
EB0	TUBR	54.43	48.96	43.36	30.74	29.82	29.87
EB7	TUBR	56.03	47.10	43.81	31.53	31.31	32.41
ViT-32	TUBR	57.12	46.34	41.39	30.48	28.49	31.19
ResNet18	FCBR	47.42	48.03	43.85	44.40	44.34	42.11
EB0	FCBR	43.48	47.43	46.76	70.03	50.46	49.34
EB7	FCBR	30.76	32.08	33.66	73.31	53.22	55.82
ViT-32	FCBR	17.28	18.38	17.43	29.58	26.11	28.06
ResNet18	TUBR + FCBR	37.99	35.12	32.18	32.72	30.05	32.87
EB0	TUBR + FCBR	60.31	53.64	50.24	60.04	51.20	48.92
EB7	TUBR + FCBR	70.76	59.37	59.37	51.90	41.09	40.17
ViT-32	TUBR + FCBR	47.97	38.93	35.75	37.62	31.52	33.88

7 Conclusions

The integration of newly annotated breast tissue data from FCDP and TUDP strengthens the robustness and generalizability of machine learning models in digital pathology. These datasets enable training and validation on diverse tissue structures, disease stages, and pathological features. A dual-dataset approach fosters the development of resilient AI models capable of handling image variability, annotation styles, and disease presentations, ultimately enhancing diagnostic precision and adaptability in pathology.

In this chapter we have introduced the science behind the annotation process so that ML researchers can better appreciate the challenges with this data. Improving the performance of an ML system and making that system clinically relevant often requires developing a good understanding of the basic science. We have also established baseline performance of some well-known algorithms. Even though performance has improved dramatically in recent years, it is clear that we have a way to go before the technology will be clinically acceptable.

Going forward, we plan to develop transformer-based architectures that encode long-term spatial context. There are numerous computational challenges associated with this since the data must be streamed into the trainer because it is impossible to fit all the data in computer memory in one time. Hence, we will need to rethink the architecture and workflows so that we can train on the entire corpus using context lengths of 100,000 steps. The amount of memory required to hold sufficient context for each frame is extremely large, so we need to explore architectures that efficiently encode context.

For more information about the resources presented here, please refer to our project website: www.nedcdata.org.

Acknowledgements This material is based on work supported by several organizations over the years, including the National Science Foundation (grants nos. CNS-1726188 and 1925494), the Temple University Catalytic Collaborative Funding Initiative, and, most recently, by the Pennsylvania Breast Cancer Coalition Breast and Cervical Cancer Research Initiative. Any opinions, findings, and conclusions or recommendations expressed in this material are those of the author(s) and do not necessarily reflect the views of these organizations.

References

1. Bacus, J. V., & Bacus, J. W. (1995). Method and apparatus for automated assay of biological specimens (United States Patent US-5473706-A). url: https://patents.google.com/patent/US5473706A.
2. Kusta, O., Rift, C. V., Risør, T., Santoni-Rugiu, E., & Brodersen, J. B. (2022). Lost in digitization – A systematic review about the diagnostic test accuracy of digital pathology solutions. *Journal of Pathology Informatics*, 13, 100136. https://doi.org/10.1016/j.jpi.2022.100136.
3. Xu, H., & Remick, D. G. (2016). Pathology: A Satisfying Medical Profession. *Academic Pathology*, 3, 2374289516661559. https://doi.org/10.1177/2374289516661559.
4. Metter, D. M., Colgan, T. J., Leung, S. T., Timmons, C. F., & Park, J. Y. (2019). Trends in the US and Canadian Pathologist Workforces From 2007 to 2017. *JAMA Network Open*, 2(5), Article 5. https://doi.org/10.1001/jamanetworkopen.2019.4337.
5. Walsh, E., & Orsi, N. M. (2024). The current troubled state of the global pathology workforce: A concise review. *Diagnostic Pathology*, *19*(1), Article 1. https://doi.org/10.1186/s13000-024-01590-2.
6. Jhala, N. (2017). Digital Pathology: Advancing Frontiers. *IEEE Signal Processing in Medicine and Biology Symposium (SPMB)*. https://doi.org/10.1109/SPMB.2017.8257013.
7. Holloman, A. M., Berg, M. P., Bryant, B., Dixon, L. R., George, M. R., Karp, J. K., Knollmann-Ritschel, B. Ec., Prieto, V. G., Timmons, C. F., Childs, J. M., Lofgreen, A., Johnson, K., & McCloskey, C. B. (2023). Experiential exposure as the key to recruiting medical students into pathology. *Academic Pathology*, 10(2), Article 2. https://doi.org/10.1016/j.acpath.2023.100074.
8. *The Paige Prostate Suite: Assistive Artificial Intelligence for Prostate Cancer Diagnosis: Emerging Health Technologies* (EH0123; p. 23). (2024). Canadian Agency for Drugs and Technologies in Health. url: https://www.ncbi.nlm.nih.gov/books/NBK608438/.
9. Matthews, G. A., McGenity, C., Bansal, D., & Treanor, D. (2024). Public evidence on AI products for digital pathology. *Npj Digital Medicine*, 7(1), Article 1. https://doi.org/10.1038/s41746-024-01294-3.
10. Bellis, M., Metias, S., Naugler, C., Pollett, A., Jothy, S., & Yousef, G. M. (2013). Digital Pathology: Attitudes and practices in the Canadian pathology community. *Journal of Pathology Informatics*, 4(1), Article 1. https://doi.org/10.4103/2153-3539.108540.
11. Clarke, E., Doherty, D., Randell, R., Grek, J., Thomas, R., Ruddle, R. A., & Treanor, D. (2023). Faster than light (microscopy): Superiority of digital pathology over microscopy for assessment of immunohistochemistry. *Journal of Clinical Pathology*, *76*(5), 333. https://doi.org/10.1136/jclinpath-2021-207961.
12. McGenity, C., Clarke, E. L., Jennings, C., Matthews, G., Cartlidge, C., Freduah-Agyemang, H., Stocken, D. D., & Treanor, D. (2024). Artificial intelligence in digital pathology: A systematic review and meta-analysis of diagnostic test accuracy. *Npj Digital Medicine*, *7*(1), Article 1. https://doi.org/10.1038/s41746-024-01106-8.
13. Shawki, N., Shadhin, M. G. M., Elseify, T., Jakielaszek, L., Farkas, T., Persidsky, Y., Jhala, N., Obeid, I., & Picone, J. (2020). The Temple University Digital Pathology Corpus. In I. Obeid, I. Selesnick, & J. Picone (Eds.), *Signal Processing in Medicine and Biology: Emerging Trends in Research and Applications* (1st ed., pp. 67–104). https://doi.org/10.1007/978-3-030-36844-9.

14. Doshna, B., Wevodau, Z., Jhala, N., Akhtar, I., Obeid, I., & Picone, J. (2021). The Temple University Digital Pathology Corpus: The Breast Tissue Subset. In I. Obeid, I. Selesnick, & J. Picone (Eds.), *Proceedings of the IEEE Signal Processing in Medicine and Biology Symposium (SPMB)* (pp. 1–3). IEEE. https://doi.org/10.1109/SPMB52430.2021.9672275.
15. Shalamzari, S. S., Bagritsevich, M., Melles, Anne-Mai, Obeid, I., Picone, J., Connolly, D., Wu, C., Brown, B., James, J., Gong, Y., & Wu, H. (2023). Big Data Resources for Digital Pathology. *Proceedings of the IEEE Signal Processing in Medicine and Biology Symposium*, 1–19. https://doi.org/10.1109/SPMB59478.2023.10372721.
16. Bagritsevich, M., Hackel, D., Obeid, I., & Picone, J. (2024). Annotation of the Fox Chase Cancer Center Digital Pathology Corpus. *Proceedings of the IEEE Signal Processing in Medicine and Biology Symposium*, 1–4. https://doi.org/10.1109/SPMB62441.2024.10842255.
17. Rolls, G. (2019). An Introduction to Specimen Preparation. In *Leica Biosystems* (p. 1). url: https://www.leicabiosystems.com/pathologyleaders/an-introduction-to-specimen-preparation.
18. Anderson, J. (2019). An Introduction to Routine and Special Staining. In *Leica Biosystems* (p. 1). url: https://www.leicabiosystems.com/pathologyleaders/an-introduction-to-routine-and-special-staining.
19. Niyas, S., Bygari, R., Naik, R., Viswanath, B., Ugwekar, D., Mathew, T., Kavya, J., Kini, J. R., & Rajan, J. (2023). Automated Molecular Subtyping of Breast Carcinoma Using Deep Learning Techniques. *IEEE Journal of Translational Engineering in Health and Medicine*, 11, 161–169. https://doi.org/10.1109/JTEHM.2023.3241613.
20. Madabhushi, A., & Lee, G. (2016). Image analysis and machine learning in digital pathology: Challenges and opportunities. *Medical Image Analysis*, 33, 170–175. https://doi.org/10.1016/j.media.2016.06.037.
21. Wu, H., Phan, J. H., Bhatia, A. K., Cundiff, C. A., Shehata, B. M., & Wang, M. D. (2015). Detection of blur artifacts in histopathological whole-slide images of endomyocardial biopsies. *2015 37th Annual International Conference of the IEEE Engineering in Medicine and Biology Society (EMBC)*, 727–730. https://doi.org/10.1109/EMBC.2015.7318465.
22. Leica Biosystems. (2008). *ScanScope SVS file format specifications*. url: https://www.leicabiosystems.com/products/whole-slide-imaging/scanscope-svs-file-format/.
23. Goode, A., Gilbert, B., Harkes, J., Jukic, D., & Satyanarayanan, M. (2013). OpenSlide: A vendor-neutral software foundation for digital pathology. *Journal of Pathology Informatics*, 4(1), Article 1. https://doi.org/10.4103/2153-3539.119005.
24. Obeid, I., & Picone, J. (2016). The Temple University Hospital EEG Data Corpus. In M. A. Lebedev (Ed.), *Augmentation of Brain Function: Facts, Fiction and Controversy. Volume I: Brain-Machine Interfaces* (1st ed., Vol. 10, pp. 394–398). Frontiers Media S.A. https://doi.org/10.3389/fnins.2016.00196.
25. Simons, J., Wevodau, Z., Doshna, B., Obeid, I., & Picone, J. (2021). *The Temple University Hospital DPATH Corpus: Annotation Guidelines* (p. 18). Temple University. url: https://isip.piconepress.com/publications/reports/2021/tuh_dpath/annotations/.
26. Hoda, S. A., Rosen, P. P., Brogi, E., & Koerner, F. C. (2020). *Rosen's Breast Pathology*. Wolters Kluwer Health. url: https://shop.lww.com/Rosen-s-Breast-Pathology/p/9781496398918.
27. Gurcan, M. N., Boucheron, L. E., Can, A., Madabhushi, A., Rajpoot, N. M., & Yener, B. (2009). Histopathological Image Analysis: A Review. *IEEE Reviews in Biomedical Engineering, 2*, 147–171. https://doi.org/10.1109/RBME.2009.2034865.
28. Raghavendra, A. S., Bassett, R., Damodaran, S., Barcenas, C. H., Mouabbi, J. A., Layman, R., & Tripathy, D. (2025). Clinical Characteristics and Survival Outcomes of Metastatic Invasive Lobular and Ductal Carcinoma. *JAMA Network Open*, *8*(4), e251888. https://doi.org/10.1001/jamanetworkopen.2025.1888.
29. Cireşan, D. C., Giusti, A., Gambardella, L. M., & Schmidhuber, J. (2013). Mitosis detection in breast cancer histology images with deep neural networks. *International Conference on Medical Image Computing and Computer-Assisted Intervention*, 16(Pt 2), 411–418. https://doi.org/10.1007/978-3-642-40763-5_51.

30. Condrescu, M., Opuni, K., Hantash, B. M., & Reeves, J. P. (2002). Cellular regulation of sodium-calcium exchange. *Annals of the New York Academy of Sciences*, 976, 214–223. https://doi.org/10.1111/J.1749-6632.2002.TB04744.X.
31. Cross, B. M., Breitwieser, G. E., Reinhardt, T. A., & Rao, R. (2014). Cellular calcium dynamics in lactation and breast cancer: From physiology to pathology. *American Journal of Physiology-Cell Physiology*, 306(6), Article 6. https://doi.org/10.1152/ajpcell.00330.2013.
32. Dumitrescu, C., Hackel, D., Obeid, I., & Picone, J. (2025). Crystallization signatures as predictive biomarkers in histopathology. *Proceedings of the IEEE Signal Processing in Medicine and Biology Symposium (SPMB)*, 1, 1–6. https://doi.org/10.1109/SPMB67169.2025.11345371
33. Sapino, A., & Kulka, J. (Eds.). *Breast Pathology* (1st ed. 2020). Cham : Springer International Publishing : Imprint: Springer, 2020. url: *https://link.springer.com/referencework/*https://doi.org/10.1007/978-3-319-62539-3.
34. Tozbikian, G., Brogi, E., Vallejo, C. E., Giri, D., Murray, M., Catalano, J., Olcese, C., Van Zee, K. J., & Wen, H. Y. (2017). Atypical Ductal Hyperplasia Bordering on Ductal Carcinoma In Situ. *International Journal of Surgical Pathology*, 25(2), 100–107. https://doi.org/10.1177/1066896916662154.
35. Elmore, J. G., Longton, G. M., Carney, P. A., Geller, B. M., Onega, T., Tosteson, A. N. A., Nelson, H. D., Pepe, M. S., Allison, K. H., Schnitt, S. J., O'Malley, F. P., & Weaver, D. L. (2015). Diagnostic Concordance Among Pathologists Interpreting Breast Biopsy Specimens. *Journal of the American Medical Association*, 313(11), 1122–1132. https://doi.org/10.1001/JAMA.2015.1405.
36. Lambert, B., Forbes, F., Doyle, S., Dehaene, H., & Dojat, M. (2024). Trustworthy clinical AI solutions: A unified review of uncertainty quantification in Deep Learning models for medical image analysis. *Artificial Intelligence in Medicine*, 150, 102830. https://doi.org/10.1016/j.artmed.2024.102830.
37. Yin, L., Duan, J.-J., Bian, X.-W., & Yu, S. (2020). Triple-negative breast cancer molecular subtyping and treatment progress. *Breast Cancer Research*, 22(1), 61. https://doi.org/10.1186/s13058-020-01296-5.
38. Nong, L., Zhang, Z., Xiong, Y., Zheng, Y., Li, X., Li, D., He, Q., & Li, T. (2019). Comparison of next-generation sequencing and immunohistochemistry analysis for targeted therapy-related genomic status in lung cancer patients. *Journal of Thoracic Disease*, 11(12), 4992–5003. https://doi.org/10.21037/jtd.2019.12.25.
39. "ICD-10-CM, Official Guidelines for Coding and Reporting", Centers for Medicare & Medicaid Services (CMS), January 01, 2020, url: https://www.hhs.gov/guidance/sites/default/files/hhs-guidance-documents/ICD-10-CM_Guidelines-FY2020_final.pdf.
40. Li, X., Sun, X., Meng, Y., Liang, J., Wu, F., & Li, J. (2020). Dice Loss for Data-imbalanced NLP Tasks. *Proceedings of the 58th Annual Meeting of the Association for Computational Linguistics*, 465–476. https://doi.org/10.18653/v1/2020.acl-main.45.
41. Jadon, S. (2020). A survey of loss functions for semantic segmentation. *2020 IEEE Conference on Computational Intelligence in Bioinformatics and Computational Biology (CIBCB)*, 1–7. https://doi.org/10.1109/CIBCB48159.2020.9277638.
42. Bagritsevich, M., Picone, J., & Obeid, I. (2024). *The TUH Digital Pathology Corpus*. url: https://isip.piconepress.com/projects/nedc/html/tuh_dpath/.
43. Thai, B., McNicholas, S., Shalamzari, S. S., Meng, P., & Picone, J. (2023). Towards a More Extensible Machine Learning Demonstration Tool. *Proceedings of the IEEE Signal Processing in Medicine and Biology Symposium*, 1–4. https://doi.org/10.1109/SPMB59478.2023.10372731
44. Thundiyil, S. C., & Picone, J. (2025). Time Series Analysis from Classical Methods to Transformer-Based Approaches: A Review. In *Signal Processing in Medicine and Biology: Applications of Artificial Intelligence in Medicine and Biology* (Vol. 1, p. 56). Springer. url: https://isip.piconepress.com/publications/book_sections/2024/springer/transformers/. (in publication).
45. He, K., Zhang, X., Ren, S., & Sun, J. (2016). Deep Residual Learning for Image Recognition. *Proceedings of the IEEE/CVF Conference on Computer Vision and Pattern Recognition* (CVPR), 770–778. https://doi.org/10.1109/CVPR.2016.90

46. Khalkhali, V., Shawki, N., Shah, V., Golmohammadi, M., Obeid, I., & Picone, J. (2021). Low Latency Real-Time Seizure Detection Using Transfer Deep Learning. In I. Obeid, I. Selesnick, & J. Picone (Eds.), *Proceedings of the IEEE Signal Processing in Medicine and Biology Symposium (SPMB)* (pp. 1–7). IEEE. https://doi.org/10.1109/SPMB52430.2021.9672285.
47. Alexandrov, D., & Picone, J. (2024). The Impact of ECG Channel Reduction on Multi-Label Cardiac Diagnosis. *Proceedings of the IEEE Signal Processing in Medicine and Biology Symposium (SPMB)*, 6. https://doi.org/10.1109/SPMB62441.2024.10842233
48. Tan, M., & Le, Q. V. (2019). EfficientNet: Rethinking Model Scaling for Convolutional Neural Networks. In K. Chaudhuri & R. Salakhutdinov (Eds.), *Proceedings of the International Conference on Machine Learning (ICML)* (Vol. 97, pp. 6105–6114). PMLR. url: http://proceedings.mlr.press/v97/tan19a.html.
49. Dosovitskiy, A., Beyer, L., Kolesnikov, A., Weissenborn, D., Zhai, X., Unterthiner, T., Dehghani, M., Minderer, M., Heigold, G., Gelly, S., Uszkoreit, J., & Houlsby, N. (2021). An Image is Worth 16x16 Words: Transformers for Image Recognition at Scale. *Proceedings of the International Conference on Learning Representations (ICLR)*, 1–21. url: https://iclr.cc/virtual/2021/oral/3458.
50. Paszke, A., Gross, S., Massa, F., Lerer, A., Bradbury, J., Chanan, G., Killeen, T., Lin, Z., Gimelshein, N., Antiga, L., Desmaison, A., Köpf, A., Yang, E., DeVito, Z., Raison, M., Tejani, A., Chilamkurthy, S., Steiner, B., Fang, L., … Chintala, S. (2019). PyTorch: An imperative style, high-performance deep learning library. *Proceedings of the 33rd International Conference on Neural Information Processing Systems.* Curran Associates Inc. https://doi.org/10.5555/3454287.3455008.
51. Atabansi, C. C., Nie, J., Liu, H., Song, Q., Yan, L., & Zhou, X. (2023). A survey of Transformer applications for histopathological image analysis: New developments and future directions. *BioMedical Engineering OnLine*, *22*(1), 96. https://doi.org/10.1186/s12938-023-01157-0.
52. Nguyen, C., Asad, Z., Deng, R., & Huo, Y. (2022). Evaluating transformer-based semantic segmentation networks for pathological image segmentation. *Proceedings of the Society of Photo-Optical Instrumentation Engineers (SPIE)*, 12032, 120323N. https://doi.org/10.1117/12.2611177.
53. Gonzalez, R., & Woods, R. (2017). *Digital Image Processing*. Pearson Deutschland. url: https://elibrary.pearson.de/book/99.150005/9781292223070.
54. Janz, A., Jakimow, B., Thiel, F., Goswami, A., van der Linden, S., & Hostert, P. (2025, June 15). *Generic Filter (Majority)*. EnMAP-Box 3 Documentation. url: https://enmap-box.readthedocs.io/en/latest/usr_section/usr_cookbook/generic_filter.html.
55. ArcGIS Pro 3.5. (2025, June 15). *Smoothing zone edges with Boundary Clean and Majority Filter*. ArcGIS Pro. url: https://pro.arcgis.com/en/pro-app/latest/tool-reference/spatial-analyst/smoothing-zone-edges-with-boundary-clean-and-majority-filter.htm.
56. Nghiem, Y., Berman, L., & Bulik, A. (2024). Machine Learning in Digital Pathology. *Senior Design I, College of Engineering, Temple University*, 1–37. url: https://isip.piconepress.com/publications/presentations_misc/2024/senior_design/mladp.
57. Agkland, B. D., & Weste, N. H. (1981). The edge flag algorithm—A fill method for raster scan displays. *IEEE Transactions on Computers*, C–30(1), 41–48. https://doi.org/10.1109/TC.1981.6312155.
58. Wang, J., Zhang, M., Zhang, J., Wang, Y., Gahlmann, A., & Acton, S. T. (2021). Graph-theoretic post-processing of segmentation with application to dense biofilms. *IEEE Transactions on Image Processing*, 30, 8580–8594. https://doi.org/10.1109/TIP.2021.3116792.
59. Nobile, M. S., Cazzaniga, P., & Ashlock, D. A. (2019). Dilation functions in global optimization. *2019 IEEE Congress on Evolutionary Computation (CEC)*, 2300–2307. https://doi.org/10.1109/CEC.2019.8790247
60. Miao, L., & Qi, H. (2006). The design and evaluation of a generic method for generating mosaicked multispectral filter arrays. *IEEE Transactions on Image Processing*, 15(9), 2780–2791. https://doi.org/10.1109/TIP.2006.877315.

61. Picone, J. (1990). Continuous Speech Recognition Using Hidden Markov Models. *IEEE Acoustics, Speech, and Signal Processing Society (ASSP)*, *7*(3), 26–41. https://doi.org/10.1109/53.54527.
62. Liu, Z., Lin, Y., Cao, Y., Hu, H., Wei, Y., Zhang, Z., Lin, S., & Guo, B. (2021). Swin Transformer: Hierarchical Vision Transformer using Shifted Windows. *2021 IEEE/CVF International Conference on Computer Vision (ICCV)*, 9992–10002. https://doi.org/10.1109/ICCV48922.2021.00986.
63. Liu, Z., Hu, H., Lin, Y., Yao, Z., Xie, Z., Wei, Y., Ning, J., Cao, Y., Zhang, Z., Dong, L., Wei, F., & Guo, B. (2022). Swin Transformer V2: Scaling Up Capacity and Resolution. *2022 IEEE/CVF Conference on Computer Vision and Pattern Recognition (CVPR)*, 11999–12009. https://doi.org/10.1109/CVPR52688.2022.01170.
64. Ronneberger, O., Fischer, P., & Brox, T. (2015). U-Net: Convolutional Networks for Biomedical Image Segmentation. In N. Navab, J. Hornegger, W. M. Wells, & A. F. Frangi (Eds.), *Proceedings of the International Conference on Medical Image Computing and Computer-Assisted Intervention (MICCAI)* (pp. 234–241). Springer International Publishing. https://doi.org/10.1007/978-3-319-24574-4_28.
65. Cui, J., Guo, H., Wang, H., Chen, F., Shu, L., & Li, L. C. (2020). Fully-automatic segmentation of coronary artery using growing algorithm. *Journal of X-Ray Science and Technology*, *28*(6), 1171–1186. https://doi.org/10.3233/XST-200707.
66. Asadi-Aghbolaghi, M., Darbandsari, A., Zhang, A., Contreras-Sanz, A., Boschman, J., Ahmadvand, P., Köbel, M., Farnell, D., Huntsman, D. G., Churg, A., Black, P. C., Wang, G., Gilks, C. B., Farahani, H., & Bashashati, A. (2024). Learning generalizable AI models for multi-center histopathology image classification. *NPJ Precision Oncology*, *8*(1), 151. https://doi.org/10.1038/s41698-024-00652-4.

MIX
Papier aus verantwortungsvollen Quellen
Paper from responsible sources
FSC® C105338

If you have any concerns about our products,
you can contact us on
ProductSafety@springernature.com

In case Publisher is established outside the EU,
the EU authorized representative is:
Springer Nature Customer Service Center GmbH
Europaplatz 3, 69115 Heidelberg, Germany

Printed by Libri Plureos GmbH
in Hamburg, Germany